Modern Manufacturing: Materials, Processes and Systems

Modern Manufacturing: Materials, Processes and Systems

Edited by
Casan Anderson

Willford Press
www.willfordpress.com

Published by Willford Press,
118-35 Queens Blvd., Suite 400,
Forest Hills, NY 11375, USA

ISBN: 978-1-68285-398-6

Cataloging-in-Publication Data

Modern manufacturing : materials, processes and systems / edited by Casan Anderson.
p. cm.
Includes bibliographical references and index.
ISBN 978-1-68285-398-6
1. Production engineering. 2. Manufacturing processes.
3. Manufacturing industries--Technological innovations. I. Anderson, Casan.
TS183 .M63 2017
670--dc23

For information on all Willford Press publications
visit our website at www.willfordpress.com

WILLFORD PRESS

Printed in China.

Contents

Preface VII

Chapter 1 **A method of meta-mechanism combination and replacement based on motion study** 1
Yadong Fang

Chapter 2 **Managing uncertainty through supply chain flexibility: reactive vs. proactive approaches** 15
Reina Angkiriwang, I Nyoman Pujawan and Budi Santosa

Chapter 3 **An investigation on the roll force and torque fluctuations during hot strip rolling process** 36
Mahdi Bagheripoor and Hosein Bisadi

Chapter 4 **Influence of subcontracting constraints on the performance of manufacturing industries in Nigeria** 50
Victor Chukwunweike Nwokocha and Iganatius Ani Madu

Chapter 5 **Lean thinking for a maintenance process** 62
Sherif Mostafa, Sang-Heon Lee, Jantanee Dumrak, Nicholas Chileshe and Hassan Soltan

Chapter 6 **A review of few unconventional machining processes based on the concept of velocity shear instability in plasma** 99
R.K. Tyagi

Chapter 7 **Production scheduling for continuous manufacturing systems with quality constraints** 111
Jalel Ben Hmida, Jim Lee, Xinchun Wang and Fathi Boukadi

Chapter 8 **Influence of surface pretreatment in resistance spot welding of aluminum AA1050** 128
Ihsan K. Al Naimi, Moneer H. Al Saadi, Kasim M. Daws and Niels Bay

Chapter 9 **Sequencing jobs in an engineer-to-order engineering environment** 144
Douglas H. Grabenstetter and John M. Usher

Chapter 10 **Criteria definition and approaches in green supplier selection – a case study for raw material and packaging of food industry** 161
Narges Banaeian, Hossein Mobli, Izabela Ewa Nielsen and Mahmoud Omid

Chapter 11 **A review of turning of hard steels used in bearing and automotive applications** 181
Suha Karim Shihab, Zahid A. Khan, Aas Mohammad and Arshad Noor Siddiquee

Chapter 12 **Stabilizing wave attenuation effects in turning process** 207
Chigbogu Godwin Ozoegwu

Chapter 13 **End mill tools integration in CNC machining for rapid manufacturing processes: simulation studies** 216
Muhammed Nafis Osman Zahid, Keith Case and Darren Watts

Permissions

List of Contributors

Index

Preface

Manufacturing of products is a primary function of industries across the globe. It encompasses an array of processes such as procurement of raw materials, packaging, supply, etc. This book unfolds the innovative aspects of manufacturing which will be crucial for the progress of this field in the future. The topics included herein are of utmost significance and bound to provide incredible insights to readers. It elucidates the concepts and innovative models around prospective developments with respect to manufacturing. With state-of-the-art inputs by acclaimed experts of this field, this book targets students and professionals alike. It will help new researchers by foregrounding their knowledge in this discipline.

All of the data presented henceforth, was collaborated in the wake of recent advancements in the field. The aim of this book is to present the diversified developments from across the globe in a comprehensible manner. The opinions expressed in each chapter belong solely to the contributing authors. Their interpretations of the topics are the integral part of this book, which I have carefully compiled for a better understanding of the readers.

At the end, I would like to thank all those who dedicated their time and efforts for the successful completion of this book. I also wish to convey my gratitude towards my friends and family who supported me at every step.

Editor

1

A method of meta-mechanism combination and replacement based on motion study

Yadong Fang*

Department of Mechanical and Electrical Engineering, Xi'an Technological University, Xi'an, China

Lacking the effective methods to reduce labor and cost, many small- and medium-sized assembly companies are facing with the problem of high cost for a long time. In order to reduce costs of manual operations, the method of meta-mechanism combination and replacement is studied. In this paper, we mainly discuss assembling motion analysis, workpieces position information acquisition, motion library construction, assembling motion analysis by Maynard's operation sequence technique, meta-mechanism database establishment, and match of motion and mechanism. At the same time, the principle, process, and system realization framework of mechanism replacement are introduced. Lastly, problems for low-cost automation of the production line are basically resolved by operator motion analysis and meta-mechanism combination and match.

Keywords: production line; low-cost automation; motion analysis

1. Introduction

Motion Study, also known as motion research, element analysis, or work study, its main content is seeking the most economical and effective working method through a variety of analytical methods. It is an effective way to study and formulate correct and reasonable motion to shorten hours and improve efficiency by increasing valuable movement and shorten or eliminate waste motion (Du, 2013). Frank Bunker Gilbreth, an American engineer, established therbligs theory in 1912. The American Society of Mechanical Engineers added 'Discovery' motion and summarized 18 kinds of basic therbligs (Hua, 2012). In the process of analysis of therbligs, operation is observed and divided finely, and motion sequence is linked with hands, feet, eyes, and head movements in detail (Hirosea, Dokia, & Kondoa, 2013). Furthermore, operation is recorded and classified by therblig symbols, and the problems about operating sequence and method, such as single-hand waiting, unreasonable motion, and waste motion, are found and improved. The application researches in the country about motion study are mainly focused on the following four aspects (Wang, 2010):

*Email: fangyadong@xatu.edu.cn

- Motion analysis applied in production line balance.

Improvement measures based on motion analysis for production line in equipment or electronics manufacturing enterprise are proposed, and it improves the cycle time and balance ratio of production line (Gu & Zhou, 2009).

- Motion analysis applied in experimental teaching and assembly line study.

Wen Wei, Ding Wen-ying, and Dong Shao-hua choose reducer assembly line as the subject investigated, and eliminate bottleneck process of production line to improve its balance ratio. Gao Qi and Wang Shao-Hua research assembly line based on threbligs, and find out stasis, invalid, waste, imbalance, and disorder phenomenon in production process (Liao & Sun, 2009).

- Operation improvement and design by motion analysis.

Liu Shao-shan and Liu Feng-shan research therblig motion analysis and work improvement under lean production system to improve resource utilization rate. Lv Xiao-Ji and Chen Cang-ming take one automobile enterprises, for example, and eliminate various waste and imbalance problem in enterprise production process by motion analysis and line balance (Li & Gui, 2008).

- Operation process optimization by motion analysis.

Jiang De-yi and Liu Hong-xiao take snatch motion as instance, construct body mathematical model, and give optimal motion of snatch object in production activities. According to motion analysis for efficiency principle, Liu Hong-xiao optimizes male shirt shoulder sewing process (Yang, Wu, Zhu, Bao, & Wei, 2013).

However, there are few reports for the automation of the assembly line, especially the motion analysis and meta-mechanisms.

Since the automation technology appears, it develops crosswise with mathematics and computer technology. Increasingly mature hardware technologies, especially the rapidly developing computer technology, lay the foundations of automation technology (Wei, 1994). However, at the present stage, the automation technology not only solves the problems of production control for the single production cell, but also solves the problems of the entire production process as well as operation and management of business. The development of automation technology is divided into 4 stages, and it is shown in Table 1.

So far, for the automation of assembly enterprises all over the world, the four stages coexist, and they are information- and system-oriented (Jiang & Sui, 2012). But we know, at the fourth stage, computer-integrated manufacturing system (CIMS) costs too much, and the manufacturing period is too long; so in some small- and medium-sized enterprises, for the simple automation production lines, we can find a technology-intensive and low-cost way.

Therefore, the replacement of assembly operation with meta-mechanism and LCA have great potential and extensive foreground. Besides, they meet the requirements currently with advantages of low-cost investment, short period, quick effectiveness, and high automation.

Table 1. The development of automation technology.

The stage	Automatic level	Industrial mode of production	Technical means	Type of manufacturing: Discrete manufacturing	Type of manufacturing: Process manufacturing
First stage	Manual control	Labor intensive	Manual operation	Simple facility	Simple facility
Second stage	Unit automation	Equipment-intensive	CNC technology	CNC machine	Unit combination instrument
Third stage	Information automation	Information intensive	Computer-aided technology	CAD/CAM/ FMS	DCS
Fourth stage	Decision automation	Knowledge intensive	Computer integrated technology	CIMS	CIPS

2. Common problems in the manual production lines

There are usually a lot of labor-intensive enterprises in the manufacturing production lines, in which the staffs are engaged in the simple and repetitive assembly motions, and easy to feel fatigued (Meng, 1994). Since the production efficiency and production quality are easily affected by workers, once lack of personnel or training, it will not be able to guarantee the delivery cycle (Son, 2005). In recent years, especially in the eastern coastal area, with the declining of the domestic manufacturing costs of parts and the increasing of the labor costs, reducing the costs of product's assembly become more and more important and evident, particularly in the industries of household appliances and electronics industries. The economic benefits brought by improving assembly efficiency are far more obviously increase than that brought by simply reducing manufacturing costs of parts. The applications of low-cost automation will make some small- and medium-sized assembly enterprises (or some large enterprises) run their own business with lower costs, and in the process, it can almost get the same profits as the profits of advanced automation enterprises. If we combine the low-cost automation with each production line, which can realize the combined operation of machines and operators, and the achievements will be much more valuable than trying to carry out the pure automation in a short term.

Although many enterprises pay much attention to the development and application of low-cost automation all over the world in recent years, but the executing process of low-cost automation in the production line is still faced with many problems (Yang, 2010). Mostly, three issues must be considered: firstly, the effective analysis of assembly motions; secondly, the replacement of similar movements by automation; and thirdly, the mechanisms must be easily replaced and just need low costs. In general, low-cost automation should have the following features:

(1) *Small investment*: The investment of computers and automation equipment generally accounts for 7–12% of all equipment investment, or less. The upper limit to the investment just likes this: the percentage of investment in a new factory should be lower than the general provisions. The percentage of some special projects should be far lower than the funds, which are approved by provinces and cities, and less than the enterprise's affordability. However, this part should account for more than 40% in CIMS and computer-integrated process system (CIPS).

(2) *Targeted investment*: Investment should increase output, improve quality, raise labor productivity, and reduce the consumption in the short term. Man should get benefits in a relatively short time (usually 1–2 years), and the lifetime of equipment should be 5–8 years.
(3) *The automation*: The most necessary part is automation. It includes reasonable classification between men's and machines' functions, and compact structures. Besides, the rate of function should be above 95%.
(4) Reliable and easy to maintain.
(5) Low-cost automation should be gradually improved, eventually progressed in the direction of the CIM and CIP.
(6) Use the mature technology as much as possible, invest least with quick effect, and reduce the investment risks.
(7) The maintenance cost must be low; the work should be reliable, and the online rate should be above 95%.

3. The principle and process of meta-mechanism replacement

Assembly enterprises are usually faced with these problems: whether input low-cost automation equipment in production line or not; how much automation equipment should be input; and what kind of low-cost automation equipment should be input, etc. But, even if enterprises have the ability to put simple automation into effect, they do not; or they spent a lot on expensive automation equipment, but it is difficult to get benefits in a short time, which will cause huge losses. In this article, through the analysis of operator motion in production line, man can establish a matching relation between operator motions and meta-mechanism combinations or automation equipment, implement low-cost automation in a production line, and then save a large part of the expenses.

The principle and operating steps about motion analysis of assembly and mechanism replacement have shown in Figure 1.

The process to the motion analysis and the method of meta-mechanism replacement includes the following aspects:

(1) Analyze the assembly motions. According to the category of products, use a video camera to do some real-time recording of the assembly operation, assembly operation is divided into parts by motion display devices, storage devices, and processing devices. At the same time, analyze the positional information about workpieces. Use the laser displacement sensor to collect the 3D data of workpieces in the assembling process from different positions.
(2) Use the algorithm of point cloud reconstruction and registration to rebuild the 3D data model of the workpieces (Wang, 2012), and then obtain the positional information of the workpieces.

The algorithm of point cloud reconstruction is shown as following:

Step1: Point cloud data to be spliced is inputted.
Step2: Initial triangle is determined.
2.1 Any point is selected randomly as first vertex P_1 of initial triangle.
2.2 Closet point P_2 from P_1 is searched, edge P_1P_2 is constructed.
2.3 Optimum point is searched by traversing other points.

2.4 If $min(|P_1P_3|+|P_2P_3|)$, then goes to next step, else returns to 2.1.
2.5 If minimum interior angle of triangle$\Delta P_1P_2P_3$ isn't less than 35 degree, then goes to step 2.6, else returns to 2.1.
2.6 If maximum interior angle of triangle$\Delta P_1P_2P_3$ isn't more than 90 degree, then goes to next step, else returns to 2.1.
2.7 Initial triangle is determined, and point P_1, P_2 and P_3 are set as dead point. Besides, $\Delta P_1P_2P_3$ and three edges of initial triangle are put into directly triangle table and edge table.
Step3: The process of triangulation is realized.
3.1 Any edge E_1 is selected randomly (point P_4 and P_5).
3.2 Minimum envelope rectangle for triangles including edge E_1 is solved.
3.3 Optimum point P_6 is searched from active point set in minimum envelope rectangle.
3.4 If $min(|P_4P_5|+|P_4P_6|)$, then goes to next step, else returns to 3.1.
3.5 If minimum interior angle of triangle$\Delta P_4P_5P_6$ isn't less than 35 degree, then goes to step 2.6, else returns to 3.1.
3.6 If maximum interior angle of triangle$\Delta P_4P_5P_6$ isn't more than 90 degree, then goes to next step, else returns to 3.1.
3.6 P_6 is set as dead point, and P_4P_5 and P_4P_6 are added into edge table. Besides, edge E_1 is deleted, and$\Delta P_4P_5P_6$ is put into triangle table.

The algorithm of point cloud registration is shown as following (Zhang, 2013):

Step1: Original point cloud data sampling.
1.1 Workpiece is scanned progressively by laser displacement sensor.
1.2 Reflected light of workpiece is collected by laser collection device.
1.3 Laser signal is converted into workpiece 3D data.
Step2: Bilateral filtering denoising of point cloud is realized.
Step3: Point cloud is simplified.
3.1 Minimum enveloping hexahedron of point cloud is calculated.
3.2 By data traversing point, grid cell is divided, and data points are classified as grid cell.
3.3 Grid cell spatial location is set as mean value of data point set, and point cloud quantity and coordinate are reset.
Step4: Corresponding closest point of reference point cloud set X_1 is calculated, according to registration point cloud set X_2 to be matched.
Step5: Minimum optimal model for corresponding point mean distance is constructed.
Step6: Translation transformation and rotation transformation matrix are determined, and registration point coordinate value X_3 in new coordinate system are calculated.
Step7: If condition of convergence is satisfied, then point cloud registration is finished, else $X_2 = X_3$, and returns to 4.

(3) Set up the motion database. According to the products' category, store each motion of the operating personnel who assembles a product, the required time, and the variation of position of ready-to-assemble parts.
(4) Express the motion with Maynard's operation sequence technique (MOST). After each motion has been divided into therbligs, we can express the assembling operations by MOST. Then, classify and integrate it, such as moving, grabbing, body movements, pressure, pull, and rotation. We can inspect strictly, delete redundant motion, and change the existing procedure by canceling, merging, rearrangement, and simplification. Last, we can adjust the working methods to get the optimal operation.
(5) Set up the meta-mechanism repository. Sort the application of the existing meta-mechanism of assembly production line and automation equipment. Forming the meta-mechanism repository of the assembly line, such as object-clamping

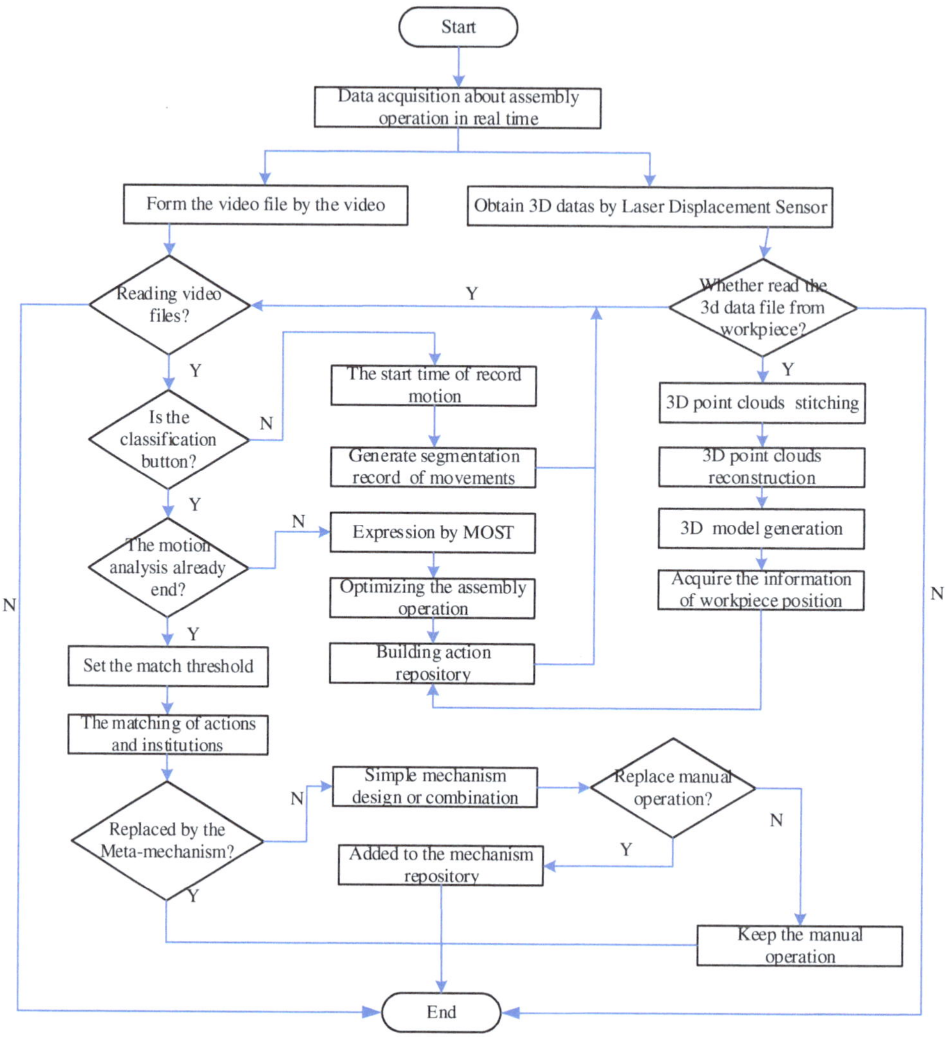

Figure 1. The principle of assembly motion analysis and mechanism substitution.

mechanism, moving mechanism, fixture-clamping mechanism, switches, positioning mechanism, locking mechanism, automatic halt mechanism, overload protecting mechanism, and so on, which usually includes 2D diagrams, key parameters, suitable operation, serial number, etc. The mechanism repository requires a lot of mechanisms and their combinations. The simple mechanisms and combinations which are commonly used can be divided into forms just like Figure 2.

(6) Match assembly motions with specific mechanisms. In view of each motion expressed by MOST in motion repository, search meta-mechanism repository and calculate the semantic similarity. If the similarity ratio is less than the setting threshold, combine mechanisms to form the project that can meet the requirements of assembly motion, which can contribute to evaluating and decision-making, and refer to the positional information about workpieces in the Step 2, then get the key technical parameters of mechanisms. The system interface has shown in Figure 3.

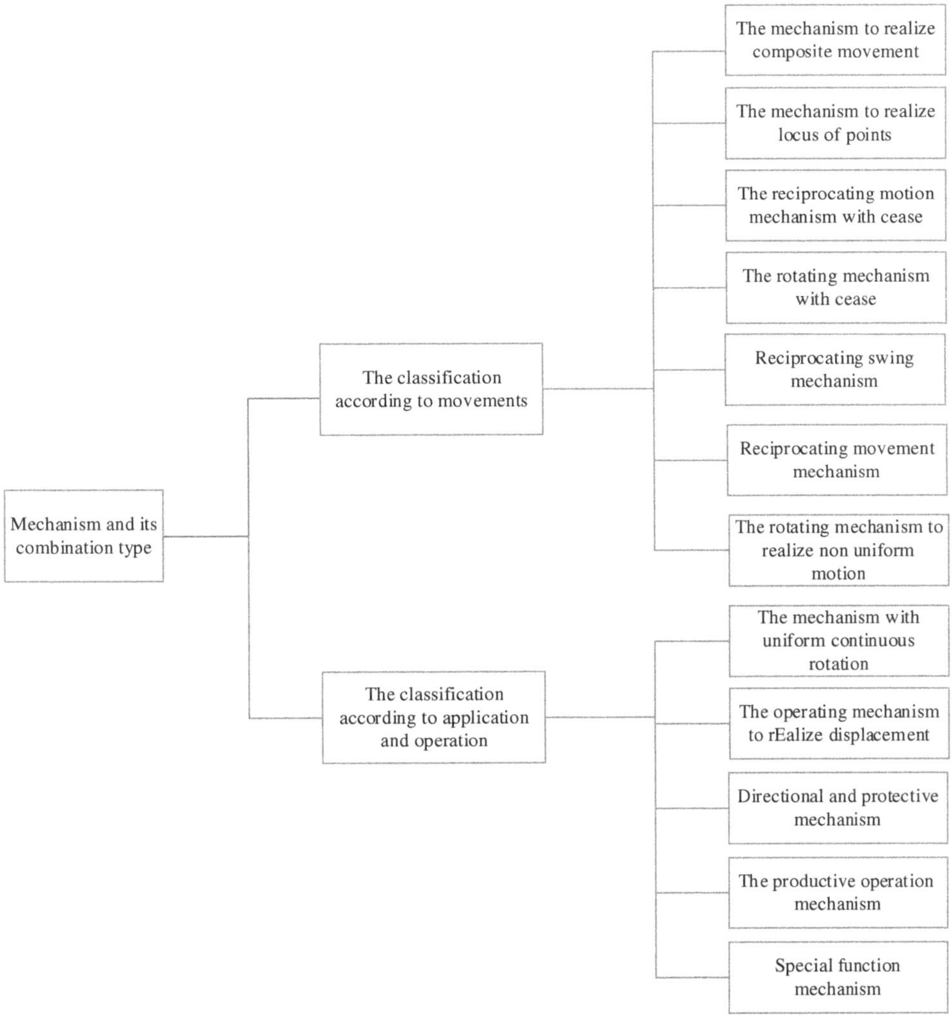

Figure 2. The classification of mechanisms and their combinations.

The realization of the above functions is based on the J2EE (Java 2 Platform Enterprise Edition) framework. Figure 4 shows that the system is divided into five layers: client layer, request to accept layer, application service layer, perception layer, and data storage layer from top to bottom. The function and effect of each layer are shown.

3.1. The client layer

When engineering technicians need data interaction, they use the server interface, which is provided by world web service and JavaBean (Enterprise Java Beans, EJB), to visit the system, and start the functions. The client layer includes two kinds of clients. One client is the thin client of Web service; the other is the specific client, which is based on Java and downloaded from the platform.

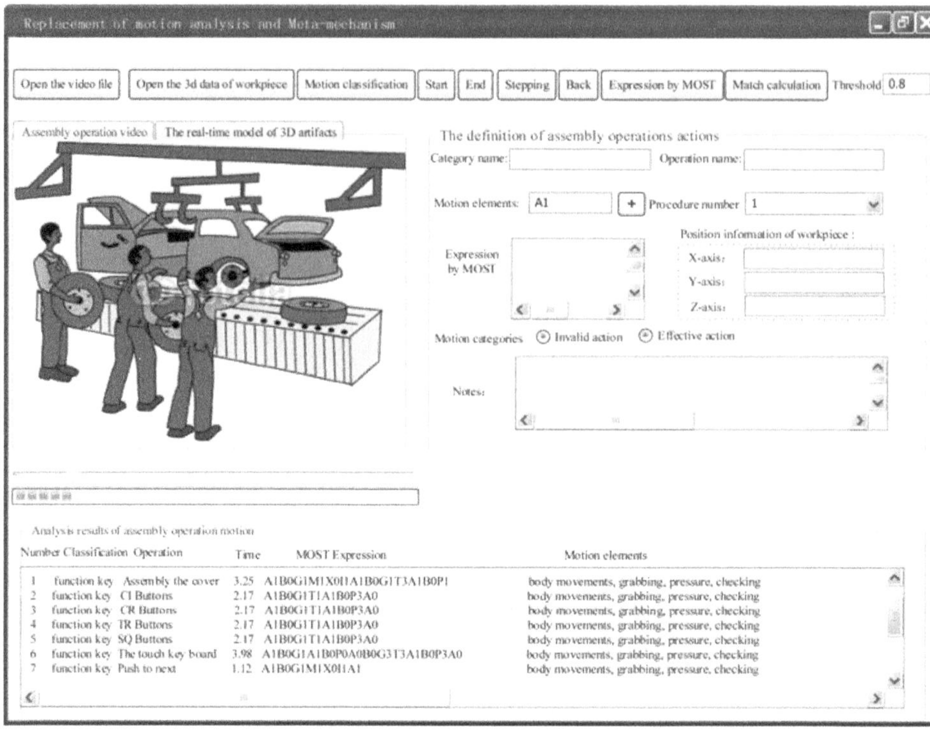

Figure 3. The interface of assembling motion analysis and meta-mechanisms.

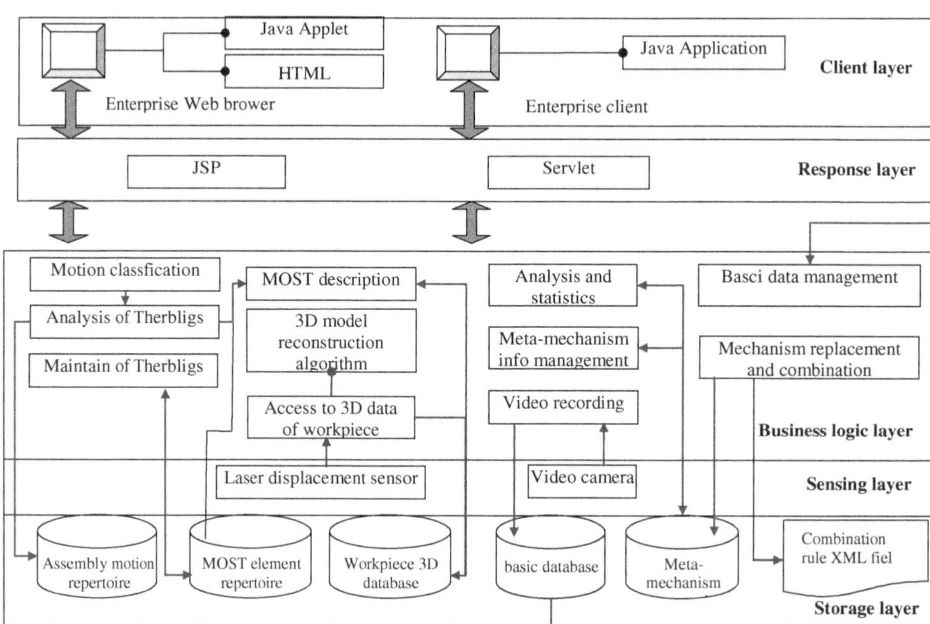

Figure 4. The framework of assembling motion analysis and meta-mechanisms.

3.2. Response layer

Response layer is used to receive the requests from the browser and give it to the bottom layer to handle, and send the result to the browser. The progress is made up of Java Server Page and Servlet that shows the information about the client layer. The request to accept layer can handle the data logically, for example, data calibration and the check of client browser. This layer cannot deal with complicated logical processing.

3.3. The business logic layer

The business logic layer is the core of the platform. It runs on the application server of Java and all logical processes are sealed in the EJB elements. The application server provides EJB elements with an optimized execution environment. The business processes are: engineering technicians use video cameras to record the assembling operation. They get 3D data of workpieces with a laser displacement sensor. Then, they use the algorithm of 3D reconstruction to reconstruct the 3D model of workpieces and get the changed information about positions of instruments. Changed positional information and the video files are synchronous. They utilize the classification of motions Bean to determine the sort of operations. They define and disassemble the assembly operations by the means of motion analysis. Then, get the standard database of MOST-elements. They use human–computer interaction to get the corresponding MOST-expression and its validity, and optimize the assembly operations. The standard database of MOST-elements can be controlled by the maintenance of therbligs. The information management module of meta-mechanism is in charge of the information maintenance of meta-mechanism in general assembling lines. Of course, this information is saved in meta-mechanism repository. The meta-mechanism combination and replacement are responsible for calculating the matched-degree between assembly operations and the typical meta-mechanisms. If the matched-degree meets the requirements, the mechanism will be replaced. After that, the operator will get changed positional 3D information of workpieces and the technical parameters. Otherwise, the operator will analyze XML, which is the rule of meta-mechanism combination, then the manual assembling operation is substituted by the valid combination of meta-mechanisms. Basic data management is responsible for personnel management, role management, user management, organizational management, and management of video files. The analytic and statistical model gets information of the assembly motion repository, foundation database, and meta-mechanism repository, so that it can analyze the operation of the assembling line and get reports and statistics of meta-mechanism replacement.

3.4. Sensing layer

The sensing layer involves two kinds of hardware: the video camera and laser displacement sensor. Video cameras record the assembly operation and send it to the database by Wi-Fi. The laser displacement sensor sends the collected 3D data to the server synchronously and stores the positional information of workpieces by the means of the reconstruction of a 3D model.

3.5. The data storage layer

The data storage layer provides the assembling motion analysis and the meta-mechanism combination and replacement with infrastructural support. It comprises assembling motion repository, MOST-element repository, 3D database of workpieces, foundation database, meta-mechanism repository, and XML files (the rule of meta-mechanism combination). The data storage layer is responsible for the storage of data, inquiring data, and backups. It maintains consistency and safety and provides the application service layer with data services. Assembling motion repository stores the motions' classifications, names, numbers, operating time, MOST-expressions, and therbligs. MOST-element repository stores the basic data table and consists of general motion table, controlling motion table, and instrument table. The 3D database of workpieces stores the changed positional information of workpieces. Fundamental database stores users, staff, roles, and motion video files. Meta-mechanism repository stores types, names, application, key parameters, and 2D images of the meta-mechanism that assembly line usually uses.

4. The mechanism replacement of a station in PS4 handles assembly line

There are some preconditions to realize meta-mechanism replacement of 'Installing R1/ R2 buttons' workstation in PS4 handles line:

(1) Realize arrangement and classification of the meta-mechanism repository of productive operation in assembly line.

The meta-mechanism combination in an assembly line can be divided into five categories according to different ways of working: feeding mechanism, selecting mechanism, carrying mechanism, grasping mechanism, and assembly mechanism. The letters in parentheses express the corresponding acronym of each mechanism.

(2) Analyze the scope of application about meta-mechanism combination.

Analyze and research the way of motion and applicable object of meta-mechanism combination, thereby mainly confirm the range of application of meta-mechanism.

(3) Complete the matching process in MOST of commonly used productive mechanism.

Achieve the matching of MOST expression to each of the operating mechanism.

(4) The MOST expression of key motions.

Extract key motions of MOST from the each mechanism, and delete invalid motions, which can be easy to filtrate the meta-mechanism combination later.

(5) The index extracting of key motion of the operation mechanism.

Extract the key motion from the MOST expression (or MOST formula), which can be easy to improve the accuracy of matching.

After the above procedures, we can begin to replace the motion of 'Install R1/R2 buttons' workstation with corresponding meta-mechanism. The process can be achieved according to Table 2, and its detail description is shown as following steps:

(1) Observe the movement of objects.

Observing the operator's working process on 'Install R1/R2 buttons' workstation, and analyze the motion of the workpieces carefully, which can narrow down the range of selected mechanisms.

(2) Express by MOST.

The 'R1/R2 buttons' station is mainly composed of six motions, whose specific names and the MOST formulas are shown in Table 3.

(3) Select the key motions from MOST formulas.

The method of selecting key motions from MOST formulas is deleting the invalid motion sequences, and selecting the key motions, then we will get the key MOST motions as shown in Table 4.

(4) Extract the index from the key motions.

When the process of selecting key motions is finished, we can begin to extract the key index from these key motions. The results are shown in Table 5.

(5) Select the meta-mechanism.

When the above steps are completed, we can find the corresponding meta-mechanism or its combinations to replace the motions. Through the analysis we can obtain that six motions can be replaced by four different mechanisms, and the specific results are shown in Table 6.

Table 2. The replacement process of 'Install R1/R2 buttons' station.

Steps	Step name	Operation process
1	Observe the movement of objects	Observe the movement of objects, then grasp the mode of motion
2	Express by MOST	Analyze the operator's MOST motion on the 'Install R1/R2 buttons' workstation, and write down the working process
3	Select key motions from MOST formulas	Select key motions from MOST formulas, and must ensure they are effective ones
4	Extract the index from the key motion	Extract the index from the key motion which has got from the MOST formulas
5	Select the meta-mechanism	According to the MOST formulas, key MOST motion, its index and the motion-object, consult the corresponding meta-mechanism combination, then select the suitable mechanism-combination to replace them
6	Sort the mechanism combination	More than one motions often correspond to the different sets of mechanisms, so sometimes need to combine the multiple mechanisms to reduce input costs

Table 3. Express the motions of 'Install R1/R2 buttons' by MOST formulas.

Station name	No.	Motion name	MOST formula
Install R1/R2 buttons	1	Get a qualified product from the front workstation	A1B0G1T3A1B0P1A0
	2	Install the button 'R1' (right)	A1B0G1A1B0P3A0B0G0M1X0I3A0
	3	Get the button 'R2' and install a button spring in it (right)	A1B0G1(A1B0G3)A1B0P3A0
	4	Install the button 'R2' in the main holder(right)	A0B0G1A0B0P3A0B0G0M1X0I3A0
	5	Use adsorption-fixture to absorb a loudspeaker and install it	A1B0G1A1B0P1A1B0G3A1B0P3A1B0P1A1
	6	Push the product to the next workstation	A1B0G1M1X0I1A1(A1B0G1A1B0P1A1)

Table 4. Select the key motions of 'Install R1/R2 buttons' workstation.

Station name	No.	Motion name	Key motions
Install R1/R2 buttons	1	Get a qualified product from the front workstation	ABG/ABP
	2	Install the button 'R1' (right)	ABG/MXI
	3	Get the button 'R2' and install a button spring in it (right)	ABG/ABP
	4	Install the button 'R2' in the main holder (right)	ABG/MXI
	5	Use adsorption-fixture to absorb a loudspeaker and install it	ABG/ABP
	6	Push the product to the next workstation	ABG/MXI

It is observed that there are four different replacing mechanisms, and two of them are repeated, which can be achieved through a simple transformation on the primary mechanisms.

(6) Sort the mechanism combination.

Because the mechanism replacement involves four mechanisms, therefore, we can renew to combine and simplify them.

Table 5. Extract the index from key motions of 'Install R1/R2 buttons' work station.

Station name	No.	Motion name	Index of key motions
Install R1/R2 buttons	1	Get a qualified product from the front workstation	G1/T3
	2	Install the button 'R1' (right)	M1/X0/I3
	3	Get the button 'R2' and install a button spring in it (right)	G1/P3
	4	Install the button 'R2' in the main holder (right)	M1/X0/I3
	5	Use adsorption-fixture to absorb a loudspeaker and install it	G1/G3/P1/P3
	6	Push the product to the next workstation	M1/X0/I1

Table 6. The mechanism replacement of 'Install R1/R2 buttons' workstation.

Station name	No.	MOST formula	Replacement of meta-mechanism
Install R1/R2 buttons	1	A1B0G1T3A1B0P1A0	The stepping feeder-mechanism on the conveyor
	2	A1B0G1A1B0P3A0B0G0M1X0I3A0	The feeder-mechanism with cam-and-lever mechanism used by conveying billet
	3	A1B0G1(A1B0G3)A1B0P3A0	The feeder-mechanism with cam-and-lever mechanism used by conveying billet
	4	A0B0G1A0B0P3A0B0G0M1X0I3A0	The intermittent rotary conveying mechanism used in packaging cardboard
	5	A1B0G1A1B0P1A1B0G3A1B0P3A1B0P1A1	The intermittent rotary conveying mechanism used in packaging cardboard
	6	A1B0G1M1X0I1A1(A1B0G1A1B0P1A1)	The conveying mechanism with slider-crank

Note: Sometimes, in order to achieve the replacement of similar motions, we can simply change the mechanism structure.

5. Conclusions

Though the motion analysis is increasingly applied in enterprise production lines, it is still limited in follow aspects: relieving operational fatigue, improving operating process, measuring, and controlling the labor-hour. The motion analysis and the meta-mechanism replacement will improve the theory of motion analysis, expand the ranges of application of motion analysis, and find a good way to save manpower and costs.

Disclosure statement

No potential conflict of interest was reported by the author.

Funding

This work was supported by China Shaanxi Provincial Non-traditional Machining Key Laboratory Open Fund (ST-12010).

References

Du, Z. (2013). Motion analysis of application research in China. *Journal of Mechanical Management Development, 1* (1), 1–2.

Gu, T., & Zhou, B. (2009). The application of motion analysis in cylinder production line. *Journal of Mechanical Manufacturing, 8* (540), 1–2, 47.

Hirosea, K., Dokia, H., & Kondoa, A. (2013). Dynamic analysis and motion measurement of ski turns using inertial and force sensors. *Procedia Engineering, 60*, 355–360. Retrieved from http://www.sciencedirect.com/science/article/pii/S1877705813011338

Hua, J. (2012). *Optimization in the idea of lean production to the fixed telephone Bit Electronics.* Shandong province: Shandong University of Science and Technology.

Jiang, J., & Sui, R. (2012). Automatic assembly of combined checking fixture for auto-body components based on fixture elements libraries. *College of Mechanical and Electrical Engineering, 6* (30), 2–6.

Li, J., & Gui, W. (2008). The applications of methods research in G company operating standards decision. *Journal of Guilin University of Electronic Science and Technology, 28*, 2–10.

Liao, Y., & Sun, H. (2009). The application research of drug packaging assistive technology. *Journal of Packaging Engineering, 30*, 199–201.

Meng, X. (1994). *The modern mechanism manual* (pp. 152–229). Beijing: The Press of Mechanical Industry.

Son, C. (2005). Comparison of intelligent control planning algorithms for robot's part micro-assembly task. *Engineering Applications of Artificial Intelligence, 7*(19), 1–4.

Wang, F. (2010). *The technology and prospect of low cost automation concept*. Beijing: Tsinghua University.

Wang, J. (2012). *Technology research to the splicing and fusion of mass point-clouds model* (p. 10). Zhejiang: Zhejiang University of Technology.

Wei, C. (1994). Intelligent machine design with object-oriented design and constraints satisfaction mechanism. In *Proceedings of the third international conference on automation technology*, Taipei, China.

Yang, P. (2010). Drug packaging operation process analysis and optimization on the basis of motion research. *Journal of Packaging Engineering, 7*, 31.

Yang, Q., Wu, D. L., Zhu, H. M., Bao, J. S., & Wei, Z. H. (2013). Assembly operation process planning by mapping a virtual assembly simulation to real operation. *Computers in Industry, 64*, 869–879.

Zhang, F. (2013). *Research of point-clouds data processing with geometric method*. Xian: Northwestern University.

2

Managing uncertainty through supply chain flexibility: reactive vs. proactive approaches

Reina Angkiriwang, I Nyoman Pujawan* and Budi Santosa

Department of Industrial Engineering, Sepuluh Nopember Institute of Technology, Kampus ITS Sukolilo, Surabaya 60111, Indonesia

The purpose of this paper is to obtain insights into the typology of uncertainty and relevant strategies adopted by manufacturing companies to achieve better supply chain flexibility. Strategies are classified into reactive (buffering) and proactive (redesigning). We develop a framework that links supply chain uncertainty, the two types of strategies for achieving supply chain flexibility and the relevant objectives to be achieved. Four case studies are compared in terms of uncertainty typology and strategies being adopted to improve supply chain flexibility. We present three propositions as a result of the study. First, to achieve better flexibility, companies have been focusing more on buffering than on proactive or redesign strategies. Second, companies tend to focus on internal operations rather than collaborating with external parties, and third, the power structure in a supply chain governs the type and configuration of supply chain flexibility.

Keywords: supply chain flexibility; uncertainty; strategy; proactive; reactive

1. Introduction

To remain competitive in business, manufacturers are required to produce products of a quality acceptable to customers and to deliver those products at competitive cost with highly reliable delivery times. Achieving high quality levels, timeliness of deliveries, and efficient processes along the supply chain cannot be reliant on a single organization, but should be ensured through collaboration and coordination with trading partners.

Increasing uncertainty has made the task of satisfying customers more challenging. Efficiency along the supply chain is important to maintain acceptable product prices, but flexibility to deal with time-varying or dynamic demand could be even more important nowadays. With the high probability that customers will suddenly increase, reduce, cancel, or move forward or backward their orders, supply chain players need to be more flexible in many respects. This may include the need to change capacity levels, to use different transportation modes, to switch supplier, to deal with small lot sizes and to have short or even negligible changeover times. Researchers have considered that flexibility, along with quality, cost, and speed of delivery, is critical for competitiveness (Avittathur & Swamidass, 2007; Bernardo & Mohamed, 1992; Gong, 2008; Shin, Collier, & Wilson, 2000) and should be viewed from the supply chain perspective in a collaborative way (Fantazy, Kumar, & Kumar, 2009; Kumar, Fantazy, Kumar, & Boyle, 2006; Pujawan, 2004; Zhang, Vonderembse, & Lim, 2006). As pointed out by Vickery,

*Corresponding author. Email: pujawan@ie.its.ac.id

Calantone, and Dröge (1999), supply chain flexibility has a positive impact on the bottom line of the companies. Their empirical study also reveals that volume flexibility is highly related to market share and market share growth.

Researchers have paid a great deal of interest to flexibility. Slack (1988) and Gupta (Gupta & Goyal, 1989; Gupta & Somers, 1992) are among the early authors that have brought up the issue of flexibility in the context of internal operations systems. Following their works, many authors have published papers on flexibility. Currently, flexibility is viewed not only from the perspective of internal operations systems; more and more authors are trying to link it to a supply chain context. Vickery et al. (1999), Duclos, Vokurka, and Lummus (2003), and Pujawan (2004) are among the authors that have extended flexibility from internal operations to the supply chain context. Seebacher and Winkler (2013) provide a comprehensive review and citation analysis of works on manufacturing as well as supply chain flexibility. Although there has been continuing interest in the flexibility issue within the context of the supply chain, further debate and discussion are required.

Although it is important to remain competitive in the market, supply chain flexibility is costly (Gupta & Goyal, 1989; Koste & Malhotra, 1999; Merschmann & Thonemann, 2011; Pujawan, 2004) and flexibility dimensions are not equally important for every supply chain (Sánchez & Pérez, 2005; Stevenson & Spring, 2007). Therefore, a supply chain should carefully assess how much flexibility is really needed (Pujawan, 2004) because increased flexibility does not always result in greater economic income (Bernardo & Mohamed, 1992; Gong, 2008). Hence, it is necessary to decide the right degree of supply chain flexibility and the appropriate strategy that should be adopted to respond to the need for flexibility. In this respect, companies may just reactively build up safety stock or add safety buffers to the lead time, or pursue more proactive paths such as redesigning products or supply chain networks. In spite of its importance, there have been very few works that attempt to classify flexibility strategies and to relate the typology of uncertainties to those strategies. The objective of this paper is to present a conceptual model and case studies that address uncertainty typology, strategies to improve supply chain flexibility, and the relevant objectives that the companies in a supply chain aim to achieve. In particular, this paper aims to address the following questions in an explorative manner:

(1) What strategies have companies pursued to improve flexibility in the supply chain?
(2) In dealing with uncertainty, do companies tend to reactively buffer themselves or use more proactive strategies?

2. Literature review

The subject of flexibility has been present in the literature for more than two decades. This topic is highly related to uncertainty, both internal and external to a company or a supply chain. Supply chain flexibility is defined as the ability of a system or a chain to respond to unexpected and unpredictable changes due to uncertain environments to meet a variety of customer needs or requirements, while still maintaining customer satisfaction without adding significant cost. The type of strategies used to increase flexibility should be in line with the typology of uncertainty.

There have been changes in the direction of flexibility research. In the early stages, at the end of the eighties, flexibility was often discussed in the context of internal manufacturing systems. Cox (1989) stated that flexibility was a new concept with no acceptable measurement. Consequently, this triggered several researchers to develop an instrument to measure or assess flexibility (Bernardo & Mohamed, 1992; Gupta & Goyal, 1989; Gupta & Somers, 1992; Slack, 1988). Sethi and Sethi (1990) mentioned that flexibility is a complex, multi-dimensional, and hard-to-capture concept leading to several studies seeking to establish dimensions of flexibility during the 90s (Koste & Malhotra, 1999; Vokurka & O'Leary-Kelly, 2000). In the last few decades, intensifying market competition forced companies to change their orientation from internal operation systems to a supply chain context. Likewise, people extended the viewpoint of flexibility from the manufacturing to the supply chain context. Since then, a number of researchers have contributed to the literature on supply chain flexibility.

The major stream of work on flexibility has been related to the development of flexibility measures, as well as the conceptual framework that defines the relationships between various flexibility dimensions. Duclos et al. (2003) and Lummus, Duclos, and Vokurka (2003) developed a conceptual model and extended flexibility dimensions to the supply chain context. A framework for assessing supply chain flexibility has been developed by Pujawan (2004). The aim was to capture the relationship between supply chain flexibility drivers and flexibility dimensions, and ultimately to identify which dimensions have to be prioritized for improvement. Sánchez and Pérez (2005) built a hierarchy of supply chain flexibility dimensions based on bottom-up classification and tried to find a relationship between supply chain flexibility and firm performance. Kumar et al. (2006) developed a framework which recommended steps for implementing supply chain flexibility. It gives an overview of the relationship between each flexibility dimension in a supply chain.

As mentioned, a number of authors have tried to develop a framework to assess the importance of flexibility dimensions or relationships between flexibility dimensions and firm performance. However, little has been done to address and classify strategies to achieve the appropriate level of supply chain flexibility. In particular, no attention has been paid to relationships between the typology of uncertainty and the type of strategies chosen for better supply chain performance. Among the few authors to have addressed flexibility, Tachizawa (2007) and Tachizawa and Thomsen (2007) and Tachizawa and Gimenez (2010) developed models that relate between supply flexibility drivers and strategies to improve supply chain flexibility. However, the focus of their study is the purchasing or supply side, not the supply chain.

3. The conceptual model

In this paper, a conceptual model that links supply chain uncertainty, supply chain flexibility strategies, and the objectives pursued by companies along the supply chain is proposed. The elements of the conceptual model were synthesized from the literature. Figure 1 depicts the model, showing uncertainties as the factors that drive the need for supply chain flexibility (left), the two categories of strategy, i.e. four types of buffering (above), eight proactive strategies (below), and three objectives (right). More details will be given below.

3.1. Supply chain uncertainty

Uncertainty has been considered as a major factor behind the need for supply chain flexibility. We classify uncertainty in the supply chain context as upstream (supply) uncertainty, internal (process) uncertainty, and downstream (demand) uncertainty.

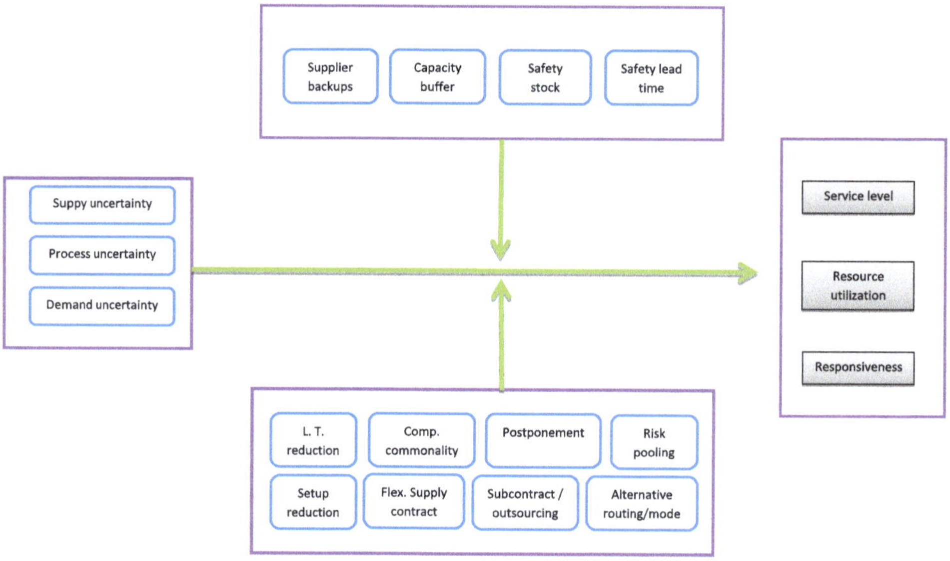

Figure 1. The conceptual model.

- *Supply uncertainty*. This is related to the uncertainty of materials' supply. Supply uncertainty may be manifested in the form of uncertainty regarding material availability/supply capacity, material price (Tang, 2006; Zhang, Shang, & Li, 2011), alternative sourcing availability (Pujawan, 2004), and supply lead time (Mohebbi & Choobineh, 2005; Osman & Demirli, 2012; Ramasesh, 1991; Thomas & Tyworth, 2006). When supply is uncertain, a higher level of flexibility is required to achieve better customer service levels. For example, when the price of a certain material suddenly increases, the ability to change the production schedule in order to produce alternative products or to use a substitute material is helpful in maintaining the efficiency and effectiveness of the supply chain.
- *Process uncertainty*. Process uncertainty is related to the probabilistic nature of machine availability (Ho, 1989; Van Kampen, van Donk, & van der Zee, 2010), yield (Gurnani & Gerchak, 2007; Ho, 1989; Schmitt & Snyder, 2012; Tang, 2006), quality (Ho, 1989; Murthy & Ma, 1991; Wu, Blackhurst, & O'grady, 2007), and processing times (Cao, Patterson, & Bai, 2005; Schmitt, 1984). Uncertainty in internal operations may also be related to such factors as labor issue, instable availability of working capital, and problems with information technology. The more uncertain the internal processes, the higher the level of flexibility needed. For example, when the reliability of a manufacturing process is low, a capacity buffer is necessary to keep the system flexible.
- *Demand uncertainty*. This refers to the probabilistic nature of demand quantity, types, timing, and locations. Demand uncertainty could be in the form of errors in the demand forecast (Schmitt, 1984), changes in customer orders (Pujawan & Smart, 2012; Van Kampen et al., 2010; Wong, Boon-Itt, & Wong, 2011), uncertainty about the product specification/mix that the customers will order (Li, Sikora, Shaw, & Tan, 2006; Van Donk & van der Vaart, 2005), and competitor actions regarding

marketing promotion (Wong et al., 2011). An empirical study involved 106 manufacturing companies by Pujawan and Smart (2012) suggested that most manufacturing companies experience order volatility from customers. Manufacturing companies, especially those producing innovative products, are facing shrinkage in product life cycle and increasing competition in the market. This ultimately creates demand uncertainty and forcing supply chain players to markdown product prices. Supply chain flexibility is important to cope with the dynamic nature of demand. For example, when demand for a product is highly uncertain, companies may have to increase the inventory buffer so that the sudden increased demand can be satisfied. Alternatively, the use of multiple transportation modes to cope with differing degrees of demand urgency may be important to improve customer service levels under highly uncertain demand.

3.2. Classification of strategies for supply chain flexibility

Flexibility is a strategy to deal with uncertainty facing the supply chain. As mentioned in the literature review, various strategies to increase flexibility have been evaluated by different authors. Koste and Malhotra (1999) mentioned that flexibility can be either reactive or proactive in nature. The reactive nature of flexibility addresses the environmental uncertainty, both internal and external, faced by an organization, while the proactive nature of flexibility allows an organization to redefine market uncertainties or influence what customers have come to expect from a particular industry. In the following subsections, we will list the strategies that belong to the above two classifications.

3.2.1. Reactive or buffering strategies

Basically, companies make no attempt to influence the level of uncertainty but, rather, react to it in an attempt to maintain their service level to customers or to maintain efficiency through, for example, better capacity utilization. These are the type of strategies which companies use to buffer themselves:

- *Safety stock*. Safety stock is one of the most common approaches to increase flexibility under the existence of demand and supply uncertainty. With safety stock, a company can reduce the probability of inventory shortage to an acceptable level. According to Van Kampen et al. (2010), safety stock also increases responsiveness. We argue that safety stock is a reactive strategy because this approach merely responds to the current level of uncertainty, without any attempt to proactively reduce it.
- *Capacity buffer*. One way to cope with uncertainty is to have flexible capacity. However, in many situations it is costly or even impossible to dynamically adjust the capacity level. Hence, to achieve flexibility, companies may set capacity higher than average demand so that they can avoid substantial shortages during peak periods. A capacity buffer could be used as a substitute of or a complement to safety stock. As mentioned by Manuj and Sahin (2011), some companies prefer to maintain surge capacity in the form of extra assembly lines that are staffed when necessary rather than maintain buffer inventory.
- *Supplier backups*. Working with a single supplier is risky. Companies often maintain multiple suppliers, which will guarantee availability but in most situations

will increase costs. The use of multiple suppliers to improve supply flexibility has been discussed in the literature, including for example Das (2011) and Manuj and Sahin (2011).

- *Safety lead times*. In dealing with uncertainty, companies often add a safety lead time to the actual cycle time (Raturi & Jack, 2004). While increasing inventory and adding costs, safety lead time enables companies to increase materials' availability and thus become more flexible in responding to demand. Van Kampen et al. (2010) suggested that adding safety lead time increases mix flexibility.

3.2.2. Proactive or redesign strategies

These are the type of strategies in which companies attempt to increase supply chain flexibility through proactively redesigning products, processes, and the supply chain network as well as proactively negotiating more effective relationships with trading partners. The following strategies belong to this classification:

- *Component commonality*: one of the obvious strategies related to product design that can improve flexibility is the use of component commonality. There has been wide discussion of the possibility that use of common component could increase efficiency as well as responsiveness when a manufacturing company offers a large product variety or when uncertainty in demand exists (Mohebbi & Choobineh, 2005). Large component commonality also increases supply chain flexibility. The use of component commonality to increase supply chain flexibility has been discussed by Pujawan (2004).
- *Postponement*: design or redesign of processes, both in terms of the manufacturing shop floor and in administrative processes, could significantly improve flexibility. Manufacturing or logistics postponement are examples of process design that improves supply chain flexibility. Increased component commonality and the use of postponement as strategies for better flexibility have been discussed in the literature (Barad & Sapir, 2003; Das, 2011; Kumar, Shankar, & Yadav, 2008; Pujawan, 2004; Tachizawa, 2007).
- *Risk pooling*: When demand is highly uncertain and involves multiple sales regions, a supply chain often designs a network for risk pooling. Reducing facilities or centralizing stocks to fewer facilities reduces risks as well as improves flexibility in allocating stock to multiple destinations or sales regions. In line with this idea, Stevenson and Spring (2009) argued that consortium purchasing among several independent firms or the use of centralized procurement among firms in the same group are among strategies to overcome the problem of buyers having minimum power and having to accept large minimum order quantities.
- *Subcontracting/outsourcing*: The use of external capacities through subcontracting and outsourcing is also a strategy used to be flexible in dealing with uncertainty. As suggested by Stevenson and Spring (2009), outsourcing is a tactic to obtain purchasing flexibility because it can reduce the risks of capacity utilization and amortization, especially when demand is uncertain, irregular, low, and/or temporary.
- *Flexible supply contract*: proactively negotiating supply contracts to alleviate minimum order quantities or to obtain a commitment from suppliers to supply materials or services in the case of a significant increase in demand are among the

strategies that could improve supply chain flexibility. Flexible procurement contracts can provide supply flexibility (Tang & Tomlin, 2008), ensure stability for the supplier, and help the buyer respond to demand fluctuations (Stevenson & Spring, 2007).

- *Lead time reduction*: reducing lead time is an important strategy to enable better flexibility. With a shorter lead time, companies will be able to better respond to demand uncertainty. Proactively reducing lead time may be done through redesigning procurement processes, changing supplier selection criteria from cost focus to speed focus, or developing suppliers for better lead time management.
- *Setup time reduction*: long production setup is a major cause of inability to create volume and mix flexibility. Significantly reducing setup time requires systematic effort within a manufacturing company. Vonderembse, Uppal, Huang, and Dismukes (2006) suggest that reduction of setup time will allow the economic production of small quantities, thereby achieving cost reductions, flexibility, and internal responsiveness. Having suppliers with short setup times would create supply flexibility and thus should be considered as an important criteria when selecting suppliers (Wu & Shen, 2006).
- *Alternative routing/mode*: when a customer places a sudden order, companies may have to use alternative transportation modes to speed up shipment. In another situation, when problems occur on a certain route, the company's flexibility to use alternative routes is important. The use of alternative transportation routing/modes to create flexibility has been mentioned in the literature (see e.g. Kumar et al., 2008; Morlok & Chang, 2004; Pujawan, 2004).

3.2. *Objectives of supply chain flexibility*

Although supply chain flexibility is often costly to achieve, it is rewarding to have appropriate levels of flexibility for companies working under uncertain environments. A study conducted by Merschmann and Thonemann (2011) involving German companies suggests that in uncertain environments companies with highly flexible supply chains perform better than companies with less flexible supply chains, while the opposite holds true under certain environments. As widely discussed in the literature, flexibility is viewed as a system's ability to cope with internal and external variation (Gong, 2008) or to cope with environmental uncertainty in an attempt to improve market responsiveness (Das, 2011), maintain customer service levels (Kumar et al., 2008), and achieve better resource utilization. Hence, we define here three objectives of supply chain flexibility that companies aim to achieve:

- *Higher service level*. Service level is a measure of a company's ability to satisfy customers' demand. It is normally measured in terms of percentage of customer demand satisfied without a backlog. When demand and supply are both uncertain, a flexible supply chain system is necessary to maintain a high level of service. As pointed out by Kumar et al. (2008), flexibility is the strategy by which the supply chain aims to maintain customer service levels by absorbing disturbances in supply and sudden changes in demand. There are many other works that address the linkage between flexibility and service level, including for example Alfredsson and Verrijdt (1999), Mohebbi and Choobineh (2005), and Van Hoek (2001).
- *Resource utilization*. Supply chain activities normally involve a variety of resources, such as those used for production, storage, material handling,

transportation, and administrative activities. Resource utilization is generally a measure of how much of the available capacity of resources is being used for productive outputs. Under uncertain operating environments, it is often difficult to obtain high levels of resource utilization. However, when appropriate supply chain flexibility is in place, there is an opportunity to better utilize the resources along the supply chain. Some earlier publications mention how flexibility may improve resource utilization. Ndubisi, Jantan, Hing, and Ayub (2005) stated that there is a relationship between resource utilization and volume flexibility. Pramod and Garg (2006) suggest that as any flexibility level increases, the system utilization increases.

- *Responsiveness*. Responsiveness is a very important capability which companies need to have in doing business. In an uncertain environment, responsiveness can only be achieved if there is sufficient level of flexibility throughout the supply chain. Flexibility in responding to changes in demand reflects responsiveness to changing customer needs and wants (Duclos et al., 2003). Kumar et al. (2006) stated that there is a need to make operations more flexible in an environment with unpredictable changes, and thereby increase responsiveness. Further, Yi, Ngai, and Moon (2011) suggest that better supply chain responsiveness can be achieved in two ways: reducing uncertainties and improving supply chain flexibility.

4. Case studies

We conducted a case study involving four manufacturing companies in Indonesia. According to Kähkönen (2011), case study is an appropriate research method for gathering rich empirical data and thereby to gain a deep understanding of the phenomenon in question. The objective of the case study was to gain more insights into supply chain flexibility in practice, and particularly the strategies which companies adopt to achieve better flexibility. We selected case companies with different supply chain characteristics, i.e. one is representing the engineering-to-order model, the other one is representing make-to-stock model, and the other two are somewhere in between of the above two models. A list of open-ended questions were prepared before interacting with each case company. In each case, we interviewed 3–5 people in 2–4 meetings involving those in charge of procurement, planning, manufacturing, product development, engineering, and marketing functions. In addition to the interview, we also collected and analysed pertinent data that could support our understanding of supply chain flexibility. We will provide details of each case in this section. Cross-case analysis will be conducted to compare uncertainty typology facing each case company and strategies they applied to create better supply chain flexibility.

4.1. Case 1: company A

This company is in the business of producing new ships and repairing old ones. As a project-based company, design, procurement, and production respond to customer orders. The company serves many domestic customers, but only three of them are considered major customers. The production facility is located on the coast of East Java. The size of the production facility limits its capacity. They can produce up to two new ships and repair up to 100 ships each year. The company employs around 550 permanent and 600 subcontracted workers.

For every product, customers will be involved at the early stage of developing the product specification. The raw material will not be ordered until the design and specifications for the end product are agreed on by the customer. Each product is composed of more than 10,000 components. However, there are only four major types of component: plate, pipe, engine, and electricity. In each of these major types, there are about five different variants. The majority of materials, about 80%, are obtained from overseas suppliers with 2–6 months' lead time. To avoid lateness of material supply, the company sets a lead time buffer of about four weeks. The company has about three suppliers for each type of component. The relationship with customers and suppliers for each project is contractual in nature rather than a partnership type. If a supplier is unable to supply the quantity ordered, an alternative supplier should be used.

The production process consists of five stages. It begins with body construction, pipe fitting, installation of machinery or engine, electrification and, finally, running tests. The duration of a project building a new ship is somewhere between 4 and 20 months, depending on the product type, size, and complexity. Substantial changes in the design at the production stage rarely happen. There are usually only minor revisions which do not significantly affect the production schedule and costs.

Unavailability of material will lead to production rescheduling. Interestingly, delays in the supply of materials are not because of problems with the suppliers, but quite often because of poor cash availability at the company. For ship production, capacity constraints can be relaxed by hiring more subcontracted workers, although it is not always easy to adjust the level of the workforce. However, for ship repair, space availability is a major constraint which cannot be adjusted. The uncertain nature of internal operations and materials supply leads to poor levels of on-time delivery to customers.

4.2. Case 2: company B

Company B is a supplier of packing items such as plastic bottles and tubes for the consumer goods industry. The company operates three factories in Indonesia with fixed production capacities, employing around 1000 people in total. Currently, they supply packing items to about 10 major customers. They have two types of product, exclusive and V-item, with about 100 variants. Exclusive products are those specific to a certain customer, whose design or patent is owned by the customer. V-item is a nonpatented product and can be sold to any customer. However, most of the products produced by this company are exclusive to certain customers. For exclusive products, the customers are involved in product design and innovation. All changes are based on customer requests.

The production process is fairly simple, with only a few steps, including molding, printing, and stamping. Even though the process is very simple, the level of process uncertainty is high due to machine breakdowns, high setup time, and defective products. Materials can be categorized into resin, master bed, and art work. There are around 20 varieties for resin, 50 for master bed, and 100 for art work. Supply of materials is not considered a problem since they can be obtained easily from local or overseas suppliers. The company has about 10 active suppliers with three to eight weeks' lead time and 80–90% on-time delivery. The relationship with the customer is generally transactional in nature. However, the company collaborates with mold suppliers to ensure that new models are built according to the customer's requirements.

Customers have high bargaining power. Demand uncertainty is passed on to the company, where orders could fluctuate by up to 40% from the initial figures they submitted to the company. This is the major cause of uncertainty facing the company and triggers much of the production rescheduling. Some customers even change their orders one day before the delivery due date. The high flexibility needed with regard to customers' rescheduling or changing orders is possible because the company keeps high levels of inventory buffers and has relatively short supply lead times for most materials.

4.3. Case 3: company C

This company produces circuit breakers for several industry sectors, such as telecommunication, transportation, watercraft and special vehicles, and industrial equipment. The company operates two factories in East Java with around 250 employees. The majority of products are sold to four major customers, all of which are located overseas. Customers' geographic area influences product variety. The majority of the materials are supplied by one main supplier. The company produces up to 4500 product variants; all of them belong to nine major product types. Variety is created through different product attributes such as ampere, mounting, number of poles, thread neck design, hardware, types of terminal, push button, color of push button, and auxiliary contact.

A representative of the company stated that this type of industry has very few competitors, because they deal in a very special product which needs extremely high performance and accuracy. Difficulty in obtaining replaceable materials and alternative suppliers, coupled with uncertain supply, results in a very high inventory buffer of materials.

Production processes for one type of product are different from those for other types and the processing times vary significantly. Even variants within the same product type could have different processes. Some types of product have less than 20 variants, but some could have thousands.

Each product consists of 11–45 components, with about 10–20% common components across different finished products of the same type. To speed up production time, subassemblies are prepared before the company receives customer orders. However, the final assembly is normally done after the receipt of orders from customers. The number of machines will determine production capacity and each machine is dedicated to a single product type only. This creates a low level of production capacity flexibility. On the other hand, the company absorbs demand fluctuation by working overtime hours.

Demand is lumpy for all product variants. The company sets no constraint on minimum order quantity. The order quantity from customers could be very large for some product variants; for other variants, customers sometimes order very small quantities, and even only one unit. This lumpiness creates much idle capacity and poor on-time delivery. To avoid having a large idle capacity, the company produces some regular products whose demand is quite certain in anticipation of customer orders. However, it results in high inventory of finished goods. The company did not set time-fencing mechanism for orders. As a consequence, there is frequent production rescheduling. On the positive side, the changeover time for producing different product variants within the same type is very short.

4.4. Case 4: company D

Company D produces pharmaceutical products, operating three plants. All are located in the West Java Province of Indonesia, with about 400 workers. Every plant is dedicated to a certain type of product. Currently, they only serve the Indonesian market. The company supplies products to about 100 customers, but only three are considered major customers. There are four types of product – tablet, capsule, syrup, and dry syrup – with 64 product variants. The main product itself cannot be modified; however, the company usually changes the product package once a year.

Raw materials originate overseas but the import process is managed by a local distributor. Supply of materials from local distributors requires two to three days of lead time. It can take up to six months to receive special materials that do not have a local distributor. This company has around 30 suppliers for main materials and 28 for packaging items, many of which are considered as backup suppliers. The policy of the management is to have at least three suppliers for each type of material. Relationships with suppliers are more transactional than partnership in type. One supplier can supply several types of main materials for different finish goods.

Materials are categorized into four: main material, supporting material, primary packaging, and secondary packaging. In total, there are about 200 types of material where 50 are categorized as main materials, 100 are supporting materials, 9 are primary packaging items, and 64 are secondary packaging items. The product structure is generally very simple. A product consists of one to four main materials, one primary packaging item and one secondary packaging item. Product variations are created by different product weights, packaging, or product composition. Some materials can be used in more than one product variant within the same type.

There is high process similarity across different products. However, the batch quantity differs significantly. Production capacity for each product type is fixed, but there is a possibility to increase production volumes through the use of overtime hours. Even though the production lines are dedicated to certain product types, a changeover is needed when there is a change in product variants to be produced. The changeover time is about 2–3 h for cleaning and about half a day for changing a mold. The company responds to this long changeover time by freezing order and production schedule for one month.

In general, demand for most products is quite stable and little affected by segments and the geographical area of a market. On the other hand, there is uncertainty associated with supply of materials, mainly due to high competition to obtain them. In addition, the warehouse capacity is quite small, which limits the company's ability to do forward buying for materials. The company normally buys sufficient materials to satisfy the production schedule for a month, even though they can actually predict material requirements for three months. For materials that have long lead times, this creates a major material availability problem.

5. Cross-case comparisons and analysis

In this section, we will compare the uncertainty typology of each case company and the strategies adopted to achieve better supply chain flexibility.

5.1. *Typology of supply chain uncertainty*

Table 1 shows the comparisons between four companies in terms of uncertainty typologies. The four cases represent different types of operations: the first is a project-based or engineering-to-order company, and is followed by make-to-order, assembly-to-order, and make-to-stock companies. With the exception of Case 4, all companies are facing highly uncertain demand. Demand uncertainty may be associated with frequently revised orders

Table 1. Typology of supply chain uncertainty.

	Case 1	Case 2	Case 3	Case 4
Type of operations	Engineering to order	Make-to-order	Assembly-to-order	Make-to-stock
Level of uncertainty				
• Demand	Highly uncertain	Highly uncertain, customers frequently change orders, high mix uncertainty	Highly uncertain product mix, high product variety, demand fluctuating over time	Low demand uncertainty (orders received well in advance)
• Supply	Highly uncertain supply lead time	Low supply uncertainty, relatively short lead time	Highly uncertain supply lead time and material price	Low supply uncertainty for most materials
• Process	Moderately uncertain, very high complexity	Highly uncertain in terms of cycle time, yield, availability	Moderately uncertain: low process similarity, high variability in processing time	Stable process, low uncertainty

by customers (as in Case 2) or the inability of the company to forecast what products they are going to sell in the future (as in Cases 1 and 3).

From the supply side, Cases 1 and 3 are facing high supply uncertainty while the other two cases are fortunate to have low supply uncertainty. In Case 1, supply uncertainty is mostly due to the long lead times (4–6 months), but in Case 3, uncertainty is associated with lead time as well as the price. In both of these cases, most suppliers are located overseas, while for Cases 2 and 4 most materials are obtained from local suppliers.

In terms of operations processes, only Case 1 deals with highly complex processes; the other three involve simple production processes. On the other hand, process uncertainty in Case 2 is high where the company is unable to predict the cycle time accurately. In addition, yield and production system availability are uncertain. It is often the case that the production system fails and requires unplanned maintenance. In Case 3, process uncertainty is relatively lower, although processing time varies greatly depending on the types of product being produced.

From the typology of uncertainty of the four case companies, it is obvious that the company in Case 4 is operating under a stable and low level of uncertainty where not much flexibility is needed. The other three cases are facing different types of uncertainty and require flexibility in the supply chain.

5.2. Flexibility strategies adopted

The way in which a supply chain creates flexibility to handle demand uncertainty differs from case to case. Table 2 shows which strategies are adopted by each case company. Low, moderate, and high refer to the intensity of the strategies being used. The ratings are given subjectively by the authors after obtaining information from each case company. An 'X' sign means there is no clear evidence of the use of that strategy by the respective case company. For example, the company in Case 3 uses safety stock with a high level of intensity but there is no evidence of it using backup suppliers. From the list of strategies presented in the framework of Figure 1, all four cases are more focused on buffering rather than the proactive or redesign strategies. For example, safety stock and safety lead time are used by all four cases, although with differing levels of intensity. On the other hand, the proactive strategies are less intensively adopted to improve supply chain flexibility. Based on this fact, we arrive at a proposition:

Proposition 1: *The flexibility of companies tends to be of a more reactive than proactive nature.*

The use of buffers to increase flexibility has been discussed extensively. For example, Caputo (1996) pointed out that buffers could be considered alternative or complementary factors to technological flexibility. Any strategy to increase flexibility has cost implications. The choice of positioning a buffer as an alternative or complementary way of achieving flexibility depends on the cost benefits of the two solutions. However, as suggested by Christopher and Lee (2004), buffers are built up as a result of no confidence or low visibility across the supply chain. Manuj and Sahin (2011) also warned that the use of buffers, especially capacity buffers, could be very costly if not planned carefully.

Although all four cases are using some type of proactive strategy, they are concentrating on internal operations rather than working with supply chain partners. For example, all four cases seem to use some level of component commonality. Increasing component commonality is possible with changes in product and process design. The objective is to reduce the number of components a company has to manage while maintaining the ability to produce a large product variety. All four case companies are adopting component commonality as a strategy to achieve better flexibility. The use of common components, along with the postponement strategy, has long been considered an effective way of increasing flexibility. Gunasekaran and Ngai (2005) pointed out the commonality of components is an important strategy to achieve flexibility, especially in the context of a built-to-order supply chain. Likewise, manufacturing postponement has long been used as a hedging strategy against the proliferation of product variety (Feitzinger & Lee, 1997; Yang, Burns, & Backhouse, 2004).

Table 2. Flexibility strategies adopted.

	Case 1	Case 2	Case 3	Case 4
Type of operations	Engineering to order	Make-to-order	Assembly-to-order	Make-to-stock
Strategies/enablers				
Safety stock	Low	High	High	Moderate
Safety lead time	High	Low	Low	Low
Capacity buffer	Low	High	Low	X
Backup supplier	High	Low	X	High
Lead time reduction	X	X	X	Low
Setup time reduction	X	X	High	X
Component commonality	Moderate	High	Moderate	Low
Postponement	High	High	Moderate	X
Risk pooling	X	X	X	X
Flexible supply contract	X	X	X	X
Subcontracting/outsourcing	High	X	X	X
Alternative routing/modes of transportation	X	X	Low	X

Lead time reduction, flexible supply contracts, and subcontracting/outsourcing are not popular strategies among the four cases, although the company in Case 4 has been able to reduce supply lead time through the use of local suppliers. Obviously, when a company is undertaking any of these three strategies, the focus should be shifting from internal operations to supply chain collaboration. Supply lead time could not be effectively reduced without the involvement of suppliers. Likewise, flexible supply contracts would require the supplier to alleviate 'minimum order quantity' or to ensure security of supply in case of increasing demand and to enable speeding up delivery in case urgent deliveries are emerging. This then leads us to the following proposition:

Proposition 2: The flexibility of companies tends to focus more on internal operations rather than on collaborating with external parties.

As mentioned in the literature review, authors have recognized the importance of better coordination and collaboration to increase supply chain flexibility (Pujawan, 2004; Vokurka & O'Leary-Kelly, 2000). However, it should be noted here that Proposition 2 above suggests that efforts to increase supply chain flexibility have not been undertaken in accordance with supply chain principles. The move from manufacturing flexibility to supply chain flexibility should be interpreted as a need for collective efforts of supply chain members to create flexibility and, hence, responsiveness to end customers (Das, 2011; Kumar et al., 2008). Literature on supply chain flexibility suggests that reasons for the need for flexibility are both internal and external (Gong, 2008). Therefore, flexibility cannot be created without involving trading partners.

5.3. Flexibility attributes, problems, and objectivess affected

Table 3 presents the most obvious flexibility attributes, dominant problems that hinder flexibility and impacts of insufficient flexibility. In Case 1, as is usually the case for ETO companies, there is high flexibility in producing different products. However, the problem is mostly in the form of the supply of materials which have long lead times and the problem of cash availability that often results in delayed procurement of materials. In Case 2, the company attempts to provide very high flexibility to customers, so order revisions are allowed. With a long setup time, such volatile demand can only be satisfied with high inventory, both in the form of materials as well as finished products. In Case 3, the company also attempts to provide high flexibility to its customers, who may even order only a single unit. This is very well supported by the internal production system, where changeover time is very short on one side and there is moderate level of component commonality on the other side, but there is a big challenge from the supply side because of the long lead time. As a result, the company has to keep a high level of inventory buffers.

5.4. The impacts of power structure in supply chain relationships

Power structure in a supply chain seems to affect supply chain flexibility configuration. Although some studies have indicated that competition has shifted from company vs. company to supply chain vs. supply chain in the past two decades, each member in the chain is still exercising their power in the relationship (Crook & Combs, 2007). The power structure in a supply chain determines the parties that have to keep higher inventories, work with higher capacity flexibility and alleviate 'minimum order quantity'

Table 3. Dominant problem for each company that hinders supply chain flexibility.

	Most obvious flexibility attributes	Problems	Impacts
Case 1	• Ability to produce various types of products	• Problems with cash availability that prevents the company to obtain sufficient materials in due time	• Poor on-time delivery (only about 70%) • Idle resources because of waiting for materials
Case 2	• No frozen schedule, customers are allowed to change orders any time	• Long setup time • Frequent breakdown	• Large inventory buffer (up to 6 weeks demand) • Fluctuated capacity utilization
Case 3	• No limit on minimum order quantity, some customer orders have a quantity of one unit • Ability to produce a very large number of products variants	• There is only one very dominant supplier that limits supply flexibility • Long supply lead time	• Poor on-time delivery reliability to customers • High inventory of raw materials (about 3–4 months)
Case 4	• None looks obvious in Case 4	• Limited warehouse capacity that prevents forward buying for materials	On-time delivery of about 90% which is quite low for highly predictable demand Lost sales

restrictions. The party that holds more power may require its trading partners to share information but reciprocate (Mishra, Raghunathan, & Yue, 2007), and hence tends to pose more uncertainties for others.

Differences in power structure and flexibility configuration are also exhibited by the cases observed in this study. In Case 2, customers are demanding high flexibility for changing orders, but the case company is unable to pass this risk on to the suppliers. The uncertainty on the part of the customers has to be absorbed by the case company. This is worsened by the fact that the case company is facing long setup times and uncertain processes. Although the supply lead time is relatively short, there is a need for the company to keep high material inventories and maintain a high level of capacity buffers. Figure 2 illustrates the power structure and flexibility configuration of Case 2. The solid arrows connecting the company with the suppliers and the customers indicate the power direction. The single direction arrow from customers to the company indicates the dominating power of the customers, while the two-way directions between suppliers and the company indicates balanced power between the two. The dashed

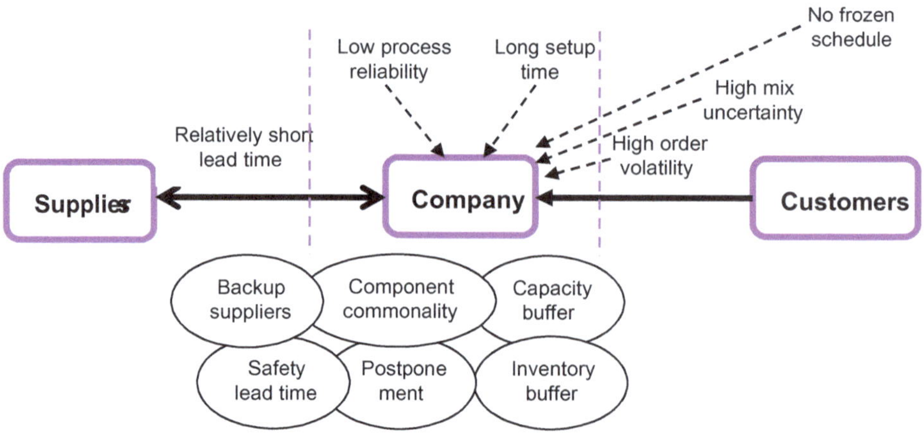

Figure 2. Illustration of power structure and flexibility configuration for Case 2.

arrows represent challenges that foster the need for better flexibility and the circles are strategies for achieving better flexibility.

The company in Case 3 seems to be positioned between dominant, inflexible suppliers and demanding customers. As a consequence, the company has to absorb uncertainties regarding inventory of materials. As presented in the two tables above, the company in Case 3 is facing customers that require high product variety and a very wide range of order quantities, from as small as one unit to as large as thousands of units. Facing this uncertainty, the case company has to be flexible in producing different lot sizes and different product types. The company has been responding to this need by reducing setup costs, postponing the assembly process until orders are received, and increasing component commonality. On the other hand, the company is very much dependent on one very large, powerful supplier with long lead times, and hence there is little choice but to keep high inventory buffers, add buffers to the lead time and switch the transportation mode to air freight in case urgent delivery is needed. Figure 3 illustrates the power structure and flexibility configuration of Case 3. Hence, we suggest the following proposition:

Proposition 3: *The power structure and typology of uncertainty in the supply chain will govern the type and configuration of supply chain flexibility.*

6. Concluding remarks

6.1. Summary

In this paper, we have presented a conceptual model that links the uncertainties, the two classifications of strategies for improving supply chain flexibility and the pertinent objectives. We classified the strategies into reactive or buffering and proactive or redesign strategies. Four case studies have been analysed to obtain insights into uncertainty typology and the strategies each case company is adopting to have better supply chain flexibility. Our case studies suggest that each company may be facing different uncertainty typology and choosing different flexibility strategies.

We suggest three main propositions that need further testing. First, companies tend to use reactive rather than proactive strategies. This is somewhat understandable as

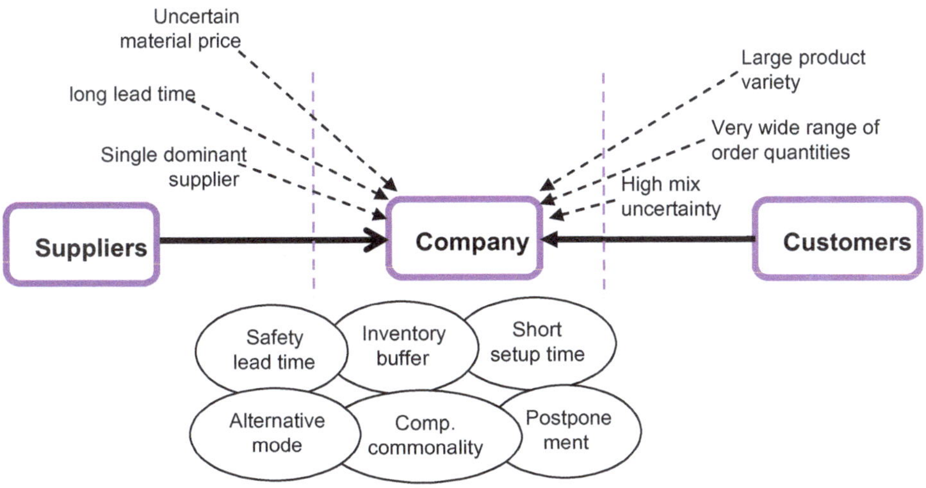

Figure 3. Illustration of power structure and flexibility configuration for Case 3.

reactive strategies are easier to carry out and require less time and resource investment, although it may not be better in the long term. We suspect that such a pattern is quite common in the manufacturing industry in general, and hence a follow-up study involving a larger number of companies is necessary. Second, companies seem to be more inward than outward-looking when there is a need for higher flexibility. There is not much evidence of collaborative improvement of flexibility across the supply chain network through, for example, flexible supply contracts, outsourcing, or supply lead time reduction. Companies tend to work more on their internal systems. However, this might be largely dependent on the power structure in a supply chain. A manufacturing company positioned between powerful customers and suppliers is normally unable to influence trading partners, and thus uncertainty has to be absorbed internally through safety stock, safety lead times, extra capacity, and other internal reactions.

6.2. Research implications

Obviously, there are a number of opportunities to extend this work. First, there is a need to do a larger scope study to confirm if the above three propositions hold true for a wider range of samples. In addition, uncertainty typology may affect the strategy applied. An interesting research question would be if any relationship exists between uncertainty typology and the flexibility strategies adopted by manufacturing companies. Finally, the strategies applied might be affected by common factors, leading to the following research question: would the choice between buffering and redesigning strategies be affected by any common factor?

Another avenue for future research is to find the balance between flexibility and schedule instability. Schedule instability is a subject that has been extensively discussed in the literature since as early as the 1970s (see e.g. an early paper by Steele, 1975) but remains a topic of ongoing research (see e.g. Pujawan & Smart, 2012 for a recent paper). A company offering high flexibility would not impose such policies as a long 'frozen schedule' and a large 'minimum order quantity'. However, being flexible with customers may also mean much disruption to the production schedule. Such disruptions are normally called schedule instability. Quantitative models or empirical research trying to find the balance between schedule instability and supply chain flexibility is an important topic for future research.

6.3. Managerial implications

Managers would benefit from our findings in a number of ways. First, the conceptual model can be used as a general guideline for achieving better supply chain flexibility when dealing with uncertainties. Managers could adopt various strategies to achieve supply chain flexibility, including four buffering strategies and eight redesign strategies. With this model, managers should be aware of various options, but the appropriateness of each strategy would depend on each specific situation.

Second, it should be quite obvious that too much focus on buffering strategies may be counterproductive in the long term. Buffering strategies are normally easier to execute and require less effort and risk. The move toward more proactive or redesigned strategies needs not only time and money, but also leadership that focuses on long-term benefits and willingness to take up short-term risks. Managers should be motivated to initiate a more long-term focus on flexibility rather than reactively responding to uncertainties with buffering strategies.

Finally, managers need to be motivated to explore possibilities to work with external parties to create better supply chain flexibility. The ideas behind supply chain management that fosters better collaboration with supply chain partners are much easier to say than to do. It is likely that most supply chain managers understand the importance of having close relationships with trading partners, but the reality, at least from our case study, suggests that such close collaboration is not easy to operate. Managers are more internal rather than external-looking. Potentials to improve flexibility together as a chain are little realized. This again implies the need to explore possibilities to work more closely with trading partners to improve supply chain flexibility.

References

Alfredsson, P., & Verrijdt, J. (1999). Modeling emergency supply flexibility in a two-echelon inventory system. *Management Science, 45*, 1416–1431.

Avittathur, B., & Swamidass, P. (2007). Matching plant flexibility and supplier flexibility: Lessons from small suppliers of US manufacturing plants in India. *Journal of Operations Management, 25*, 717–735.

Barad, M., & Sapir, D. E. (2003). Flexibility in logistic systems – Modeling and performance evaluation. *International Journal of Production Economics, 85*, 155–170.

Bernardo, J. J., & Mohamed, Z. (1992). The measurement and use of operational flexibility in the loading of flexible manufacturing systems. *European Journal of Operational Research, 60*, 144–155.

Cao, Q., Patterson, J. W., & Bai, X. (2005). Reexamination of processing time uncertainty. *European Journal of Operational Research, 164*, 185–194.

Caputo, M. (1996). Uncertainty, flexibility and buffers in the management of the firm operating system. *Production Planning & Control, 7*, 518–528.

Christopher, M., & Lee, H. (2004). Mitigating supply chain risk through improved confidence. *International Journal of Physical Distribution & Logistics Management, 34*, 388–396.

Cox, T., Jr. (1989). Toward the measurement of manufacturing flexibility. *Production and Inventory Management Journal, 30*, 68–72.

Crook, T. R., & Combs, J. G. (2007). Sources and consequences of bargaining power in supply chains. *Journal of Operations Management, 25*, 546–555.

Das, K. (2011). Integrating effective flexibility measures into a strategic supply chain planning model. *European Journal of Operational Research, 211*, 170–183.
Duclos, L. K., Vokurka, R. J., & Lummus, R. R. (2003). A conceptual model of supply chain flexibility. *Industrial Management & Data Systems, 103*, 446–456.
Fantazy, K. A., Kumar, V., & Kumar, U. (2009). An empirical study of the relationships among strategy, flexibility, and performance in the supply chain context. *Supply Chain Management: An International Journal, 14*, 177–188.
Feitzinger, E., & Lee, H. L. (1997). Mass customization at Hewlett-Packard: The power of postponement. *Harvard Business Review, 75*, 116–123.
Gong, Z. (2008). An economic evaluation model of supply chain flexibility. *European Journal of Operational Research, 184*, 745–758.
Gunasekaran, A., & Ngai, E. W. (2005). Build-to-order supply chain management: A literature review and framework for development. *Journal of Operations Management, 23*, 423–451.
Gupta, Y. P., & Goyal, S. (1989). Flexibility of manufacturing systems: Concepts and measurements. *European Journal of Operational Research, 43*, 119–135.
Gupta, Y. P., & Somers, T. M. (1992). The measurement of manufacturing flexibility. *European Journal of Operational Research, 60*, 166–182.
Gurnani, H., & Gerchak, Y. (2007). Coordination in decentralized assembly systems with uncertain component yields. *European Journal of Operational Research, 176*, 1559–1576.
Ho, C. (1989). Evaluating the impact of operating environments on MRP system nervousness. *International Journal of Production Research, 27*, 1115–1135.
Kähkönen, A. K. (2011). Conducting a case study in supply management. *Operations and Supply Chain Management, 4*, 31–41.
Koste, L. L., & Malhotra, M. K. (1999). A theoretical framework for analyzing the dimensions of manufacturing flexibility. *Journal of Operations Management, 18*, 75–93.
Kumar, V., Fantazy, K. A., Kumar, U., & Boyle, T. A. (2006). Implementation and management framework for supply chain flexibility. *Journal of Enterprise Information Management, 19*, 303–319.
Kumar, P., Shankar, R., & Yadav, S. S. (2008). Flexibility in global supply chain: Modeling the enablers. *Journal of Modelling in Management, 3*, 277–297.
Li, J., Sikora, R., Shaw, M. J., & Tan, G. W. (2006). A strategic analysis of inter organizational information sharing. *Decision Support Systems, 42*, 251–266.
Lummus, R. R., Duclos, L. K., & Vokurka, R. J. (2003). Supply chain flexibility: Building a new model. *Global Journal of Flexible Systems Management, 4*(4), 1–13.
Manuj, I., & Sahin, F. (2011). A model of supply chain and supply chain decision-making complexity. *International Journal of Physical Distribution & Logistics Management, 41*, 511–549.
Merschmann, U., & Thonemann, U. W. (2011). Supply chain flexibility, uncertainty and firm performance: An empirical analysis of german manufacturing firms. *International Journal of Production Economics, 130*, 43–53.
Mishra, B. K., Raghunathan, S., & Yue, X. (2007). Information sharing in supply chains: Incentives for information distortion. *IIE Transactions, 39*, 863–877.
Mohebbi, E., & Choobineh, F. (2005). The impact of component commonality in an assemble-to-order environment under supply and demand uncertainty. *Omega, 33*, 472–482.
Morlok, E. K., & Chang, D. J. (2004). Measuring capacity flexibility of a transportation system. *Transportation Research Part A, 38*, 405–420.
Murthy, D. N. P., & Ma, L. (1991). MRP with uncertainty: A review and some extensions. *International Journal of Production Economics, 25*, 51–64.
Ndubisi, N. O., Jantan, M., Hing, L. C., & Ayub, M. S. (2005). Supplier selection and management strategies and manufacturing flexibility. *Journal of Enterprise Information Management, 18*, 330–349.
Osman, H., & Demirli, K. (2012). Integrated safety stock optimization for multiple sourced stockpoints facing variable demand and lead time. *International Journal of Production Economics, 135*, 299–307.
Pramod, M., & Garg, S. (2006). Analysis of flexibility requirements under uncertain environments. *Journal of Modelling in Management, 1*, 196–214.
Pujawan, I. N. (2004). Assessing supply chain flexibility: A conceptual framework and case study. *International Journal of Integrated Supply Management, 1*, 79–97.

Pujawan, I. N., & Smart, A. U. (2012). Factors affecting schedule instability in manufacturing companies. *International Journal of Production Research, 50*, 2252–2266.
Ramasesh, R. V. (1991). Procurement under uncertain supply lead times – A dual-sourcing technique could save costs. *Engineering Costs and Production Economics, 21*, 59–68.
Raturi, A. S., & Jack, E. P. (2004). Creating a volume-flexible firm. *Business Horizons, 47*, 69–78.
Sánchez, A. M., & Pérez, M. P. (2005). Supply chain flexibility and firm performance: A conceptual model and empirical study in the automotive industry. *International Journal of Operations & Production Management, 25*, 681–700.
Schmitt, T. G. (1984). Resolving uncertainty in manufacturing systems. *Journal of Operations Management, 4*, 331–345.
Schmitt, A. J., & Snyder, L. V. (2012). Infinite-horizon models for inventory control under yield uncertainty and disruptions. *Computers & Operations Research, 39*, 850–862.
Seebacher, G., & Winkler, H. (2013). A citation analysis of the research on manufacturing and supply chain flexibility. *International Journal of Production Research, 51*, 3415–3427.
Sethi, A. K., & Sethi, S. P. (1990). Flexibility in manufacturing: A survey. *The International Journal of Flexible Manufacturing Systems, 2*, 289–328.
Shin, H., Collier, D. A., & Wilson, D. D. (2000). Supply management orientation and supplier/buyer performance. *Journal of Operations Management, 18*, 317–333.
Slack, N. (1988). Manufacturing systems flexibility – An assessment procedure. *Computer Integrated Manufacturing Systems, 1*, 25–31.
Steele, D. C. (1975). The nervous MRP system: How to do battle. *Production and Inventory Management, 16*, 83–89.
Stevenson, M., & Spring, M. (2007). Flexibility from a supply chain perspective: Definition and review. *International Journal of Operations & Production Management, 27*, 685–713.
Stevenson, M., & Spring, M. (2009). Supply chain flexibility: An inter-firm empirical study. *International Journal of Operations & Production Management, 29*, 946–971.
Tachizawa, E. M. (2007). *Uncertainty, integration and supply flexibility* (Unpublished dissertation). Universitat Pompeu Fabra, Barcelona, Spain.
Tachizawa, E. M., & Gimenez, C. (2010). Supply flexibility strategies in Spanish firms: Results from a survey. *International Journal of Production Economics, 124*, 214–224.
Tachizawa, E. M., & Thomsen, C. G. (2007). Drivers and sources of supply flexibility: An exploratory study. *International Journal of Operations & Production Management, 27*, 1115–1136.
Tang, C. S. (2006). Perspectives in supply chain risk management. *International Journal of Production Economics, 103*, 451–488.
Tang, C., & Tomlin, B. (2008). The power of flexibility for mitigating supply chain risks. *International Journal of Production Economics, 116*, 12–27.
Thomas, D. J., & Tyworth, J. E. (2006). Pooling lead-time risk by order splitting: A critical review. *Transportation Research Part E: Logistics and Transportation Review, 42*, 245–257.
Van Donk, D. P., & van der Vaart, T. (2005). A case of shared resources, uncertainty and supply chain integration in the process industry. *International Journal of Production Economics, 96*, 97–108.
Van Hoek, R. I. (2001). The rediscovery of postponement a literature review and directions for research. *Journal of Operations Management, 19*, 161–184.
Van Kampen, T. J., van Donk, D. P., & van der Zee, D.-J. (2010). Safety stock or safety lead time: Coping with unreliability in demand and supply. *International Journal of Production Research, 48*, 7463–7481.
Vickery, S. N., Calantone, R., & Dröge, C. (1999). Supply chain flexibility: An empirical study. *The Journal of Supply Chain Management, 35*, 16–24.
Vokurka, R. J., & O'Leary-Kelly, S. W. (2000). A review of empirical research on manufacturing flexibility. *Journal of Operations Management, 18*, 485–501.
Vonderembse, M., Uppal, M., Huang, S., & Dismukes, J. (2006). Designing supply chains: Towards theory development. *International Journal of Production Economics, 100*, 223–238.
Wong, C. Y., Boon-Itt, S., & Wong, C. W. Y. (2011). The contingency effects of environmental uncertainty on the relationship between supply chain integration and operational performance. *Journal of Operations Management, 29*, 604–615.
Wu, T., Blackhurst, J., & O'grady, P. (2007). Methodology for supply chain disruption analysis. *International Journal of Production Research, 45*, 1665–1682.

Wu, I. L., & Shen, Y. C. (2006). A model for exploring the impact of purchasing strategies on user requirements determination of e-SRM. *Information & Management, 43*, 411–422.

Yang, B., Burns, N. D., & Backhouse, C. J. (2004). Management of uncertainty through postponement. *International Journal of Production Research, 42*, 1049–1064.

Yi, C. Y., Ngai, E. W. T., & Moon, K.-L. (2011). Supply chain flexibility in an uncertain environment: Exploratory findings from five case studies. *Supply Chain Management: An International Journal, 16*, 271–283.

Zhang, G., Shang, J., & Li, W. (2011). Collaborative production planning of supply chain under price and demand uncertainty. *European Journal of Operational Research, 215*, 590–603.

Zhang, Q., Vonderembse, M. A., & Lim, J.-S. (2006). Spanning flexibility: Supply chain information dissemination drives strategy development and customer satisfaction. *Supply Chain Management: An International Journal, 11*, 390–399.

3

An investigation on the roll force and torque fluctuations during hot strip rolling process

Mahdi Bagheripoor* and Hosein Bisadi

Department of Mechanical Engineering, Iran University of Science and Technology, Tehran, Iran

Accurate prediction of the roll separating force and roll torque is critical to assuring the quality of the product. Fluctuation of these parameters during the rolling process is even more important because of its effects on rolling setups, geometrical accuracy (especially flatness in strip rolling), and the uniformity of mechanical and microstructural properties of the rolled material. The main approach of the present study is a precision analysis of the roll force and roll torque and their instabilities during the rolling process. In doing so, a coupling multi-variable simulation model for hot strip rolling was built using non-linear thermo-viscoplastic finite element method. Fluctuations of roll force and torque about the steady-state operating point are derived and the effects of main process parameters such as rolling speed and strip reduction on these instabilities are investigated. To check the validity of the employed model, predicted results are compared with experimental data.

Keywords: hot rolling process; finite element method; roll force; roll torque; aluminum alloys

1. Introduction

Considering the market requirements for flat-rolled aluminum products, an increasing demand can be observed for tight dimension tolerance, especially for thickness distribution along the strip length and profile (thickness distribution along the strip width). At the same time, the increased running speed brings out possible vibration problems in the rolling process, especially in a cold strip mill. If the thickness variation of the hot-rolled aluminum strip can be reduced, it will be possible to increase the production speed of the cold strip mill. These claims set new demands on the acceptable rolling forces and torques which, together with the roll pass design, have a significant influence on gage variations, the shape of the product, and power consumption.

Important pieces of information such as separating force, required torque, their time history and variations during the rolling process, and causes of these variations can be utilized for developing and optimizing pass schedules and providing setup data to the mills renting powerful process tools to satisfy the requirement of product development.

Throughout this century, much research has been devoted to the roll force and torque modeling and prediction using different techniques. Conventional models which are mostly based on thermo-mechanical theory were aimed at relating the roll force, roll torque, and power or energy consumption to the rolling parameters such as the roll diameter, initial temperature of the strip, strip thickness, reduction ratio, number of

*Corresponding author. Email: m_bagheripoor@mecheng.iust.ac.ir (M. Bagheripoor)

passes, speed of rolling, and interface friction (Ford & Alexander, 1964; Sims, 1954). Arnold and Whitton in 1975 proposed a formula for roll-separating force based on Sims' (1954) hot flat rolling theory, which included modifications for projected area of contact and empirical factors (Biswas, 2003). Joun and Hwang (1992) presented a new method of roll force prediction which was based on approximate solutions to the velocity, strain rate and stress distributions in the roll gap. Said, Lenard, Ragab, and Elkhier (1999) determined the roll separating forces and torques via low carbon steel rolling as a function of the area reduction at entry temperatures of 900, 950, 1000, 1050, and 1100 °C. They examined four empirical, mathematical models of the process for their ability to predict the roll separating forces. Yanagimoto, Morimoto, Kurahashi, and Chikushi (2002) proposed a mathematical model for prediction of rolling force and microstructure evolution in hot strip rolling (HSR), by Orowan's theory, FDM analysis for temperature, and incremental theory for the evolution of microstructure. Bayoumil (2007) developed a kinematic analytical approach for the prediction of rolling force, rolling torque, and forward slip in the hot strip continuous rolling, based on formulating a velocity field in the roll bite zone that expresses the effect of interfacial friction on the distribution of axial velocity and longitudinal stresses across the strip thickness. Chen, Zhang, Sun, Wang, and Song (2012) developed an online rolling force model for tandem cold rolling mill by numerical integration method, whereby the plastic deformation zone was divided into small units perpendicular to the motion of the strip.

The use of the finite element (FE) technique to predict roll forces and torque has been a popular approach for many researchers, and many commercial FE software packages now exist. Kim, Lee, Shin, and Shivpuri (1992) investigated shape rolling process using computer software, TASKS, based on the finite slab-element method and their predicted roll force and torque indicated reasonable agreement with other published results. Combining the FE and boundary element methods, Shangwu, Rodrigues, and Martins (1999) carried out the three-dimensional (3-D) modeling of hot rolling process of flat strips. They predicted rolling force, rolling torque, and contact pressure on the roll for both rigid and flexible roll cases. Kwak, Lee, Hwang, and Kim (2002) used FE simulation to investigate the effect of HSR process variables on some selected non-dimensional parameters characterizing the thermo-mechanical behavior of the strip and based on those parameters derived an online model for the precise prediction of roll force and roll power. Duan and Sheppard (2002) simulated the hot flat rolling of aluminum alloy 3003 by the commercial FEM program FORGE3. They adopted an inverse analysis method to match the calculated rolling force and torque with the measured rolling force and torque by treating the friction law and the friction coefficient as free parameters. A large-deformation constitutive model applicable to the calculation of roll force and torque in heavy-reduction rolling was presented by Byon, Kim, and Lee (2004). 3-D FE analysis coupled with the proposed constitutive models has been carried out to calculate work-piece deformation and the rolling force in their investigation. An approximate model for predicting roll force and torque in plate rolling was proposed by Moon and Lee (2008). Peening effect which unavoidably occurs in plate rolling due to a small ratio of work roll radius over mean thickness of slab being deformed has been considered in their model. Wang, Peng, Xu, and Liu (2010) investigated comprehensive influences of normal stress and shear stress in longitudinal, transverse direction, and altitude to improve the accuracy of the calculation of rolling force. In their study, the rolling force was solved in 3-D finite differential method. A new model for the prediction of the roll force and tension profiles was presented by Kim, Kwak, Shin, and Hwang (2010). Their approach was based on an approximate 3-D theory of plasticity, and considered the effect of the pre-deformation of the strip. They used a rigid-plastic FE model to examine the prediction accuracy of the proposed model. Byon, Na, and Lee (2013) coupled FE method with a modified constitutive model in which the flow stress equation has been modified according to the working temperature and strain rate at four stands (passes) of the finishing block mill of an actual rod mill, and showed that their model is more precise in prediction of roll force in continuous rod rolling where strain rates are in the range of $100-400\ s^{-1}$ and temperature is in the range of 900−1050 °C.

Recently, some researchers have started exploring the possibility of utilizing soft computing techniques (particularly neural network and fuzzy set theory) for investigating the roll force and torque in the hot rolling process. As an improvement, Poliak, Shim, Kim, and Choo (1998) examined a proposed linear regression model to predict the roll force in hot plate rolling. Lee and Lee (2002) proposed a long-term learning method using neural network to improve the accuracy of rolling-force prediction in hot rolling mill. They combined neural network method with the conventional learning algorithm in the pre-calculation stage to reduce the thickness error at the head-end part of the strip. Neural networks were used by Yang, Linkens, Talamantes-Silva, and Howard (2003) to develop a rolling force and torque prediction model, without the requirement of a physically based or an empirical model. Lee and Choi (2004) adopted an online adaptable network for the rolling force setup and according to their field test results, they deduced that the prediction ability has been improved about 30% by the proposed method than the conventional method (mathematical models). Moussaoui, Selaimia, and Abbassi (2006) discussed the combination of an artificial neural network with analytical models to improve the performance of the prediction model of finishing rolling force in HSR process. Gudur and Dixit (2008) used a radial basis function neural network for prediction of velocity field and neutral point location in the cold flat rolling process that were further refined and post-processed in the FE code to calculate required parameters such as roll force, torque, and equivalent strain and thus provided highly accurate results with less computational time. The method of fuzzy control modification and Elman dynamic recursive network-combined prediction was used by Jia, Shan, and Niu (2008), to predict the rolling force for a tandem cold mill. A multiple radial basis function neural network model to predict rolling force based on wavelet analysis was proposed by Liu, Tong, and Lin (2012). In their model, the multi-resolution wavelet analysis method has been employed to separate the rolling force signal into several sub-signals corresponding to different factors.

In summary, many techniques have been utilized for predicting the roll separating force and torque in the hot rolling process. However, almost none of the above-mentioned efforts investigated the fluctuations of roll force and torque during the process and they have generally focused on the prediction of the mean roll force and torque in the steady-state region of each pass. The roll force and roll torque are the main parameters in controlling the flatness and geometrical accuracy, yield strength and the hardness of the product, the mill stretch and vibrations of the rolling mill, and roll flattening (Lenard, 2002). Therefore, it is worth to make a more detailed investigation on the variation of these parameters and consider their changes in the steady-state region of rolling process. The present study is aimed at a precision analysis of the roll force and torque during the rolling process and understanding the effect of process condition on the instabilities of these parameters. In doing so, the FE approach has been utilized to model HSR of aluminum alloys. The model is capable of considering temperature-dependent thermo-physical properties of rolling material and 3-D rolling conditions as it occurs during HSR. To verify the validity of the analysis, the predicted results are compared with the experimental data.

2. Mathematical model

2.1. *Deformation in the roll gap*

The thermo-viscoplastic FEM based on the flow formulation of the penalized form of the incompressibility was used to analyze the HSR process. According to the variational principle, the basic equation for the FE formulation is expressed as Equation (1) (Kobayashi, Oh, & Altan, 1989):

$$\int_V \bar{\sigma}\delta\dot{\bar{\varepsilon}}\mathrm{d}V + \int_V K\ \dot{\varepsilon}_v\delta\dot{\varepsilon}_v\mathrm{d}V - \int_S F_i\delta u_i\mathrm{d}s = 0 \qquad (1)$$

Here, $\bar{\sigma}$ is the effective stress which is a function of strain, strain rate and temperature, $\dot{\bar{\varepsilon}}$ the effective strain rate, F_i the surface tractions, u_i the velocity field, V the volume, S the boundary surface of subdomain, K the penalty constant, and $\dot{\varepsilon}_v$ is the volumetric stain rate.

The metal starts flowing or deforming plastically when the applied stress reaches the value of yield stress. For investigating the metal plastic deformation behavior, the material yield stress under uniaxial conditions, as a function of strain, strain rate, and temperature, should also be considered as flow stress. This was done using a hyperbolic sine equation, shown in Equation (2), which relates the steady-state flow stress of the material to the strain rate and temperature under which it is deformed (Brown, Kim, & Anand, 1989):

$$\dot{\varepsilon} = A\exp\left(\frac{-Q_{\mathrm{def}}}{R\theta}\right)\left[\sin h\left(\xi\frac{\bar{\sigma}}{s}\right)\right]^{\frac{1}{m}} \qquad (2)$$

where A, ξ, and m are material constants, $\dot{\varepsilon}$ is the mean equivalent plastic tensile strain rate, $\bar{\sigma}$ the equivalent flow stress, Q_{def} the activation energy for deformation, R the universal gas constant, θ the absolute temperature, and s is a scalar internal variable with dimensions of stress, called the deformation resistance. The coefficients of the hyperbolic sine equation for AA1100 deformation are summarized in Table 1 (Brown et al., 1989).

The flow behavior in the metal interior follows Equation (1). However, the effect of contact friction on metal flow should be considered when the interaction among metal surfaces exists. Contact friction between strip and roll is solved using Coulomb model, which is used mostly for forming simulations. Interfacial friction for the contact area is proportional to the normal force as shown in Equation (3):

$$\tau_{\mathrm{crit}} = \mu p \qquad (3)$$

where τ_{crit} is the critical shear stress, μ the coefficient of friction, and P is the contact pressure. The friction coefficient μ varies in wide ranges due to different rolling conditions and different options of the authors.

2.2. *Heat transfer in the roll gap*

Hot rolling is a thermo-mechanical process analyzed by solving the heat conduction equation coupled to a deformation analysis. During the rolling process, the temperature

Table 1. Summary of hyperbolic sine constants for AA1100 (Brown et al., 1989).

Material	A	Q (kJ/mol)	ξ	m
AA1100	1.91E+07	175.3	7	.23348

distribution in the strip can be calculated using the governing partial differential equation shown in Equation (4) (Wang, Yang, & He, 2008):

$$\frac{\partial}{\partial x}\left(k\frac{\partial T}{\partial x}\right)+\frac{\partial}{\partial y}\left(k\frac{\partial T}{\partial y}\right)+\frac{\partial}{\partial z}\left(k\frac{\partial T}{\partial z}\right)+\dot{Q}=\rho c\frac{\partial T}{\partial t} \tag{4}$$

where ρ is the density of the rolled strip, c the specific heat of the rolled metal, k the thermal conductivity of the rolled strip, and $\dot{Q}$ represents the volumetric rate of heat generation arising from the deformation. The heat generation term $\dot{Q}$ is calculated using Equation (5):

$$\dot{Q}=\eta\sigma\dot{\varepsilon} \tag{5}$$

where η is the fraction of plastic work converted into heat. A distributed surface flux, q_{fric} is also generated from frictional sliding and rises rapidly near the entry and exit regions along the arc of contact inducing a dramatic change in the relative slip. Its overall contribution to the thermal balance in the hot rolling process is low, but if it is considered, its value is determined as follows (Shahani, Setayeshi, Nodamaie, Asadi, & Rezaie, 2009):

$$q_{\text{fric}}=|\tau\upsilon| \tag{6}$$

where τ is the shear stress and υ is the sliding velocity. Since the heat lost from the strip is gained by the work roll in the roll gap, a simultaneous solution of the governing equations of both the strip and the work roll is required. To do so, the heat transfer in the work roll can be written as (Serajzadeh, Karimi Taheri, & Mucciardi, 2002):

$$\frac{1}{r}\frac{\partial}{\partial r}\left(k_r\frac{\partial T}{\partial r}\right)+\frac{\partial}{\partial z}\left(k_r\frac{\partial T}{\partial z}\right)=\rho_r c_r\frac{\partial T}{\partial t} \tag{7}$$

where ρ_r, c_r, and k_r are the density, the specific heat, and the thermal conductivity of the work roll, respectively. At the contact interface between the strip and the work roll, an interfacial heat transfer coefficient is assumed. The boundary conditions for heat transfer between the strip and work roll and surrounding temperature are applied as the equations employed by Kumar, Samarasekera, and Hawbolt (1992), and the interface heat transfer coefficient is determined based on the experimental work that has been conducted by Chun and Lenard (1997).

2.3. FE modeling

As it is observed there is a complex interconnection between governing equations. In order to solve these equations, the simulation was carried out by a 3-D coupled thermo-mechanical FE analysis using commercial FE software package Abaqus/Explicit. Combining the merits of both Lagrangian and Eulerian formulations, ALE formulation was developed to handle mesh distortion, mesh entanglement, and special boundary condition changes in hot rolling process. Employing symmetry along the center line, a quarter of the strip was considered in the model. A sensitivity analysis was employed to determine the effect of changing the mesh density on the predicted model results. The work roll geometry is limited to a 90° section with a thickness of 5 mm. In both the strip and work roll, eight-node isoparametric brick elements were employed. The chosen elements were suitable for coupled thermal-stress analysis. The steel work roll was defined as an elastic material with a Young's modulus of 200 GPa. The large differences in elastic modulus between the work roll and the strip causes the work roll to behave as a virtually rigid material. The thermo-physical properties of the strip and the steel work roll are taken from Biswas and Mandal (2011) and Shahani et al. (2009), respectively.

3. Results and discussion

3.1. Roll force and torque variations during the process

In this study, the experimental rolling programs of Hum, Colquhoun, and Lenard (1996) are used to investigate the rolling force and torque changes during HSR of commercial pure aluminum. Typical contact pressure distribution in the roll gap is given in Figure 1, in which an AA1100 strip with 6.28 mm thickness and 50 mm width is rolled at 500 °C with a 250 mm diameter work roll. The amounts of reduction and rolling speed are 30.4% and 80 rpm, respectively, and only one-quarter of the strip is considered in the model (as detailed in previous section). It can be seen from this figure that the distribution of the roll pressure is quite uniform after rising from zero rapidly following which a moderate rise is indicated until the peak at the neutral point, after which the pressure drops gradually. As Jing (2001) describes, besides work-hardening and temperature drop, the additional constraint caused by the friction forces (so called resistance to flow in rolling) is responsible for pressure build-up near the natural point. This type of pressure distribution is often referred to as a 'friction hill' in which the more homogeneous deformation pattern is expected (Dvorkin et al., 1997). According to the Li and Kobayashi (1982) findings, the amount of reduction, work roll radius, and the strip entry thickness are the most important parameters that shape the rolling pressure distribution.

Having the pressure distribution, the separating force as well as the acting torque can be calculated by integrating over the area of contact. But it is difficult to find the actual pressure distribution theoretically and working with approximate values as the average pressure will not lead to a precise analysis of rolling force and torque. Accurate

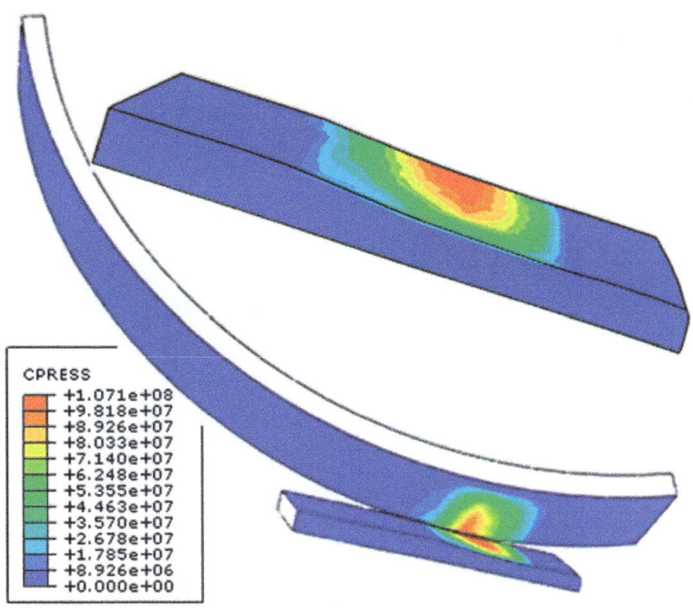

Figure 1. Typical contact pressure distribution in the roll gap.

information about the rolling force is essential when designing a rolling pass schedule. The rolling force directly determines the thickness precision of rolled products. The rolling force is also the basis for computing the rolling torque. The accurate prediction of both of these parameters is imperative if mill breakdowns are to be avoided and to ensure maximum productivity in terms of geometric and property requirements. The predicted rolling force for the mentioned rolling program, as the material traverses the roll gap, is shown in Figure 2. As expected the rolling force increases gradually when the strip feeds into the mill gap and reaches a relatively steady value when the deformation is in the steady-state regime. Since only a quarter of strip is modeled, the actual rolling force is twice the average value in the steady-state deformation zone. As it shown in Figure 2, the rolling force fluctuates in the steady-state region and it varies between 21 and 27 kN.

Perhaps, the most important feature when designing the pass schedule is the calculation of energy requirement. An underpowered unit is the most obvious risk since this will lead to a reduction in productivity due to stalling. Thus, we must ensure that the motor is more than adequate for any immediate or future workloads. In the present

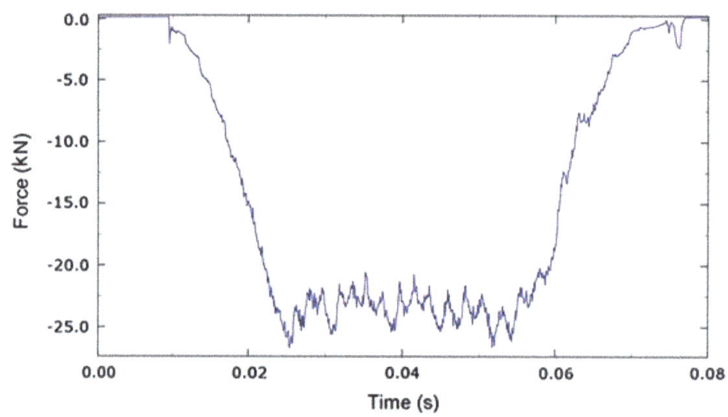

Figure 2. Rolling force variations during the time.

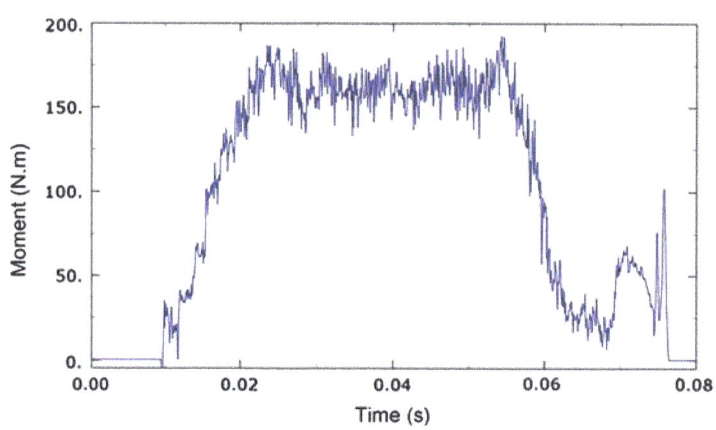

Figure 3. Rolling torque variations during the time.

work, the rolling torque is investigated vs. different condition of process. The predicted rolling torque for described rolling condition is shown in Figure 3. As the rolling force, the rolling torque reaches a relatively steady value when the deformation is in the steady-state regime. But as it can be seen from Figure 3, the values of rolling torque fluctuate in the steady-state region with a high frequency.

The plots of Figures 2 and 3 indicate that the force and torque time history (in the steady-state region) show relatively large fluctuations about the average value. Such fluctuations of rolling force around the steady-state point cause variations in roll gap profile which may lead to strip flatness problems without a work roll bending system (Hu et al., 2006; Wang, Zhang, Xu, & Zhang, 2011). And the more fluctuations in roll torque leads to the shorter service life of rolling mill components (Pyun et al., 2013). Significant amplitude of the roll force and roll torque fluctuations (than the steady-state point value) can also cause mill vibrations and inhomogeneity in mechanical and microstructural properties of the rolled material. In follows, we will try to consider the effects of main process parameters on the rolling force and torque predictions. It should be noted that in order for the results not to be mesh-dependent, the mesh of the strip was kept the same in all models.

3.2. Effects of thickness reduction

Three real rolling pass schedules that are listed in Table 2 are chosen for considering the effects of strip thickness reduction (Hum et al., 1996). A 50 mm width strip of commercial pure aluminum (AA1100) is rolled with a 250 mm diameter work roll in all passes. Figure 4(a) and (b) compared rolling force and torque variations during the time for different thickness reduction amounts. As it is observed form Figure 4(a), the amplitude of rolling force fluctuations in the steady-state region increases with increasing thickness reduction and it varies about 7.3, 8.7, and 11.1% around the average value for 21.36, 30.4, and 39.05% thickness reduction amounts, respectively. Similar results can be deduced from Figure 4(b) and it is seen that rolling torque fluctuations are increased in higher thickness reductions (the amplitude of torque variations varies about 12.5, 13.7, and 14.2% about average value for 21.36, 30.4, and 39.05% reduction amounts, respectively). The most common effect of these phenomena is that there is more thickness variations during the rolled strip length in the higher thickness reduction amounts. It can be caused by surface condition changes such as interfacial friction in the contact area that affect the contact shear stress distribution and stick slip conditions. According to Hum et al. (1996) observations, the amounts of forward slip and friction coefficient rise with increase in thickness reduction.

Figure 5 compares the predicted and published experimental measurements (Hum et al., 1996) of rolling force and torque. Since only a quarter of strip is modeled, the actual rolling forces and torques are twice the average values in the steady-state regions. As it is observed, the predicted values are all lower than the actual ones. However, the relative error for all samples is less than 13.5%, which is known as a good accuracy in

Table 2. Chosen pass schedules for investigation of thickness reduction effects.

Pass no.	Inlet thick. (mm)	Thick. red. (%)	Roll speed (rpm)	Inlet strip temp (°C)
1	6.32	21.36	80	500
2	6.28	30.4	80	500
3	6.3	39.05	80	500

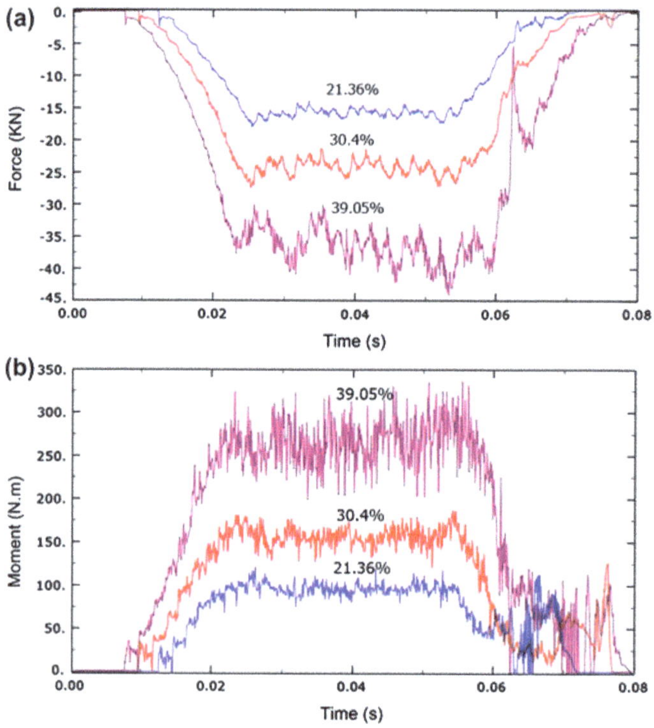

Figure 4. The effect of thickness reduction amount on the variations of (a) roll force and (b) roll torque.

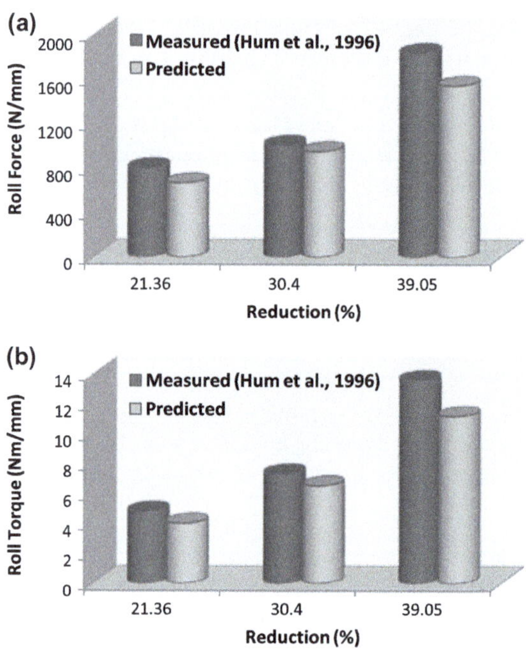

Figure 5. Comparison of: (a) roll force and (b) roll torque predictions with experimental data at different reduction amounts.

metal forming theories. As expected, rolling forces and the torques clearly increase with increasing reduction. There are several mechanisms contributing to this finding, all of which are well known. Increasing the volume of plastic deformation, and hence more mechanical work is the dominant phenomena. The higher reduction also causes the more strain hardening and affects the continuously changing strip temperature distribution and hence, the metal flow behavior during the rolling.

3.3. Effects of rolling speed

Rolling conditions that are summarized in Table 3 are chosen to consider the effects of rolling speed. Roll force and torque variations during the time are shown in Figure 6(a)

Table 3. Chosen pass schedules for investigation of rolling speed effects.

Pass no.	Inlet thick. (mm)	Thick. red. (%)	Roll speed (rpm)	Inlet strip temp (°C)
4	6.29	30.68	20	500
5	6.28	30.41	60	500
6	6.29	31.48	100	500

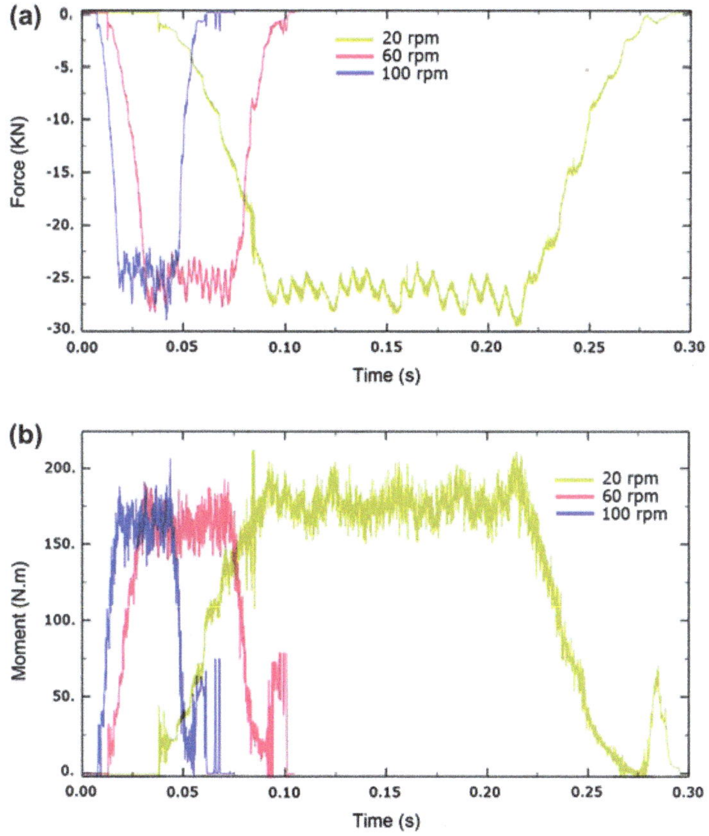

Figure 6. The effect of rolling speed amount on the variations of (a) roll force and (b) roll torque.

and (b) for different rolling speeds. The magnitude of the rolling force fluctuations in the steady-state region increases as the rolling speed increases and it varies about 7, 7.9, and 9.7% around the average value for 20, 60, and 100 rpm rolling speed amounts, respectively (Figure 6(a)). For the rolling torque, amounts of variations are about 12.7, 13.1, and 14.6% around the average value for 20, 60, and 100 rpm rolling speeds (Figure 6(b)). Hence, more fluctuations of roll force and torque is expected at higher rolling speeds. The phenomena that contribute to the changes of fluctuation amplitude with rolling speed are the strip forward slip and the contact shear stress distribution which affect stick slip condition in different rolling speeds.

A comparison between the predicted and measured (Hum et al., 1996) rolling forces and torques vs. the rolling speed is shown in Figure 7. The prediction error has a mean of 13.2% for rolling force and 13.4% for rolling torque. As observed, the dependence of the rolling force and torque on the speed of rolling is not quite clear. The lack of speed effect on the forces and torques may be attributed to the balance between the strain rate dependence of the material's resistance to deformation which is increasing with the speed and the lower pass times (which decreases heat flow from the strip to the roll and environment) and higher internal heat generation (due to strain rate rises), both of which result in softening of the material and cause lower forces and torques. Therefore, effect of rolling speed on the rolling force and torque is completely related to the material flow behavior, heat transfer condition, and other process parameters such as initial strip thickness and the amount of reduction which affect the heat generation of plastic work and changing the strip material and process condition will affect the relation of rolling force and torque to the rolling speed.

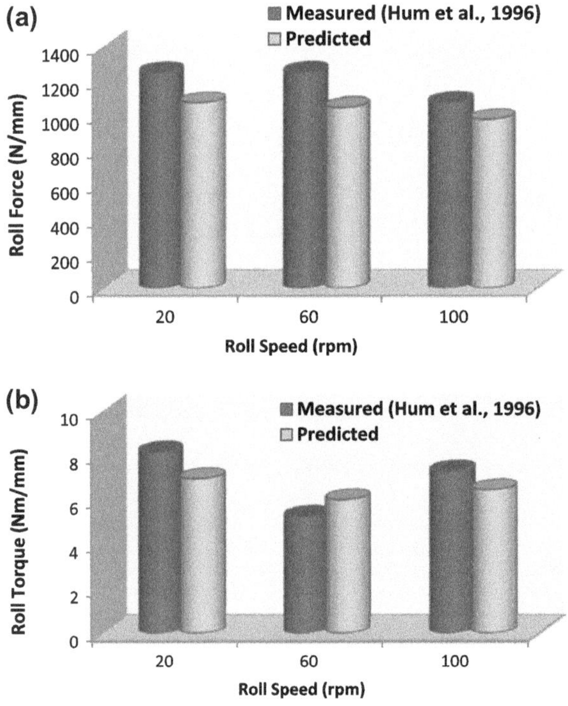

Figure 7. Comparison of: (a) roll force and (b) roll torque predictions with experimental data at different rolling speeds.

4. Conclusions

A mathematical model based on the FE method has been developed to simulate hot rolling of commercial pure aluminum strips. The major innovation of this work is a precision analysis of the roll force and torque and their fluctuations during the rolling process. The effects of main process parameters such as rolling speed and strip reduction are considered. By comparing the model predictions with the experimental data, the performance of the model was proved with a reasonable accuracy.

The results show that the rolling force and torque values are not stable in the steady-state region and fluctuate around the average value. These fluctuations are increased in higher thickness reduction amounts that can be caused by contact condition changes. Furthermore, the average roll separating force and torque are clearly increased with increasing reduction. Predicted results indicate that rolling force and torque fluctuations rise as the rolling speed increases. However, the dependence of the average rolling force and torque on the speed of rolling is not quite clear.

References

Bayoumil, S. (2007). A kinematic analytical approach to predict roll force, rolling torque and forward slip in thin hot strip continuous rolling. *Ironmaking & Steelmaking, 34*, 444–448.

Biswas, P., & Mandal, N. R. (2011). Effect of tool geometries on thermal history of FSW of AA1100. *Welding Journal, 90*, 129–135.

Biswas, S. (2003). *Simulation of thermo-mechanical deformation in high speed rolling of long steel products*. Worcester, MA: Worcester Polytechnic Institute.

Brown, S. B., Kim, K. H., & Anand, L. (1989). An internal variable constitutive model for hot working of metals. *International Journal of Plasticity, 5*, 95–130.

Byon, S. M., Kim, S. I., & Lee, Y. (2004). Predictions of roll force under heavy-reduction hot rolling using a large-deformation constitutive model. *Proceedings of the Institution of Mechanical Engineers, Part B: Journal of Engineering Manufacture, 218*, 483–494.

Byon, S. M., Na, D. H., & Lee, Y. S. (2013). Flow stress equation in range of intermediate strain rates and high temperatures to predict roll force in four-pass continuous rod rolling. *Transactions of Nonferrous Metals Society of China, 23*, 742–748.

Chen, S. Z., Zhang, D. H., Sun, J., Wang, J. S., & Song, J. (2012). Online calculation model of rolling force for cold rolling mill based on numerical integration. In *Proceeding of the control and decision conference*, Taiyuan, (pp. 3951–3955).

Chun, M. S., & Lenard, J. G. (1997). Hot rolling of an aluminum alloy using oil/water emulsions. *Journal of Materials Processing Technology, 72*, 283–292.

Duan, X., & Sheppard, T. (2002). Three dimensional thermal mechanical coupled simulation during hot rolling of aluminium alloy 3003. *International Journal of Mechanical Sciences, 44*, 2155–2172.

Dvorkin, E. N., Goldschmit, M. B., Cavaliere, M. A., Amenta, P. M., Marini, O., & Stroppiana, W. (1997). 2D finite element parametric studies of the flat-rolling process. *Journal of Materials Processing Technology, 68*, 99–107.

Ford, H., & Alexander, J. M. (1964). Simplified hot rolling calculations. *Journal of the Institute of Metals, 92*, 397–404.

Gudur, P. P., & Dixit, U. S. (2008). A neural network-assisted finite element analysis of cold flat rolling. *Engineering Applications of Artificial Intelligence, 21*, 43–52.

Hu, X. L., Zhang, Q. S., Zhao, Z., Tian, Y., Liu, X. H., & Wang, G. D. (2006). Application of approximation full-load distribution method to pass scheduling on plate mill with hydro-bending system. *Journal of Iron and Steel Research, International, 13*, 22–26.

Hum, B., Colquhoun, H. W., & Lenard, J. G. (1996). Measurements of friction during hot rolling of aluminum strips. *Journal of Materials Processing Technology, 60*, 331–338.

Jia, C. Y., Shan, X. Y., & Niu, Z. P. (2008). High precision prediction of rolling force based on fuzzy and nerve method for cold tandem mill. *Journal of Iron and Steel Research, International, 15*, 23–27.

Jing, L. (2001). *Rolling mill roll design*. Durham: Durham University.

Joun, M. S., & Hwang, S. M. (1992). An approximate analysis of hot-strip rolling – A new approach. *International Journal of Mechanical Sciences, 34*, 985–998.

Kim, N., Lee, S. M., Shin, W., & Shivpuri, R. (1992). Simulation of square-to-oval single pass rolling using a computationally effective finite and slab element method. *Transactions of ASME Journal of Engineering for Industry, 114*, 329–335.

Kim, Y. K., Kwak, W. J., Shin, T. J., & Hwang, S. M. (2010). A new model for the prediction of roll force and tension profiles in flat rolling. *ISIJ International, 50*, 1644–1652.

Kobayashi, S., Oh, S. I., & Altan, T. (1989). *Metal Forming and the finite-element method.* Oxford: Oxford University Press.

Kumar, A., Samarasekera, I. V., & Hawbolt, E. B. (1992). Roll-bite deformation during the hot rolling of steel strip. *Journal of Materials Processing Technology, 30*, 91–114.

Kwak, W. J., Lee, J. H., Hwang, S. M., & Kim, Y. H. (2002). A precision on-line model for the prediction of roll force and roll power in hot-strip rolling. *Metallurgical and Materials Transactions A, 33*, 3255–3272.

Lee, D. M., & Choi, S. G. (2004). Application of on-line adaptable neural network for the rolling force set-up of a plate mill. *Engineering Applications of Artificial Intelligence, 17*, 557–565.

Lee, D. M., & Lee, Y. (2002). Application of neural-network for improving accuracy of roll-force model in hot-rolling mill. *Control Engineering Practice, 10*, 473–478.

Lenard, J. G. (Ed.). (2002). *Metal forming science and practice*. Oxford: Elsevier.

Li, G. J., & Kobayashi, S. (1982). Rigid-plastic finite-element analysis of plane strain rolling. *Journal of Engineering for Industry, 104*, 55–63.

Liu, Y., Tong, C., & Lin, F. (2012). Rolling force prediction based on wavelet multiple-RBF neural network. In *Proceeding of the IEEE international conference on information and automation*, Shenyang, (pp. 941–944).

Moon, C. H., & Lee, Y. (2008). Approximate model for predicting roll force and torque in plate rolling with peening effect considered. *ISIJ International, 48*, 1409–1418.

Moussaoui, A., Selaimia, Y., & Abbassi, H. A. (2006). Hybrid hot strip rolling force prediction using a Bayesian trained artificial neural network and analytical models. *American Journal of Applied Sciences, 3*, 1885–1889.

Poliak, E. I., Shim, M. K., Kim, G. S., & Choo, W. Y. (1998). Application of linear regression analysis in accuracy assessment of rolling force calculations. *Metals and Materials, 4*, 1047–1056.

Pyun, Y. S., Cho, I. H., Suh, C. M., Park, J., Rogers, J., Kayumov, R., & Murakami, R. (2013). Reducing production loss by prolonging service life of rolling mill shear pin with ultrasonic nanocrystal surface modification technology. *International Journal of Precision Engineering and Manufacturing, 14*, 2027–2032.

Said, A., Lenard, J. G., Ragab, A. R., & Elkhier, M. A. (1999). The temperature, roll force and roll torque during hot bar rolling. *Journal of Materials Processing Technology, 88*, 147–153.

Serajzadeh, S., Karimi Taheri, A., & Mucciardi, F. (2002). Unsteady state work-roll temperature distribution during continuous hot slab rolling. *International Journal of Mechanical Sciences, 44*, 2447–2462.

Shahani, A. R., Setayeshi, S., Nodamaie, S. A., Asadi, M. A., & Rezaie, S. (2009). Prediction of influence parameters on the hot rolling process using finite element method and neural network. *Journal of Materials Processing Technology, 209*, 1920–1935.

Shangwu, X., Rodrigues, J. M. C., & Martins, P. A. F. (1999). Three-dimensional simulation of flat rolling through a combined finite element-boundary element approach. *Finite Elements in Analysis and Design, 32*, 221–233.

Sims, R. B. (1954). The calculation of roll force and torque in hot rolling mills. *ARCHIVE: Proceedings of the Institution of Mechanical Engineers 1847–1982 (vols 1–196), 168*, 191–200.

Wang, P. F., Zhang, D. H., Xu, L., & Zhang, W. X. (2011). Research and application of dynamic substitution control of actuators in flatness control of cold rolling mill. *Steel Research International, 82*, 379–387.

Wang, X., Peng, Y., Xu, L., & Liu, H. (2010). A 3-D differential method for solving rolling force of PC hot strip mill. *Journal of Iron and Steel Research, International, 17*, 36–39.

Wang, X., Yang, Q., & He, A. (2008). Calculation of thermal stress affecting strip flatness change during run-out table cooling in hot steel strip rolling. *Journal of Materials Processing Technology, 207*, 130–146.

Yanagimoto, J., Morimoto, T., Kurahashi, R., & Chikushi, I. (2002). Mathematical modelling for rolling force and microstructure evolution and microstructure controlling with heavy reduction in tandem hot strip rolling. *Steel Research, 73*, 56–62.

Yang, Y. Y., Linkens, D. A., Talamantes-Silva, J., & Howard, I. C. (2003). Roll force and torque prediction using neural network and finite element modelling. *ISIJ International, 43*, 1957–1966.

4

Influence of subcontracting constraints on the performance of manufacturing industries in Nigeria

Victor Chukwunweike Nwokocha* and Iganatius Ani Madu

Department of Geography, University of Nigeria Nsukka, Nigeria

In this work, an attempt has been made to show the influence of subcontracting constraints on firm performance in Nigeria. The study in line with the literature identified a number of constraints hindering an effective subcontracting arrangement in the study area. While the constraints were found not to have affected the use of subcontracting in the country, low capital intensity, disclosure of commercial secrets, poor services and interest conflict were found to have restricted subcontracting arrangements in the study area to sharing of equipment and short-term contracts. These constraints however were found not have affected the performance of manufacturing industries in the study area. This paper keeping in mind the findings of this study suggested that manufacturing industries in Nigeria should invest more in machineries and tools so as to increase subcontracting co-operations among industries.

Keywords: subcontracting; firms; performance; manufacturing industries; contraints; Nigeria

1. Introduction

Industrialisation is one of the packages in the economic development process of any country that could be used to accelerate economic growth. It acts as a catalyst to transform the economic structure of countries, from simple, slow-growing, and low-value activities to more productive activities (manufacturing) that enjoy greater margins driven by technology, and have higher growth prospects (Logan, 2006).

The success of industrialisation or industrial production, however, largely depends on the establishment of useful linkages between industries (Nguyen and Khac, 2013). Industrial linkage is the interrelationship amongst various industrial activities through the input–output relationship or the economic value chain (Mayhew, 2009). Subcontracting which is a type of industrial linkage based on alliance is a work contract that externalise production processes from one firm to another through interfirm relationships (Ajayi, 2007). In developing countries, business linkages with local small and medium enterprises, including procurement, distribution and sales, offer large firms an avenue through which to address some of these concerns. These relationships can allow large firms to reduce input costs while increasing specialisations and flexibility.

In Nigeria, subcontracting; a strategic positioning of industrial activities started in the early 1960s, the post-independence period (Ajayi, 2007). The earliest stage in the adoption of production subcontracting as an industrial production technique in Nigeria

*Corresponding author. Email: chukwuweike_nwokocha@yahoo.com

was characterised by insignificant growth and rapid growth thereafter. Subcontracting became very important after the introduction of the Structural Adjustment Programme in 1986, and it is perceived by industrialist as very important in reducing the cost of production (Ajayi, 2007) and most recently following the works of Nwokocha, Madu, Ocheje, and Olerum (2015a), subcontracting is very important in enhancing the operational efficiency in industries, induce specialisation so as to gain professional resources and to strengthen cooperation amongst industries.

Firm performance on the other hand is a relevant construct in strategic management research and frequently used as a dependent variable. Despite its relevance, there is hardly a consensus about its definition, dimensionality, and measurement, what limits advances in research and understanding of the concept (Santos & Brito, 2012). The research into firm performance suffers from problems such as lack of consensus, selection of indicators based on convenience, and little consideration of its dimensionality (Combs, Crook, & Shook, 2005; Crook, Ketchen, Combs, & Todd, 2008; Richard, Devinney, Yip, and Johnson 2009). Many studies measure firm performance with a single indicator and represent this concept as one-dimensional, even while admitting its multidimensionality (Glick, Washburn, & Miller, 2005). If several dimensions exist, a researcher should choose the dimensions most relevant to his or her research and judge the outcome of this choice (Richard et al., 2009). Ray, Barney, and Muhanna (2004) stress this, warning against the difficulties of testing the resource-based theory using aggregated measures of performance and suggesting the use of indicators directly connected to the resources under analysis.

Consequently, this paper proposes to use a multidimensional measurement model of firm performance, structured in the Goal-Setting Theory (Chong, 2008; Locke & Latham, 2002) in order to assess the influence of subcontracting constraints on firm performance in Nigeria. Goal setting is the mechanism by which a firm delivers results against its strategy (e.g. subcontracting) on the extent to which there is clarity, challenge, commitment, feedback, and task complexity (Locke & Latham, 2002). It is explained using a combination of both financial and non-financial indicators. The financial indicators are: sales growth, growth in profits, changes in assets by gross value plant and machinery, return on assets to measure capital efficiency. The non-financial indicators were: growth in market share; product success; increase in number of employees and; labour productivity (Gakure, Kimemia, & Waititu, 2014; Hu, Zheng, & Wang, 2011; Kongmanilaa & Takahashib, 2009; Marimuthu, Arokiasamy, & Ismail, 2009; Ongonga & Abeka, 2011; Tuan and Yoshi, 2010).

Empirical researches on subcontracting have received growing attention in the academia during the past few years. This is evidenced by an increasing number of publications and studies on the topic (Ajayi, 2001, 2002, 2003, 2007; Berry, 1997; Diaz-Mora & Triguero-Cano (2007); Dahane, Clementz, & Rezg, 2008; Dahane, Dellagi, Clementz & Rezg, 2011; Deardorff & Djankov, 2000; Furlan, Grandinetti, & Camuffo, 2007; Grossman & Helpman, 2005; Gubik, 2005; Hajej, Rezg, & Gharbi, 2014; Kongmanila and Takahashi, 2009; Nwokocha et al. 2015a; Nwokocha, Madu, Ocheje, Olerum, & Nwosu, 2015b; Taymaz and Kiliçaslan, 2002; Tuan and Yoshi, 2010; Watanabe, 1971). According to these studies, subcontracting helps small and medium scale enterprises as well as large scale firms to stimulate, reduce, and control operation cost in manufacturing, improve company's focus, and to access world class capabilities, develop joint maintenance and production strategies amongst others.

Similarly, researches on subcontracting constraints have never fallen short in the academia. Some of these researches include the works of (Annim & Machethe, 1998;

Bbenkele, 1998; Dahane et al., 2008; Hajej et al., 2014; Kumar & Subrahmanya, 2007; Okatch, Mukulu, & Oyugi, 2011). While the works of Dahane et al. (2008), generalised subcontracting constraint in computational engineering, Hajej et al. (2014), developed optimal production plans of principal and subcontracting machines which minimises the total production and inventory cost for cases without and with returned products under service level and subcontracting transportation delay. These two studies however were structured to solve subcontracting constraint arising from transportation delay, joint maintenance, and production management. Similarly, while Annim and Machethe (1998), in a study on business linkages in Kwazulu-Natal, point out that the constraint on the expansion and improvement of subcontracting linkages is, according to suppliers: limited application of new technology, poor product quality, unreliable delivery of goods or services, and high products prices, the studies of Bbenkele (1998) and Okatch et al. (2011) posited that the constraints to subcontracting are inability of the SMEs to supply quality products to schedule, lack of local suppliers for certain parts, the proliferation of makes and models, and competition from imported second hand vehicles from Japan and Europe.

Conceding from the above, there is a widespread consensus on the constraints of subcontracting on industrial activities. These empirical findings, however, did not state the influence or contributions of these constraints to the performance of manufacturing industries. Following the findings of Berry (1997), Yasuda (2005), Marsall, Mcivor, and Lamming (2007), and Gakure et al. (2014)), which suggested that subcontracting has a positive impacts on firm's performance and the works of Hu et al. (2011), which observed that there is no significant relationship between subcontracting and firm performance, it will be difficult to make an informed assessment on the influence of subcontracting constraints on firm performance. This work is thus oriented to investigate the influence of subcontracting constraints on the performance of manufacturing industries in Nigeria.

1.1. Theoretical framework

The theoretical framework for analysing subcontracting is hinged on three theories. These theories are the dualistic approach, development approach, and networking and clustering approach. While dualistic approach considers subcontracting as an unequal power relationship whereby large contractors realise benefit at the expense of small contractors, development approach formulated by development economists considers subcontracting as a relationship between large and small firms, but emphasises a positive role of it.

The networking and clustering approach, largely regarded as the modern theoretical foundation of subcontracting, supports networking initiatives and the development of industrial cluster (Pyke, 1992; UNCTAD, 1994). In the case of the cluster approach, subcontracting itself is one of the main types of networking on which clusters could be established. This approach looks at a group of firms cooperating (and competing) within a complex web of supportive institutions. Externalities, linkages, and economics of scale generated by this form of cooperation and competition are internalised by the network so that the collective efficiency and flexibility of the industry is enhanced (Ceglie & Dini, 1999).

From the fore going, the approaches to subcontracting have shown that subcontracting is a form of relationship established by firms not only to be more competitive in their production activities, but also to share mutual benefits. These approaches however failed to disclose the impact of subcontracting constraints on the performance of manufacturing industries. Based on this, three key questions are posed in this research in order to answer the above-stated question:

(1) What are the constraints hampering effective subcontracting arrangement?
(2) Have these constraints affected the use of subcontracting arrangement by firms?
(3) Have these constraints affected the performance of firms?

This work tends to explore the lacuna in the approaches to subcontracting in order to strengthen it.

1.2. Material and methodology

The purpose of this study was to establish the impact of subcontracting constraints on the performance of firms in Nigeria. The study adopted a number of methods comprising semi-structured interviews, field observations, reference to relevant literature, and questionnaire survey of 120 manufacturing industries as described elsewhere (Nwokocha et al., 2015a).

The appropriate sample size for the population-based survey was determined largely by the estimated prevalence of the variable of interest – subcontracting in this instance (estimated at 15%); the desired level of confidence, 95%; and the acceptable margin of error, 5% (standard value of .05). A pilot test on 20 firms helped to remove ambiguities and improve the instrument as well as test for its reliability and validity. Open- and close-ended questions asking respondents to rate various questionnaire items using a five-Likert scale of; 5 – very important, and 1 – not at all important. Likert-type ordinal scale representing a spectrum of subjective feelings and opinions were used to solicit specific responses from the industrialists.

1.2.1. Instrument/questionnaire validation

To establish the degree of reliability of the questions in the questionnaire, 20 manufacturing industries were made to rate each question in the questionnaire on a five-point scale. Principal Component Analysis method was used to extract the factors. The criteria states that: Cronbach's Alpha of a scale should be greater than .70 for items to be used together as a scale while factor loadings greater than .40 are considered statistically significant for studies with sample size less than 200. Therefore, in the present study, ±0.40 was used as the cut-off for loadings since the sample size of the study was less than 200. The higher the factor loadings were, the closer they were related to the variable.

1.2.2. Constraints to subcontracting

Constraints to subcontracting were the independent variable in this study. Following our field observations, fourteen (14) variables were identified as the constraints to subcontracting in the industrial sector of Nigeria. The constraints were analysed using a five-Likert scale of; 5 – very important, and 1 – not at all important. The pilot test result can be seen in Table 1.

Following the results in Table 1, the original 14 factors which constituted the constraints to subcontracting were reduced to 10 factors. This was because four factors with

Table 1. Factor analysis of subcontracting constraints.

Items	Factor loading
Proliferation of makes	0.67
Increased demand for specialised technology	0.29*
Low visibility	0.56
Poor services	0.70
Limited application of new technology	0.83
Competition from imported parts	0.86
Low capital intensity	0.81
Lack of local suppliers for certain parts	0.58
Unfulfilled order from subcontractors	0.13*
Interest conflicts with subcontracting partners	0.66
Disclosure of commercial Secrets	0.55
Legal disputes	0.11*
Weaken Culture	−0.02*
Decrease compatibility of innovation	0.75
Number of Items	14
Cronbach's Alpha	0.91

Notes: Authors Computation, 2015.
NB: The factors marked with (*) were eliminated from further analysis.

loadings less than .40 were discarded as shown in Table 1. Thus, ten (10) factors with factor loadings between .55 and .93 were subsequently considered valid as the constructs to represent the constraints to subcontracting. The Cronbach's Alpha for these 10 factors was .91, which exceeded the reliability cut-off value of .70.

1.2.3. Firm performance

Firm performance was the dependent variable in the present study and in accordance with the literature, a combination of both financial and non-financial indicators led to a balanced performance measurement. The pilot test was conducted using a five-Likert scale of; 5 – significantly increased, and 1 – significantly decreased.

Firm performance measures had a total of 8 items generated from literature comprising both the financial and non-financial indicators. The results indicate that with factor loadings of between .70 and .90, the construct of the 8 measurement items was valid for firm performance (see Table 2).

The Cronbach's Alpha coefficient for the 8 indices was =0.92, which means the instrument had an excellent level of consistency, and fit for use in data collection.

Table 2. Factor analysis of firm performance.

Growth in sales (volume)	0.90
Growth in profit	0.88
Gross value of capital (machinery)	0.81
Return on assets	0.78
Growth in market share	0.87
Product success	0.76
Labour productivity	0.71
Increase in workers	0.70
No. of indices	8
Cronbach's Alpha	0.92

Note: Author's Computation, 2015.

1.3. Data analysis

Ordinal multinomial logit model was used to test for the effect of explanatory variables on those responses that had more than two categories. Similarly, ordinary logit model was used for binary responses. Relevant statistical techniques were used to analyse the data. All analyses were carried out with the aid of Statistical Packages for Social Sciences version 17 (SPSS 17).

2. Results

The result of the study is discussed below. The focus is on the impact of subcontracting constraints on the performance of manufacturing industries in Nigeria.

2.1. Analysis of respondents

Out of the 120 questionnaires administered, 80 (66.7%) were considered valid, with no missing data. The response rate of 66.7% was, therefore, considered adequate for the study. The targeted enterprises were aged 10 years for the firms and 45 years and above for the managers on the average. About 64.5% of the firms were managed by men. The preliminary findings showed that 80.4% of the respondents had attained secondary level of education while 50.6% had attained product-related skills training through apprenticeship and learning on the job. From our observations, industrial products manufactured were Plastics and rubber items (ranging from gallons, buckets, paper and plastic bags, etc. 25%), metallic windows and doors including aluminium roofing sheets (28%) food and Beverage products (28%), and foams and chemical products (ranging from detergents, engine oil, bed forms, etc. 19%). About 85.6% of the investigated firms were engaged in subcontracting arrangements.

2.2. Constraints to subcontracting

Despite the benefits of production subcontracting, there are constraints limiting its effectiveness in the industrial sector. These constraints based on the field observation have been synthesised into ten (10) factors, and how they appeal to the industries were analysed using a five-point Likert-type scale ranging from 1 = ‘not at all important’ to 5 = ‘very important’ in relation to how they affect the industrialist (see Table 3).

The analysis shown in table 3 reveals that the majority of the industries believed that engaging in production subcontracting has a number of constraints associated with it. From the analysis, most of the industries identified: disclosure of commercial secrets, low capital intensity, interest conflicts, decrease compatibility of innovation, importation of parts and poor services with mean and standard deviation values of 4.23 (STD = .77), 4.23 (STD = .77), 4.18 (STD = .72), 3.52 (STD = .50), 3.52 (STD = .50), 3.52 (STD = .50), and 4.13 (STD = .74), respectively, as the major constraints affecting the effectiveness of subcontracting in Nigeria. Most of the industrialists interviewed have experienced one of these problems at one time or the other making them either to change their subcontracting partner or acquire the capacity to handle the activity (ies) in house. While ‘Low capital intensity’ was found to have made most of the firms in the study area to invest very little in machineries, reducing subcontracting strategies predominantly to sharing of equipment and short-term contracts, ‘Poor services, Disclosure of commercial secrets, importation of parts and interest conflict’ were found

Table 3. Mean distribution of responses to subcontracting constraints.

Variables	*N* statistic	Range statistic	Minimum statistic	Maximum statistic	Mean statistic	Std deviation
Proliferation of makes		0.00	2.00	2.00	2.00	0.00
Limited application of advance technology		0.00	2.00	2.00	2.00	0.00
Decrease compatibility of innovation		1.00	3.00	4.00	3.52	0.50
Disclosure of commercial secrets		3.00	2.00	5.00	4.23	0.77
Interest conflicts		3.00	2.00	5.00	4.18	0.72
Low visibility		0.00	2.00	2.00	2.00	0.00
Poor services		2.00	1.00	3.00	4.13`	0.74
Imported goods		1.00	3.00	4.00	3.52	0.50
Low capital intensity		3.00	2.00	5.00	4.23	0.77
Lack of local suppliers for certain parts		0.00	2.00	2.00	2.00	0.00
Valid number	80					

Source: Field Work and Author's Computation, 2015.

to have forced firms to frequently change their subcontracting partners or acquire the capacity to handle the activity (ies) in house.

2.3. *Firm performance*

Firm performance, the dependent variable in this study was measured in both financial and non-financial indicators. The respondents were asked to evaluate their firm's performance by rating various indicators of their business performance in the last 10 years on a scale of 1–5, where 1 represented – significantly decreased and 5 represented – significantly increased. Following the result in Table 4, 7.90% of the respondents had experienced significant increase while 35.31% had seen relative increase in firm performance in the last 10 years. Furthermore, while 18.12% of the manufacturing industries had experienced relative decrease in their business performance, 33.13% had experienced significant decrease. The result also revealed that 33.96% of the SMEs were static and recorded no changes (see Table 4).

The large number of firms that experienced significant decrease with reference to the little number that experience significant increase in performance is an indication that the Nigerian Industrial sector have underperformed. Similarly, the overall mean and standard deviation score of 2.38 and .71 shows that majority of the firms experienced some increase in performance. The respondents indicated that they have achieved highest performance in Profit Growth with an average mean value of 2.84 and standard deviation of .80 and least performance in return to assets with an average mean of 2.04, and standard deviation value of .54 (see Table 4).

Furthermore, in order to show the impact of these constraints on subcontracting activities in the study area, the share of subcontracting for all the industries was computed using a ten years (10 years) time frame. The share of subcontracting according to Taymaz and Kiliçaslan (2002) and Holl (2007) is subcontracted input – the share of a plant's inputs subcontracted to supplier plants, in total inputs and subcontracted output – the share of output subcontracted by plants, in total output. Subcontracted input share

Table 4. Percentage and mean distribution of responses to firm performance.

Item	Significantly decreased (1) %	Relatively decreased (2) %	Static (3) %	Relatively increased (4) %	Significantly increased (5) %	Mean	Standard deviation
Growth in sales	2.30	28.30	10.00	40.40	19.00	2.15	0.46
Profit Growth	3.50	29.30	16.60	50.60	0.00	2.82	0.80
Return on assets	5.60	16.70	43.10	24.60	10.00	2.04	0.54
Gross value of capital	5.50	19.00	34.30	21.00	20.20	2.64	0.89
Profit success	0.60	12.60	39.40	45.10	2.30	2.15	0.36
Growth in market share	10.00	13.40	25.20	50.30	1.10	2.44	0.90
Increase in workers	5.00	10.90	48.10	35.40	0.60	2.16	0.96
Labour productivity	5.00	14.90	55.00	15.10	10.00	2.64	
Total average	33.13	18.12	33.96	35.31	7.9	2.38	0.71

Note: Field work and Author's Computation, 2015.

Table 5. Subcontracted input and output share over a period of ten (10) years.

Years	Subcontracted input share (%)	Subcontracted output share (%)
2000–2001	6.31	6.56
2001–2002	5.22	5.30
2002–2003	4.54	4.80
2003–2004	4.41	4.52
2004–2005	5.38	5.40
2005–2006	4.41	4.52
2006–2007	4.41	4.00
2007–2008	4.50	4.87
2008–2009	5.36	5.50
2009–2010	5.77	5.99

Note: Fieldwork Author's Computation, 2015.

can also be referred to as the proportion of subcontracted input in total inputs while subcontracted output share refers to the proportion of subcontracted output in total output (Morrison & Yasar, 2008). These definitions were based on the survey definition of income from subcontract as income generated from the processing of materials provided by the firm offering the subcontract, see Table 5.

The results in Table 5, Figures 1 and 2 shows that the subcontracted input share and output share of the firms have been relatively stable. This relative stability shows that most firms still make use of this process despite the constraints associated with subcontracting as was noted earlier. This lays credence to our earlier finding which noted that about 85.6% of the surveyed firms were engaged in subcontracting arrangements in the study area. This result also reveals that the constraints to subcontracting have no impact or influence on the performance or underperformance of firms. This is because the subcontracted input and output share shown in Table 5 have remained relatively stable and therefore cannot account for the performance or underperformance of firms in the study area.

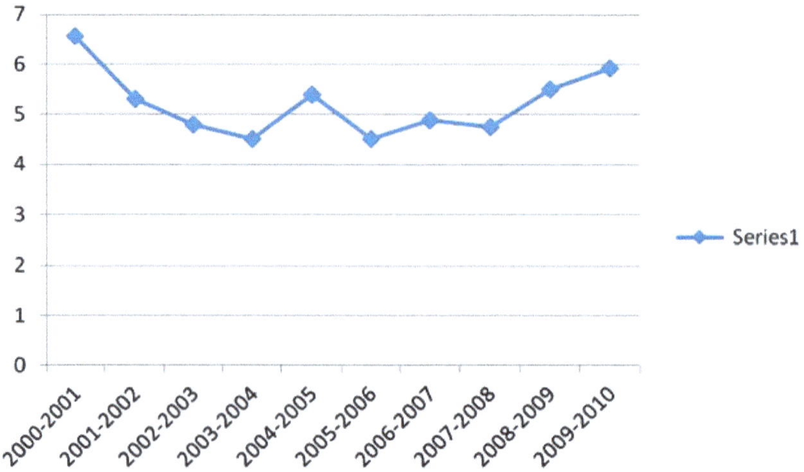

Figure 1. Subcontrated input share (time series analysis).

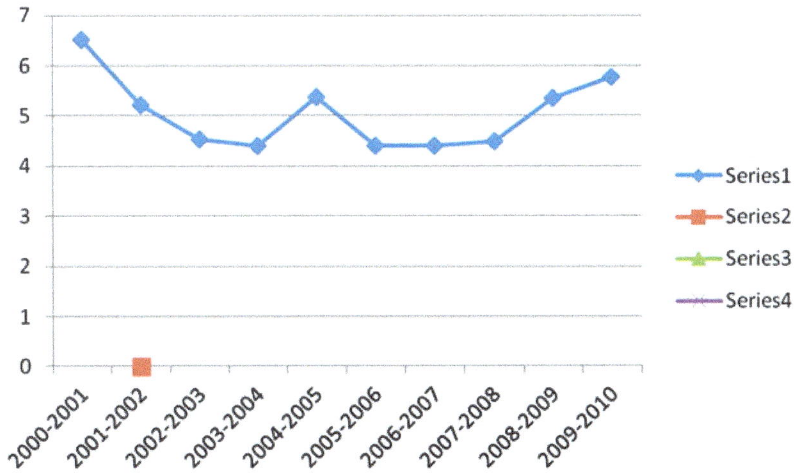

Figure 2. Subcontrated output share (time series analysis).

2.4. Summary of finding and conclusion

This paper in line with the literature identified a number of constraints to subcontracting arrangement in the study area. These constraints ranges are disclosure of commercial secrets, poor quality of service, interest conflicts, decrease compatibility of innovation, importation of parts, and unfulfilled orders from subcontractors. While 'Low capital intensity' was found to have made most of the firms in the study area to invest very little in machineries and tools thereby making their subcontracting strategies in the study area to sharing of equipment and short-term contracts, 'Poor services, Disclosure of commercial secrets, importation of parts and interest conflict' was found to have forced firms to frequently change their subcontracting partners or acquire the capacity to handle the activity (ies) in house. Similarly, the analysis of firm performance also showed that a majority of the firms have experienced some increase in performance. The firms indicated that they have achieved highest performance in Profit Growth and least performance in return to assets. Going forward, the paper also discovered that subcontracted input and output share of the firms have been relatively stable. This showed that despite the constraints associated with subcontracting, firms have continued to make use of this production process, signifying that subcontracting constraints have no impact on the use of subcontracting strategy by the firms.

Keeping in view the findings of the study and conclusion drawn, this paper recommends that manufacturing industries in Nigeria should invest more in machineries and tools so as to increase subcontracting co-operations amongst industries in the study area. This will not only help the firms to be more competitive, it will also help them to engage in subcontracting more intensely as well as building up superior capabilities in order to bypass some of the aforementioned constraints associated with subcontracting.

Disclosure statement

No potential conflict of interest was reported by the authors.

References

Ajayi, D. D. (2001). Industrial subcontracting linkages in the Lagos Region, Nigeria. *The Nigerian Journal of Economic and Social Studies (NJESS), 43*, 265–277.

Ajayi, D. D. (2002). Temporal pattern of production subcontracting in Nigeria. *Annals of the Social Science Academy of Nigeria, 14 & 15*, 67–81.

Ajayi, D. D. (2003). Nature and scope of production subcontracting in Nigeria. *Africa Development, XXVIII*, 89–111.

Ajayi, D. D. (2007). Recent trend and patterns in Nigeria's industrial development. *Africa Development, 32*, 139–155.

Annim, F. D. L., & Machethe, C. (1998). *Promoting the growth of MSEs through business linkages in the Northern Province of Kwazulu-Natal.* South African Province: Research Report: United States Agency for International Development Bureau for Africa.

Bbenkele, E. C. (1998). Enhancing economic development by fostering business linkages between the pharmaceutical companies and the traditional medicines sector. Research report. Center for Partnerships in Enterprise Research and Technology Transfer. University of Natal, Pietermaritzburg South Africa.

Berry, A. (1997). *SME competitiveness: The power of networking and subcontracting* in R. W. Gakure, P. N. Kimemia, & G. A. Waititu. Influence of Subcontract Offering on the Performance of Manufacturing Micro and Small Enterprises in Kenya. *Journal of Humanities and Social Science (IOSR-JHSS), 19*, Ver. II, 37–46.

Ceglie, G., & Dini, M. (1999). *SME cluster and networking development in developing countries: The experience of UNIDO* (Working Paper No. 2). Vienna: Private Sector Development Branch, UNIDO.

Chong, H. (2008). Measuring performance of small-and medium sized enterprises: The grounded theory approach. *Journal of Business and Public Affairs, 2*(1), 1–10.

Combs, J. G., Crook, T. R., & Shook, C. L. (2005). The dimension of organizational performance and its implications for strategic management research. In D. J. Ketchen & D. D. Bergh (Eds.), *Research methodology in strategy and management* (pp. 259–286). San Diego, CA: Elsevier.

Crook, T. R. Ketchen, D. J., Combs, J. G., & Todd, S. Y. (2008) Strategic resources and performance: A meta-analysis. *Strategic Management Journal, 29*, 1141–1154.

Dahane, M., Clementz, C., & Rezg, N. (2008). Analysis of joint maintenance and production policies under a subcontracting constraint. *International Journal of Production Research, 46*, 5393–5416.

Dahane, M., Dellagi, S., Clementz, C., & Rezg, N. (2011). Development of joint maintenance and production strategies in a subcontracting environment. *International Journal of Production Research, 49*, 6937–6961.

Deardorff, A., & Djankov, S. (2000). Knowledge transfer under subcontracting: Evidence from Czech firms. *World Development, 28*, 1837–1847.

Diaz-Mora, C., & Triguero-Cano, A. (2007). *Why do some firm contract out Production? Evidence from the firm-level panel data.* Albacete: Department of International Economics, Faculty of Juridical and Social Science, University of Castilla-La Mancha Albacete Spain.

Furlan, A., Grandinetti, R., & Camuffo, A. (2007). How do subcontractors evolve? *International Journal of Operations & Production Management, 27*, 69–89.

Gakure, R. W., Kimemia, P. N., & Waititu, G. A. (2014). Influence of Subcontract Offering on the Performance of Manufacturing Micro and Small Enterprises in Kenya. *IOSR Journal of Humanities and Social Science (IOSR-JHSS), 19*, Ver. II, 37–46.

Glick, W. H., Washburn, N. T., & Miller, C. C. (2005). *The myth of firm performance.* Proceedings of the Annual Meeting of American Academy of Management, Honolulu, Hawaii, USA, in J. B. Santos & L. A. L. Brito (Eds.), *Towards a subjective measurement model for firm performance* (pp. 95–177). Brazilian Administrative Review.

Grossman, G. M., & Helpman E. (2005). Outsourcing in a global economy. *Review of Economic Studies, 72*, 135–159.

Gubik, A. (2005). New opportunities for SMEs founded by cooperation. *European Integration Studies, Miskolc, 4*, 25–36.

Hajej, Z., Rezg, N., & Gharbi, A. (2014). Forecasting and maintenance problem under subcontracting constraint with transportation delay. *International Journal of Production Research, 52*, 6695–6716

Holl, A. (2007). *Production subcontracting and location*. Madrid: FEDEA Foundation for Applied Economics Studies.
Hu, Z., Zheng, J., & Wang, J. (2011). Impact of industrial linkages on firm performance in Chinese development zones, Yangtze River Delta, Jiangsu Province, China. *The Chinese Economy, 44*, 78–105.
Khac, N. N. (2013). *Demand creation and competition effect of export-platform FDI on backward linkages; evidence from panel data analysis of Vietnames supporting industries*. Paper provided by Centre d'Études des Politiques Économiques (EPEE), Université d'Evry Val d'Essonne in its series Documents de recherche with number 13-02, 1–42.
Kongmanilaa, X., & Takahashib, Y. (2009). Determinants of subcontracting and firm performance in Lao PDR: Evidence from a Garment industry cluster. *Asia Pacific Management Review, 15*, 97–112
Kumar, R. S., & Subrahmanya, B. (2007). Subcontracting relationships of Indian SMEs with global TNCs: Do SMEs gain, How. *Journal of Asian Economics, 5*, 2–35
Locke, E. A., & Latham, G. P. (2002). Building a practically useful theory of goal setting and task motivation: A 35-year odyssey, American Psychologist. *American Psychological Association, 57*(9), 705–717.
Logan, M. I. (2006). The shift from informal to formal: Some cost and benefits of large scale industrialization in Asian Countries. *Singapore Journal of Tropical Geography, 3*(2), 170–176
Marimuthu, M., Arokiasamy, L., & Ismail, M. (2009). Human capital development and its impact on firm performance: Evidence from developmental economics. *The Journal of International Social Research, 2/8*(Summer 2009), 265–272.
Marsall, D., Mcivor, R., & Lamming, R. (2007). Influences and outcomes of outsourcing: Insights from the telecommunications industry. *Journal of Purchasing & Supply Management, 13*, 245–260.
Mayhew, S. (2009). *A dictionary of Geography*.(4th ed.), Oxford: Oxford University Press.
Morrison, C. J., & Yasar, M. (2008). *Outsourcing, productivity, and input composition at the plant level. Canadian Journal of Economics, 42*(2), 422–439.
Nwokocha, V. C., Madu, I. A., Ocheje, J. F., & Olerum, V. N. (2015a). Production subcontracting: A strategy for the survival of small and medium scale industries in Nigeria. *Mediterranean Journal of Social Sciences, 6*, 641–651.
Nwokocha, V. C., Madu, I. A., Ocheje, J. F., Olerum, V. N., & Nwosu, I. G. (2015b). Production subcontracting: A policy issue for small and medium scale manufacturing industries in Nigeria. *Academic Journal of Interdisciplinary Studies, 4*, 375–385.
Okatch, B. A., Mukulu, E., & Oyugi, L. (2011). Constraints to subcontracting arrangements between SMEs and large firms in the motor vehicle industry in Kenya. *International Journal of Business and Social Science, 2*(15), 208–223.
Ongonga J. O., & Abeka E. O. (2011): Networking in the Kenyan informal sector: An attempt to manage the market failures. *African Journal of Business Management, 5*, 11323–11334.
Pyke, F. (1992). *Industrial development through small-firm co-operation*. Geneva: ILO.
Ray, G., Barney, J. B., & Muhanna, W. A. (2004). Capabilities, business processes, and competitive advantage: Choosing the dependent variable in empirical tests of the resource-based view. *Strategic Management Journal, 25*, 23–37.
Richard, P. J., Devinney, T. M., Yip, G. S., & Johnson, G. (2009). Measuring organizational performance: towards methodological best practice. *Journal of Management, 35*(3), 718–804.
Santos, J. B., & Brito, L. A. L. (2012). Towards a subjective measurement model for firm performance [Special issue]. *Brazilian Administrative Review, 9*, 95–117.
Taymaz, E., & Kiliçaslan, Y. (2004). *Determinants of subcontracting and regional development: An empirical study on Turkish textile and engineering industries*. Ankara: Department of Economics Middle East Technical University
Tuan, N. P. & Yoshi, T. (2010). Vertical Linkage and Firm's Performance in Supporting industries in Vietnam. *Asian Journal of Management Research, 1*(1), 1–14.
UNCTAD. (1994). *Technological dynamism in industrial districts: An alternative approach to industrialization in developing countries*. New York, NY: United Nations.
Watanabe, S. (1971). Subcontracting, industrialisation and employment creation. *International Labour Review, 104*, 51–76.
Yasuda, T. (2005). Firm growth, size, age and behaviour in Japanese manufacturing. *Small Business Economics, 24*, 1–15.

5

Lean thinking for a maintenance process

Sherif Mostafa[a]*, Sang-Heon Lee[b], Jantanee Dumrak[c], Nicholas Chileshe[a] and Hassan Soltan[d]

[a]School of Natural and Built Environments, University of South Australia, Adelaide 5000, Australia; [b]School of Engineering, University of South Australia, Mawson Lakes 5095, Australia; [c]Global Management Program, Torrens University Australia, Adelaide, SA 5000, Australia; [d]Production Engineering and Mechanical Design Department, Faculty of Engineering Mansoura University, Mansoura 35516, Egypt

The maintenance process shares significant operating costs in an organisation. Lean thinking can be incorporated into maintenance activities through applying its principles and practices/tools. Lean maintenance (LM) is a prerequisite for lean manufacturing systems. This research proposes a new structure for LM process based on a systematic literature review of a significant number of related articles that were published on LM. The process structure is designed based on the five lean principles to guide and support organisations to pursue maintenance excellence. This study establishes a scheme for LM tools that are structured into 2 level 4 bundles and 26 lean practices/tools and develops a House of Waste (HoW) to demonstrate the association between maintenance wastes and the LM tools. With a successful accomplishment of the proposed scheme, the performance of a maintenance department can create more improvement opportunities over time to reach the maintenance excellence status.

Keywords: total productive maintenance (TPM); lean maintenance; maintenance wastes and value stream mapping; lean maintenance tools; House of Waste (HoW)

1. Introduction

Maintenance function becomes a significant contributor towards to achieve strategic objectives of an organisation in today's competitive markets (Fraser, 2014). The maintenance process is to serve the production facilities of high productivity. It comprises planned and unplanned actions carried out to retain a physical asset to the acceptable operating conditions (Faccio, Persona, Sgarbossa, & Zanin, 2014). Maintenance aims at increasing the value of the reliability, safety, availability, and quality of an asset (e.g. production plant, equipment, or building) with acceptable economical costs (Márquez, 2007). Over the last decades, the maintenance has been considered as a necessary evil from the organisational management as the maintenance operation is limited to corrective functions that are usually executed under the emergency situations such as machine breakdown. However, this practice is no longer acceptable since the role of maintenance has been recognised as a strategic element of revenue generation for organisations. The maintenance with role creates a significant impact on some critical elements in production plants such as product quality, safety requirements, and operating budget levels (Khazraei & Deuse, 2011).

*Corresponding author. Email: Sherif.Mostafa@mymail.unisa.edu.au

The cost of maintenance activities could be ranged from 15 to 70% of the total production costs (Fraser, 2014; Pinjala, Pintelon, & Vereecke, 2006). This is the second largest part, after energy costs, of the operational budget. In the United States, the estimated cost of maintenance increased from $200 billion in 1979 to $600 billion in 1989 (Bevilacqua & Braglia, 2000). Maintenance activities account for an average 28 per cent of the total cost of finished goods (Blanchard, 1997). One of the reasons for such significant portion of maintenance of the total operation cost is that the machinery has become highly automated and technologically very complex. For example, usually, the modern operation systems depend on sensor-driven management systems that provide alerts, alarms, and indicators. Consequently, maintenance costs are expected to be even higher in future. Generally, the maintenance costs are proportional to the downtime (DT). The DT is the time interval when equipment/system is down for maintenance until it is back to the normal working conditions (Tinga, 2013). The increased DT is caused by the non-value added (NVA) activities or wastes within the entire maintenance process. One of waste elimination strategies is the application of lean thinking in all activities between suppliers and customers (value stream). Integrating lean thinking in maintenance is known as lean maintenance (LM). Baluch, Abdullah, and Mohtar (2012) emphasised that LM is prerequisite for success of a lean manufacturer as it provides a holistic approach to the function of maintenance. In general, the lean integration in any process is carried through adopting lean principles which begins with specifying the customer value (Bhasin, 2015). In the maintenance environment, any maintenance service could be considered as a final intangible product. The service is provided to a customer which, in this case, could be assumed as an asset (e.g. production line). Therefore, it is essential to identify the value from the asset perspective which can be improving its availability and reliability through efficient maintenance. Then, mapping the maintenance value stream which fundamentally consists of all the collective activities to deliver the maintenance service. Later, improving the maintenance value stream by abolishing the waste which assist in minimising the lead time (in this case is DT).

Investigation into the applicability of lean principles in maintenance in the existing research is still at a marginal level. Davies and Greenough (2010) emphasis on the necessity for more researches to apply lean principles to maintenance operations. Ghayebloo and Shahanaghi (2010) formulate a multi-objective decision-making (MODM) model that can determine the minimal level of maintenance requirements (i.e. labour and spare parts) which satisfies expected reliability level with the use of the lean concept. Tendayi and Fourie (2013) use a combined approach of quality function deployment (QFD) and analytic hierarchy process (AHP) to evaluate the importance of a set of maintenance excellence criteria and prioritise the lean tools against these criteria. Soltan and Mostafa (2014) introduce a framework for measuring maintenance strategies based on lean and agile components, i.e. waste removal and responsiveness. The study of McCarthy and Rich (2004) discussed lean total productive maintenance system (or lean TPM) which conceptualises the application of lean-specific techniques in TPM. The system is focused on maintaining equipment in its optimal operational state and continually improving its productivity. However, an integrative structure of lean thinking or lean TPM (e.g. principles, practices/tools, waste identification, and value stream mapping (VSM)) within the maintenance activities has not been fully established. This provides an opportunity for this study to develop and propose a process for lean thinking to be integrated to the maintenance operation. The process is formulated according to the hypothesis of Womack and Jones (2003) stating that lean principles can be applied to any sector.

This study carried out a systematic literature review following Denyer and Tranfield (2009) approach. The review was conducted to identify and understand the existing literature on LM and evaluate contributions and summarise knowledge, thereby, identifying potential directions of future research. The grand electronic databases were explored to gather the literature on LM. A total of 43 related articles published between 2000 and 2014 have been included in this study.

This study is organised as follows. Section 2 provides an overview on maintenance and lean manufacturing concepts. Sections 3 and 4 demonstrate the aim, objectives, and methodology of the proposed research. Section 5 discusses the LM and LM practices/tools. In Section 6, a LM process is proposed with the discussion of the five stages of the lean process in maintenance and introduction of the LM scheme including four bundles and 26 lean tools. The section also demonstrates the relationships between maintenance wastes and lean tools using House of Waste (HoW). Section 7 highlights benefits of the proposed LM process and Section 8 concludes the study with some directions of future research.

2. Maintenance and lean manufacturing

2.1. Lean manufacturing

The word lean in manufacturing means the efficient use of the available resources by cutting the NVA activities or wastes (Carrasqueira & Machado, 2008). Lean manufacturing represents a collection of tools that work together synergistically to create a streamlined, high-quality system that produces finished products at the same pace of the customer demand (Shah & Ward, 2007). Waste in lean manufacturing is defined as any activities that add cost to a product/service without adding values from a customer's perspective. It may be classified to three major types: unobvious waste, less obvious waste, and obvious waste (Hopp & Spearman, 2004). De Treville and Antonakis (2006) list examples of obvious waste as unnecessary inventory, unneeded processes, excessive set-up times, unreliable machines, and rework. They also argue that the less obvious waste occurs due to the various reasons such as process times, delivery times, yield rates, staffing levels, and demand rates.

In recent decades, lean system is gaining a momentum across different industrial sectors. It has been originally started as Toyota production system which describes the manufacturing philosophy of Toyota Motor Corporation (Dombrowski & Malorny, 2014; Holweg, 2007; Womack & Jones, 2003). This lean system has been successfully extended to service industries, such as maintenance service, retail banking, airlines, restaurants, public sector, education, food, and hospitals (Burgess & Radnor, 2013; George, 2003; Pedersen & Huniche, 2011; Resta, Powell, Gaiardelli, & Dotti, 2015; Smith & Hawkins, 2004; Thirkell & Ashman, 2014; Zarei, Fakhrzad, & Jamali Paghaleh, 2011). As an example, in the health care sector, hospitals achieved high-quality health care outcomes (e.g. recorded lower 30-day mortality rates) by applying lean management practices (McConnell, Lindrooth, Wholey, Maddox, & Bloom, 2013). Clearly, the implementation of lean concept in diverse sectors proves to be true Womack and Jones (2003) hypothesis stating that lean principles can be transferred to any organisation. They coined this process as lean thinking which refers to the thinking process of lean inside an organisation and its extended supply chain. This means that within the same organisation, lean thinking should be extended from the shop floor or production level to other areas such as maintenance department. The key behind this lean thinking is that service/maintenance departments and production processes are inseparable and complement each other to sustain the competitive edge of an organisation.

2.2. Maintenance strategies

Maintenance includes all activities required to keep an asset at maximum operating condition. The activities are usually carried out according to a certain maintenance strategy. The maintenance strategies may have developed accordingly with the development of manufacturing systems (Shahin, Shirouyehzad, & Pourjavad, 2012). In the early days, maintenance had been mainly concentrated around corrective maintenance, for example, repairs and replacements were conducted when needed with no optimisation strategy and there was no or little consideration of the DT. More recently, maintenance became a full-scale function, instead of a sub-function of the whole production operation. Today, maintenance management becomes a complex function, encompassing technical and management skills, while still requiring flexibility to cope with the dynamic business environment (Lee & Wang, 2008). Maintenance strategies have gradually changed from preventive maintenance (PM) (including condition-based maintenance (CBM) and time-based maintenance (TBM) to design-out maintenance (DOM) and total productive maintenance (TPM) as demonstrated in Figure 1.

The classification of the maintenance strategies shown in Figure 1 is based on the time of maintenance activities and failure that requires maintenance (Potes Ruiz, Kamsu Foguem, & Grabot, 2014). The maintenance activities are only performed after the failure occurrence in the corrective maintenance strategy. Whereas, in the PM, the intervention of maintenance activities is conducted before the failure occurrence. Maintenance strategies have been diversely used in the existing literature using similar terms such as preventive, predictive, planned, corrective, and TPM. The most common three maintenance strategies are discussed below.

2.2.1. Corrective maintenance

Corrective maintenance is known as failure-based maintenance, emergency maintenance, fire-fighting maintenance, or breakdown maintenance as the concept of corrective

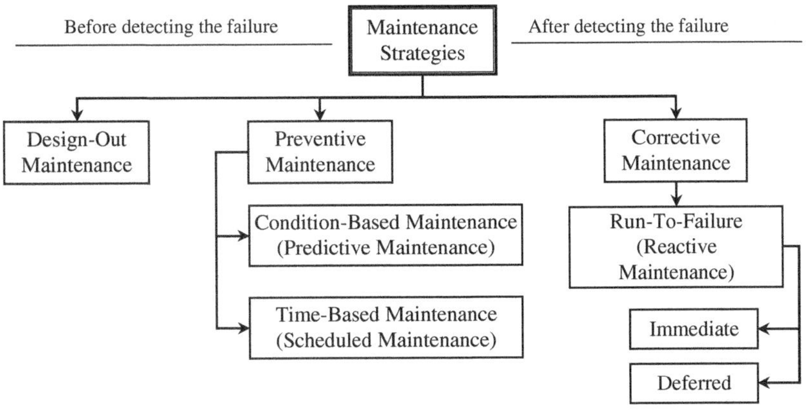

Figure 1. Types of maintenance strategies.

maintenance strategy is based on fix-it when broke (Márquez, 2007). Corrective maintenance is a conventional maintenance strategy appeared early in the industry. It has been employed in maintenance operations due to knowledge shortage on the equipment failure behaviours (Waeyenbergh & Pintelon, 2002). Corrective maintenance can be carried out immediately or deferred by appropriate maintenance technicians whom are contracted to assess the situation and fix the repairs. In situations where the failure is not critical (i.e. plenty of DT is available) and the values of the assets are not of a great concern, the corrective mode of maintenance may prove to be an acceptable option. However, the market competition, environmental and safety issues force the maintenance managers to search for more efficient maintenance strategies besides the corrective maintenance (Shahin et al., 2012).

2.2.2. Preventive maintenance

PM is carried out according to prescribed criteria. It intends to reduce the probability of failure or the degradation of the functioning of an item (Fouladgar, Yazdani-Chamzini, Lashgari, Zavadskas, & Turskis, 2012). PM can be divided into TBM and CBM. In TBM, the maintenance activities are performed based on fixed operating time interval or number of output units without considering the current condition state of the item. On the other hand, CBM is based on performance and/or parameter monitoring (e.g. corrosion and electric current monitoring, lubricant and vibration analysis, leak and crack detection, and ultrasonic testing) (Al-Najjar & Alsyouf, 2003; Khazraei & Deuse, 2011). CBM could be described as a process that integrates technology and human skills using a combination of all available diagnostic and performance data, maintenance history, operator logs, and design data to determine the likelihood of a potential failure. As a result, CBM requires a high initial cost to acquire and install the necessary sensors as well as to monitor technology (Hellingrath & Cordes, 2014; Nezami & Yildirim, 2013).

2.2.3. Design-out maintenance

DOM focuses on improving the design of a product in order to eliminate the cause to maintenance. DOM makes maintenance easier during the life cycle of a product (Waeyenbergh & Pintelon, 2004). DOM is based on the successive design corrections derived from the maintenance knowledge. It is appropriate for items with high maintenance cost, which arises because of defective design or operation outside design specifications. The DOM concept is used in some parts of motor vehicles such as permanent bearing (bearing using solid lubricant and permanently sealed) (Gopalakrishnan & banerji, 2013).

2.3. TPM and lean thinking

TPM is a concept developed in Japan, which could be defined as a productive maintenance that involves total participations meaning all employees across the whole levels of the operational hierarchy (Cua, McKone, & Schroeder, 2001; Jasiulewicz-Kaczmarek, 2013). It attempts to eliminate any losses in equipment and production efficiency through active team-based participation. Waeyenbergh and Pintelon (2002) identified the six categories of losses in TPM: (1) breakdown losses; (2) set-up and adjustment losses; (3) minor stoppage/idling losses; (4) reduced speed losses; (5) defects/rework losses; (6) start-up losses. The first two losses refer to time losses and influence equipment availability. The third and fourth losses denote speed losses and measure the equipment performance efficiency. The last two losses designate the quality of production process and are used to calculate the quality rate of the equipment (Chan, Lau, Ip, Chan, & Kong, 2005). The effectiveness of TPM strategy with regard to these six losses is measured using overall equipment effectiveness (OEE). The measurement of OEE is a function of the availability, performance efficiency, and quality. TPM entails eight main elements/pillars that can be considered as principles/tools of TPM in an organisation.

- Autonomous maintenance
- Performance improvement
- Early equipment management
- Planned maintenance
- Environment health and safety
- Office TPM
- Education and training
- Quality maintenance

The detailed description of these pillars mentioned above is presented in Ahmed, Hassan, and Taha (2005) and Chong, Chin, and Hamzah (2012). The TPM pillars can be collectively used to improve equipment availability and reliability. However, according to Baluch et al. (2012), removing or missing application of any pillars could lead to unattainable results. Likewise, the lack of comprehensive approach for the TPM implementation has resulted in a decrease of 50% TPM initiatives in US. Moreover, there are some deficiencies related to the TPM strategy. TPM concerns the operational issues that handle equipment failures rather than long-term strategic business issues. Levitt (2008) stated that TPM is no longer lean as it focuses primarily on the operational issues of equipment and it would take a long time to implement. Moreover, EPA (2015) highlighted that TPM failed to consider the environmental aspects during equipment efficiency improvement. This leaves potential waste minimisation and pollution prevention opportunities to be more researched.

While referring to major literature on TPM, it was observed that there is a link between TPM and lean as well as LM emerged as an effective maintenance strategy. Levitt (2008) mentioned that TPM and lean are broadly linked. This is clear in some of the literature that considered TPM as a subsection of lean system which mainly focused on enhancing the efficiency and availability of the manufacturing facilities. For instance, Shah and Ward (2003) postulated 21 manufacturing practices as the key features of lean system. They grouped the inter-related and internally consistent practices into bundles. TPM was one bundle which contained four practices: PM, maintenance optimisation, new process equipment, and safety improvement. These four practices sound very generic and not provide practitioners what tools to undertake the maintenance optimisation, or new process equipment. Another study by Mostafa (2011) extended the TPM practices, suggested by Shah and Ward (2003), to include housekeeping, cross-training and teams of maintenance technicians, operator involvement (autonomous maintenance), and information tracking of the work orders. Clearly, these suggested practices indicated another dimension of TPM in lean system by considering human resource perspective. Abdulmalek, Rajgopal, and Needy (2006) gathered 14 lean tools into three areas: quality, production process, and methods. TPM was located under quality area which can be considered as another perspective for TPM in lean environment. It can be concluded that TPM is considered as the foundation for the maintenance process in lean environment which must be supplemented with some lean practices. That is why some studies coined relatively new terms like lean TPM (Georgescu, 2010; McCarthy & Rich, 2004) and LM (Baluch et al., 2012; Okhovat, Ariffin, Nehzati, & Hosseini, 2012; Romano, Murino, Asta, & Costagliola, 2013; Smith & Hawkins, 2004).

LM principles take its lead from lean manufacturing through applying some new techniques to TPM concepts to render a more structured implementation path (Levitt, 2008; McCarthy & Rich, 2004; Romano et al., 2013; Smith & Hawkins, 2004). It is a prerequisite for success as a lean manufacturer that provides a holistic approach to the function of maintenance (Baluch et al., 2012; Soltan & Mostafa, 2014). As lean concept has been taking hold in the manufacturing sector, there is an increasing insight that maintenance must not be seen only from narrow operational perspective dealing with equipment failures and their consequences. Instead, maintenance must be viewed in the long-term strategic perspective and must integrate the different technical and commercial issues in an effective manner. However, LM approach cannot just be a mirror image of a lean manufacturing approach because the business dynamics of asset maintenance and those of production are fundamentally different (Baluch et al., 2012; Brown, Collins, & McCombs, 2006; Clarke, Mulryan, & Liggan, 2010). Therefore, it is clear that there is a need to develop an effective process to collectively integrate lean thinking into the maintenance with long-term strategic perspective. This study addresses such an issue to cover the major deficiencies in the existing literature.

3. Research aim and objectives

This research aims to adopt lean principles and practices/tools and collectively integrate them into the maintenance process. It has adopted recommendations of previous research works such as a study of Davies and Greenough (2010) that emphasised on the necessity of conducting more research in application of lean manufacturing principles in maintenance operations. To achieve the main aim of this research, four main objectives have been established:

(1) To review the LM concept.
(2) To develop a process for adopting lean principles into maintenance processes.
(3) To document the maintenance wastes and establish a scheme of lean practices/tools applied to maintenance.
(4) To develop the HoW that demonstrates the association of lean tools with the maintenance wastes.

4. Research methodology

To achieve the research objectives, this study employed a systematic review of the literature in order to explore major publications related to lean manufacturing and maintenance concepts, especially research works published from 2001 to 2014. Okoli and Schabram (2010) indicate that a systematic literature review is a systematic, explicit, comprehensive, and reproducible method for identifying, evaluating, and synthesising the existing body of completed and recorded work produced by researchers, scholars, and practitioners. According to Denyer and Tranfield (2009), the systematic literature review has become an essential scientific activity.

From the literature review, it has been found that exploration of the applicability of lean principles in maintenance in the existing research works is minimal. Davies and Greenough (2010) and Soltan and Mostafa (2014) state the necessity of conducting more research on applying lean principles in maintenance operations. As an initial step of this research, a seven-step systematic literature is conducted as summarised in Figure 2 to demonstrate development of the proposed LM process. It is noted that these research steps are adapted from several academic sources, such as Joffe (2011), Okoli and Schabram (2010), and Tranfield, Denyer, and Smart (2003).

4.1. Research selection criteria

The research inclusion and exclusion criteria are very critical to the quality assessment of papers (Booth, Papaioannou, & Sutton, 2012). Okoli and Schabram (2010) indicated that simplifying research using these criteria (by reviewing the title and then the abstract when needed) saves the researcher time and effort. In this study, the authors examined research articles by title and then abstracts. In the systematic review of this study, the criteria were addressed in order to clarify the selection of the research-related articles. The following criteria have been considered to include/ exclude the articles:

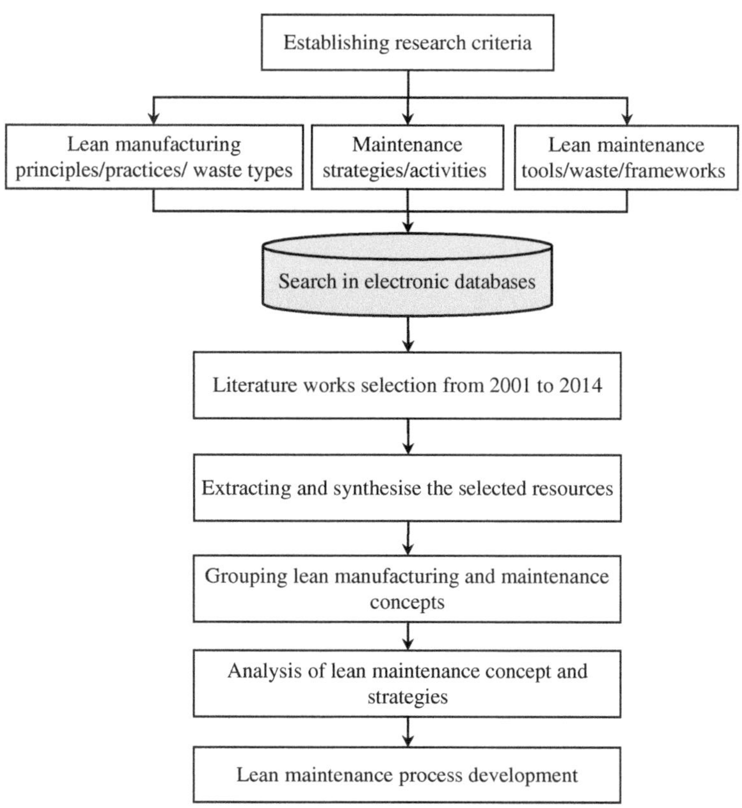

Figure 2. Research methodology framework.

- Articles published between 2001 and 2014 as the rational for considering the year 2001 is that LM as a field of research was firstly addressed by Davies and Greenough (2001)
- Search for articles published in peer-reviewed scientific journals or conference proceedings
- Ensure essential relevance by requiring that selected articles contain at least one of the search keywords in the title or abstract (see Table 1)
- Search well-known online databases which are Taylor & Francis, EBSCO, Emerald, IEEE, Inderscience, ProQuest, Sage, Science Direct, and Springer Link

5. Development of LM in the literature

The screened LM studies are dated from 2001 to 2014 with the total number of 43 retrievable publications as shown in Table 2.

The selection of published LM studies shows that most of the studies were published in the year 2014 and focused on three aspects, namely LM implementation initiative, suggested lean tools, and interrelationship within LM application as illustrated in Figure 3.

From the review of 43 publications, the focus on the LM implementation has been continuously growing since 2011 for general practice and specific industries. The suggested lean tools and interrelationship within the LM initiatives were not in parallel with the LM implementation. The LM tools were found in 18 publications. None of these publications provided a complete set of lean tools. Only two publications in 2012 and 2013 demonstrated the interrelationship within LM application. The low volume of publications indicates that more research in LM is crucially needed, especially in establishing a logical interrelationship between LM concepts and implementation and a complete set of LM tools.

5.1. Lean maintenance

LM term was coined in the last decade of the twentieth century. Smith (2004) defines LM as a proactive maintenance operation employing planned and scheduled maintenance activities through TPM practices using maintenance strategies developed through application of reliability-centred maintenance (RCM) decision logic and practiced by

Table 1. Research primary and secondary keywords.

Time period considered	Search engines used	Primary keywords	Secondary keywords
2001–2014	Taylor & Francis, EBSCO, Emerald, IEEE, Inderscience, ProQuest, Sage, Science Direct, and Springer Link	• Lean maintenance • Lean TPM	• Lean principles AND maintenance • Lean practices AND maintenance • Lean tools AND TPM

Table 2. Review of existing literature on lean maintenance.

S. no.	Author(s) and year	Key aspects covered	Application industry	Suggested lean maintenance implementation	Limitations
1	Davies and Greenough (2001)	(1) Investigate the level of presence of lean thinking within the maintenance domain and its influence and interaction with the maintenance function		No	No
2	Davies and Greenough (2003)	(1) A set of performance measures that can be used to analyse the impact of lean thinking within the maintenance function		No	No
3	Davies (2003)	(1) Explained the contribution of lean thinking to the maintenance of manufacturing system (2) An overall measure of maintenance performance comprised a number of indicators that signify change through maintenance activity has been suggested		No	No
4	Heisler (2003)	(1) Highlight the role of effective planning and scheduling of maintenance process in lean environment	General-planning and scheduling	Lean maintenance process flow	• Present the flow of maintenance processes for the lean maintenance planning and scheduling • No lean tools or principles demonstrated
5	Smith (2004)	(1) Defined the lean maintenance (2) Discussed the elements for lean maintenance success		No	

(*Continued*)

Table 2. (*Continued*).

S. no.	Author(s) and year	Key aspects covered	Application industry	Suggested lean maintenance implementation	Limitations
6	Smith and Hawkins (2004)	(1) Reviewed the history and evolution of lean (2) Explained the lean maintenance element and TPM (3) Discussed the pre-planning and mobilising lean maintenance (4) Introduced the key elements for sustain lean	General	The maintenance management pyramid	• Limited tools provided • Flow of the implementation process is limited
7	Verma and Ghadmode (2004)	(1) A model for ship fleet repair and maintenance based upon lean principles and current best practices	Ship fleet repair	Proposed implementation model for ship fleet repair	• The model contained 7 steps with no interrelationship or logic to apply them • No feedback or measuring of the performance • No lean tools suggested
8	Zwas (2006)	(1) Investigated how lean techniques can be employed in bus and rail maintenance		No	
9	Colbert (2008)	(1) Benefits of integrating lean into IR predictive maintenance programmes (2) Lean concepts/tools are able to better manage and maintain the real value of their IR programmes while drastically increasing their ROI		No	No
10	Van den Heuvel (2008)	(1) Increasing the efficiency of the activities during regular maintenance shutdowns at steel Sheet Plant using lean and six sigma approaches		No	No

11	Marks (2008)	(1) Created an accurate and enhanced CMMS database and a safe and efficient Rotable Plant storage system (2) Explored and implemented an Effective Rotable Plant Management Programme that fulfils the lean maintenance strategy (3) Demonstrated how all value adding in terms of Effective Rotable Plant Management is maintained using very little resources		No	No
12	Yile, Hang, and Lei (2008)	(1) Based on the value stream analysis (VSA), inherit 'value' relation of TBM, RCM, and remanufacture engineering is established (2) A simple layered technology framework of lean maintenance	Frictional clutch maintenance	Maintenance management pyramid	• The pyramid defined the lean maintenance as an integration of 13 practices • No limitation process flow were provided • The lean principles were not included
13	Georgescu and Militaru (2009)	(1) A lean TPM approach to the current understanding of the TPM system (2) 7Ss as a critical first step in any improvement programme; instant maintenance; improvement set-up operations		No	No
14	Thiruvengadam (2009)	(1) A method for assessing the breakdown maintenance factors using a value stream maintenance map in a clearly and precisely measurable technique		No	No

(*Continued*)

Table 2. (*Continued*).

S. no.	Author(s) and year	Key aspects covered	Application industry	Suggested lean maintenance implementation	Limitations
15	Clarke et al. (2010)	(1) One of the most important aspects of lean maintenance is developing an understanding of the maintenance processes and applying a risk-based approach in a pharmaceutical asset maintenance. This involves evaluating whether each element of maintenance practice used adds value to the product and benefits the customer		No	No
16	Davies and Greenough (2010)	(1) A lean practice template which represent activities possible within a company's maintenance department (2) Performance indicators to identify the impact of lean thinking within maintenance activities		No	No
17	Georgescu (2010)	(1) A new maintenance system called lean TPM. It is a systematic approach based on three techniques: the 7Ss as a critical first step in any improvement programme; instantaneous maintenance; improvement set-up operations		No	No
18	Ghayebloo and Shahanaghi (2010)	(1) A MODM model to determine minimum level of maintenance requirements and to satisfy expected reliability level		No	No
19	Sheng and Tofoya (2010)	(1) Four core lean rules (activities, connections, flows, and improvements) transferred in the maintenance context		No	No

20	Ayeni, Baines, Lightfoot, and Ball (2011)	(1) Application of lean within MRO in the aviation industry		No	No
21	Kolanjiappan and Maran (2011)	(1) Application of root cause failure analysis (RCFA), RCM, Failure mode and effect analysis, and maintenance procedure effectiveness analysis	Aircraft maintenance	Schematic diagram for servicing time reduction of aircrafts	• The focus of the diagram is to reduce the service time not to implement lean maintenance • process flow is missing • no lean tools provided
22	Qiang, Zhu, and Li (2011)	(1) The characteristics of the military equipment lean maintenance (MELM) and main components for improving the MELM	Military equipment	Components of MELM	• The MELM contains three stages: models of maintenance, components of MELM, and output • No implementation introduced • Lean tools has been suggested
23	Baluch et al. (2012)	(1) Lean maintenance cannot be a mirror of lean production as the business dynamics of asset maintenance and production are fundamentally different (2) TPM and lean initiatives that allowed organisations to focus on improving the efficiency of production processes		No	No
24	De Silva, Ranasinghe, and De Silva (2012)	(1) Improving maintainability of buildings can promote lean maintenance (2) Challenges of lean maintenance in the construction industry		No	No

(Continued)

Table 2. (*Continued*).

S. no.	Author(s) and year	Key aspects covered	Application industry	Suggested lean maintenance implementation	Limitations
25	Huang, Bian, and Cai (2012)	(1) Key technical characteristics of lean maintenance (RCA, FMEA, value stream analysis) (2) Realisation of lean maintenance (LM) through updating maintenance concept, highlighting the academic innovation and integration, and enhancing the culture construction of LM theory (3) A theoretical basis for the equipment maintenance reform	Military equipment maintenance	LM technical supporting frame	• The frame showed the key techniques of LM • Some lean tools have been provided • No implementation of the LM introduced
26	Okhovat et al. (2012)	(1) Developed a strategic framework to provide guidance and support to reach world-class standards both in maintenance and manufacturing processes through continual improvement (2) The framework integrated three process improvement strategies: six sigma, TPM, and lean	General-world-class manufacturing	Framework for world-class standard	• General frame for production and maintenance included TPM, lean, and 6σ • Limited lean tools documented for maintenance process
27	Rastegari (2012)	(1) A maintenance strategy in manufacturing organisation that can be linked to the company's business strategy in lean environment		No	No
28	Tinashe George Tendayi and Fourie (2012)	(1) A framework for implementing a lean in a rolling stock maintenance environment that seeks to eliminate waste, add value, and continuously improve its supply chain	Rolling stock maintenance	Lean maintenance supply chain framework	• Frame for specific case study supply chain • Not incorporated the lean principles • Lean tools are not included

29	Hou, Zhao, Meng, and Ma (2013)	(1) Lean Maintenance System Model for China's hydropower equipment maintenance enterprises by the (2) Application of the lean production theory and ideas, and the model consists of lean culture building, system file structure, maintenance project management, maintenance on-site management and safety management		No	No
30	Jahanbakhsh, Moghaddam, and Samaie (2013)	(1) Benefits of using lean approach in Teen Dairy Industry to reduce the amount of losses and rework of maintenance systems		No	No
31	Romano et al. (2013)	(1) A LRCDA method to reduce scraps and work-in-process in manufacturing context (2) LRCDA merges steps of RCFA, lean maintenance, and TPM	Power cables factory	LRCDA model	• The model is based on the RCDA not lean principles • The model contained limited lean practice within the model
32	Tendayi (2013)	(1) A framework, based on lean thinking tools and relevant performance measures to prove the applicability or otherwise, of lean thinking in an operational maintenance environment outside the traditional domain of manufacturing	Rolling stock maintenance	Framework for lean thinking in maintenance	• Insufficient implementation process flow • Limited lean tools introduced
33	Djurovic and Bulatovic (2014)	(1) Implementation of the 'S5' LEAN methods concept in maintenance, storage, as well as other elements of the business system		No	No
34	Bulatovic and Djurovic (2014)	(1) Lean production cannot be implemented without the use of lean strategy in the equipment maintenance		No	No

(Continued)

Table 2. (*Continued*).

S. no.	Author(s) and year	Key aspects covered	Application industry	Suggested lean maintenance implementation	Limitations
		(2) 5S is one of the most basic tools in the lean maintenance			
35	Dunn (2014)	(1) Focus on key elements within each of the work management framework processes that will produce improvements in maintenance productivity and remove waste (1) Three ways in which continuous improvement approaches can further improve Maintenance Productivity and eliminate waste: • Improving maintenance efficiency • Improving the predictive maintenance programme • Eliminating defects		No	No
36	Irajpour, Fallahian-Najafabadi, Mahbod, and Karimi (2014)	(1) Identify and evaluate the effectiveness of a given maintenance strategy and to rank components of maintenance system (2) Using DEMATEL method on maintenance strategy as a guideline to rank the lean components in maintenance	General-rank component of maintenance	Model of effective maintenance strategy	• The model is used to rank the effective elements of maintenance strategy • Lack of model elements dependency • Limited lean tools provided • Lean principles is not incorporated in the model
37	Jasiulewicz-Kaczmarek (2013)	(1) The common elements of the lean and green paradigms introduced for better understanding of synergy between them and overall improvement of maintenance efficiency	General-production management	Lean & green maintenance results	• The frame demonstrated the requirements and results of using lean and green in maintenance process • No presentation of how to incorporate either lean or green in the maintenance

38	Nima, Hesamodin, and Alireza (2014)	(1) Development of a practical model to define an optimised effective maintenance strategy using lean approach		No	No
39	Önder (2014)	(1) How maintenance may contribute to decreasing the environmental impact of production	General-lean green maintenance concept	Reducing environmental impacts and improving efficiency with lean green maintenance	• Generic framework contained four phases: plan, learn, action, and improve • Not comprehensive nor designed based on lean principles • Very limited lean tools were suggested
40	Shou, Wang, Wang, Hou, and Truijens (2014)	(1) Apply BIM and lean concepts to practical maintenance to improve efficiency		No	No
41	Soltan and Mostafa (2014)	(1) Lean and agile can manage the maintenance processes (2) Lean and agile maintenance should be considered as a prerequisite for any successful lean or agile application (3) A framework to measure the performance of maintenance strategies based on lean and agile factors	General	Lean and agile factors in maintenance operations	• No lean tools provided • No implementation process flow provided

(Continued)

Table 2. (*Continued*).

S. no.	Author(s) and year	Key aspects covered	Application industry	Suggested lean maintenance implementation	Limitations
42	Stuchly and Jasiulewicz-Kaczmarek (2014)	(1) The maintenance aim has been shifted from production paradigm to sustainable development that resulted in a change towards the product life cycle as well as taking into account economic, environmental, and social aspects (2) Maintenance offers numerous opportunities of decreasing influence of business processes on natural environment and more efficient resources utilisation		No	No
43	Sunjka and Murphy (2014)	(1) Determine the status of lean implementation within South African aircraft maintenance organisations (AMOs)		No	No

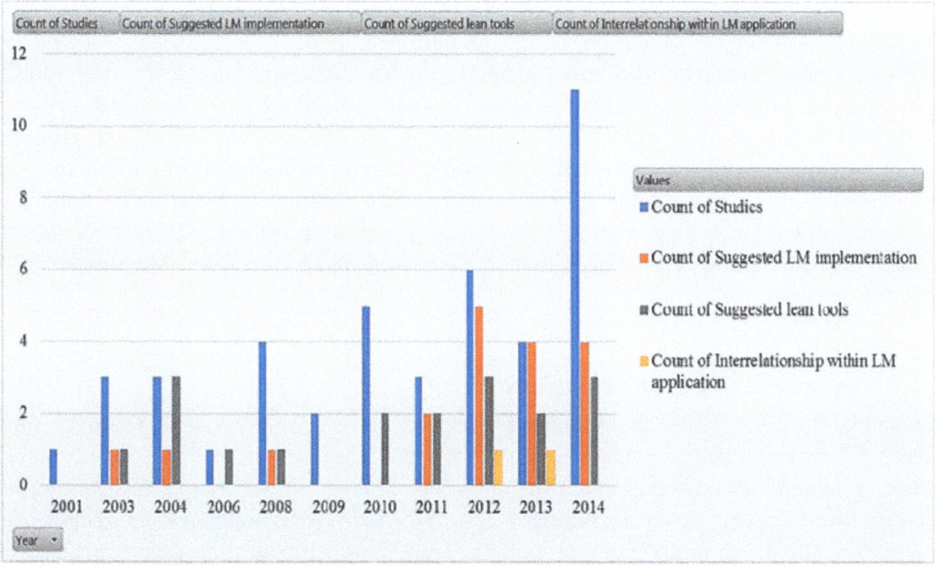

Figure 3. Distribution of lean maintenance studies.

empowered (self-directed) action teams. LM generates a desirable outcome by minimising consumption of inputs (Smith & Hawkins, 2004). LM represents adopting lean principles into the maintenance, repair, and overhaul (MRO) operations. It could reduce unscheduled DT through optimising maintenance support activities and maintenance overhead. The lean tools are representing the lean principles for the implementation process (Mostafa, Dumrak, & Soltan, 2013). To achieve LM improvement effectively, key lean tools such as VSM, 5S, and visual management need to be employed (Smith, 2004; Smith & Hawkins, 2004). A comprehensive lean tools developed for maintenance activities within an organisation include 5S, TPM, OEE, Kaizen, Poka-Yoke, process activity mapping, Kanban, computer managed maintenance system (CMMS), enterprise asset management (EAM) system, and Takt time (Davies & Greenough, 2010; Smith, 2004).

Despite the benefits of LM mentioned earlier, the review of previous studies conducted in this study found that the investigation on the applicability of lean principles into maintenance is marginal. The existing research works have been largely limited to the manufacturing environment where LM is practised as a prerequisite for lean manufacturing (Tendayi, 2013). This proposition has been mentioned in Davies and Greenough (2010) emphasising on the necessity of conducting more research on practical application of lean manufacturing principles in maintenance operations. It was discovered that the previous studies mainly focused on ranking the maintenance strategies based on some specific scope. Moreover, few initiatives have included comprehensive frameworks or models that can integrate lean thinking in operational maintenance environments outside of the manufacturing context. Ghayebloo and Shahanaghi (2010) formulate a model for determining the minimal level of maintenance requirements and satisfying reliability level through the use of the lean concept. Tendayi and Fourie (2013) use a combined approach between QFD and AHP to evaluate the importance of maintenance excellence criteria and prioritise the lean tools upon these criteria. The recent study of Soltan and Mostafa (2014) introduces a framework for measuring maintenance strategies based on lean and agile components, i.e. waste removal and responsiveness. However, the study cannot provide sufficient practical application of lean concept in the maintenance process. Romano et al. (2013) formulate lean root cause and defect analysis (LRCDA) to reduce scraps and work-in-process in manufacturing system. Nevertheless, the LRCDA model introduced is based on RCDA (not lean principles) as well as contains limited lean practice within the model. The paucity of practical application in the existing LM studies provides an opportunity for this study to expand the prevailing knowledge into a new framework for lean integration in the maintenance process.

5.2 Lean tools for maintenance activities

Reducing the NVA within maintenance activities can be accomplished through implementing lean tools (Jasiulewicz-Kaczmarek, 2013). The lean tools that suit the maintenance activities have been stated in previous studies. Smith and Hawkins (2004) identify the key lean tools including VSM, 5S, and visual management. Davies and Greenough (2010) develop a comprehensive lean tools template that represents possible lean activities within the maintenance process within an organisation. The tools are 5S, TPM, OEE, standards, mapping, inventory management, and visual management. Okhovat et al. (2012) suggest six lean tools that fit the maintenance processes of an organisation. These tools include visual control, 5S, seven wastes, single minute exchange of die (SMED), and Poka-Yoke (mistake proofing). Clarke et al. (2010) target eight LM practices as a preparation for delivering lean project objectives in a pharmaceutical organisation. A list of the references including LM tools is demonstrated in Table 3.

6. A proposed LM process

This section demonstrates an attempt to propose a process for adopting lean thinking into the maintenance activities. The process adopts the hypothesis of Womack and Jones (2003) that lean principles can be deployed to all organisations and sectors. Lean principles have being increasingly extended for industrial and service sectors. This is known as lean thinking which refers to the thinking process of lean (Holweg, 2007). The process proposed in this study is designed on the basis of the five lean manufacturing principles stated by Womack and Jones (2003). Some authors including Karim and Arif-Uz-Zaman (2013) develop lean implementation methodology based on the five lean principles for a manufacturing environment. Mostafa et al. (2013) state that lean practices/tools represent lean principles in the implementation process.

The process introduced in this study could be considered as an attempt to pave the way of applying or adopting lean principles to maintenance activities (Figure 4). The process provides guidance and support towards maintenance excellence for an organisation pursuing to extend lean practice to its maintenance department or other organisations starting introducing lean thinking to maintenance department. The process was developed through conceptual integration of five lean principles as they are the backbones of any lean initiatives (Womack & Jones, 2003). The principles specify the value, identify the value stream, flow the value, pull the value, and pursuing perfection. In addition, analysing and addressing limitations of the existing initiatives assist in developing the comprehensive LM process proposed in this study. The process is more flexible and can be adjusted according to any maintenance strategy. It could work

Table 3. Lean maintenance practices/tools reported in previous studies.

Lean maintenance tool	Previous study 1	2	3	4	5	6	7	8	9	10	11	12	13	14	15	16	17	18
TPM	*	*	*	*		*	*			*	*			*		*	*	
5S/CANDO	*	*	*	*	*		*		*		*		*	*	*	*		
Kaizen (continuous improvement)	*	*	*	*						*		*		*	*	*	*	*
CMMS/EAM	*					*	*				*			*			*	
Distributed MRO storerooms	*	*				*								*				
RCFA/FMEA	*		*	*		*		*			*						*	
PdM	*	*				*								*			*	
Autonomous maintenance		*					*			*				*			*	
SMED		*			*		*		*									
Poka-Yoke			*	*	*				*	*	*	*	*		*	*		
PDCA										*	*					*		
OEE				*														
Kanban				*				*					*		*	*	*	
Jidoka													*					*
JIT/inventory management			*	*			*						*		*	*		*
RCM						*	*	*			*			*			*	
Process mapping (VSM)				*						*			*			*		
Maintenance and reliability group				*										*				
Work standardisation	*		*	*				*				*		*	*	*		
Story boarding			*	*														
Visual control			*	*	*				*						*			
Work order system			*											*			*	
Self-audit				*														
Supplier association				*														
Open book management		*		*														
Empowered maintenance team												*		*				
Multi-skilled work team												*						
Maintenance crew training and learning							*					*						
Hoshin planning										*							*	*
A3										*								

Note: 1 – Baluch et al. (2012); 2 – Clarke et al. (2010); 3 – Davies and Greenough (2003); 4 – Davies and Greenough (2010); 5 – Djurovic and Bulatovic (2014); 6 – Huang et al. (2012); 7 – Irajpour et al. (2014); 8 – Kolanjiappan and Maran (2011); 9 – Okhovat et al. (2012); 10 – Önder (2014); 11 – Qiang et al. (2011); 12 – Romano et al. (2013); 13 – Smith and Hawkins (2004); 14 – Smith (2004); 15 – Tendayi (2013); 16 – Verma and Ghadmode (2004); 17 – Yile et al. (2008); 18 – Zwas (2006).
*Presence of the lean maintenance practice within the study.

simultaneously and complementary with previous framework developed for shop floor area as it is inclusive for the maintenance processes. The proposed process entails five stages and detailed steps within each stage.

6.1. *Stage one: specify the value*

The first stage focuses on defining an organisational maintenance system including activities, maintenance planning, strategies, and maintenance crew. In this stage, the employees training on LM wastes are assigned. Furthermore, identifying the types of wastes in maintenance processes is included. The core concept of lean manufacturing is eliminating the seven cardinal forms of waste. The seven cardinal types of waste in the maintenance process can be discussed in the same manner as in the eight waste types identified in the production system (Baluch et al., 2012; Clarke et al., 2010; Davies & Greenough, 2010).

(1) Too much maintenance: performing PM and predictive maintenance (PdM) tasks at intervals more often than optimal which results in the overproduction of maintenance work.
(2) Waiting for maintenance resources: production department is waiting for maintenance personnel to perform the maintenance service. It involves waiting for tools, parts documentation, and buys extra tools and stores them near the job location.
(3) Centralised maintenance: the centralisation of the MRO stores that are far from the job, commonly used repetitive parts that have not been kitted, documentation that must be hunted down, and work orders for machines that are not available all cause excess transportation. Therefore, maintenance personnel spend more time in motion and transportation which do not add value to the process.
(4) Non-standard maintenance: maintenance operations are normally conducted to achieve operation as soon as possible with no standard guidelines. This sometimes eliminates an opportunity to perform a higher quality repair.
(5) Excessive stock: the MRO inventory contains needed materials and spares. Additionally, work-in-process inventories may be used to ensure availability of required materials. Inventory for a maintenance operation also includes the work order backlog. Excessive maintenance work inventory results in slow response, unexpected breakdowns, and a high reactive labour percentage.
(6) Double handling: the wasted motion is usually concentrated around PM tasks. Doing inspection monthly on a pump that has not changed status in three years should be extended longer to quarterly, semi-annually, or annually depending upon the criticality of that piece of equipment.
(7) Poor maintenance: performing incorrect repair is a source of poor maintenance. Incorrect maintenance requires several repeated times to complete the repair job correctly. This affects the maintenance cost and the quality of the product. Applying proper training and detailed procedures can assist in poor maintenance elimination.
(8) Under-utilisation of maintenance crew: maintenance technicians do NVA work or do not perform as required/at the best interest of the organisation.

6.2. Stage two: identify the value stream

This stage includes all maintenance-related activities and processes. The stage starts by mapping the maintenance value stream then locating the wastes sources. This stage ends with setting equipment performance measures such as availability, OEE, and meantime between failures. VSM is used for visualising the flows of information and materials within a supply chain. VSM primarily helps an organisational management to recognise different forms of waste and its sources. One key metric of VSM is value added (VA) time percentage which measures VA activities against NVA activities (Monden, 1998). Standard icons for drawing the current and future VSM are available in Sullivan, McDonald, and Van Aken (2002). These icons should be modified to fit the maintenance activities. In this study, a new set of VSM icons fitting all the maintenance activities was introduced (see Figure 5). The new icons were designed using Edraw Max™

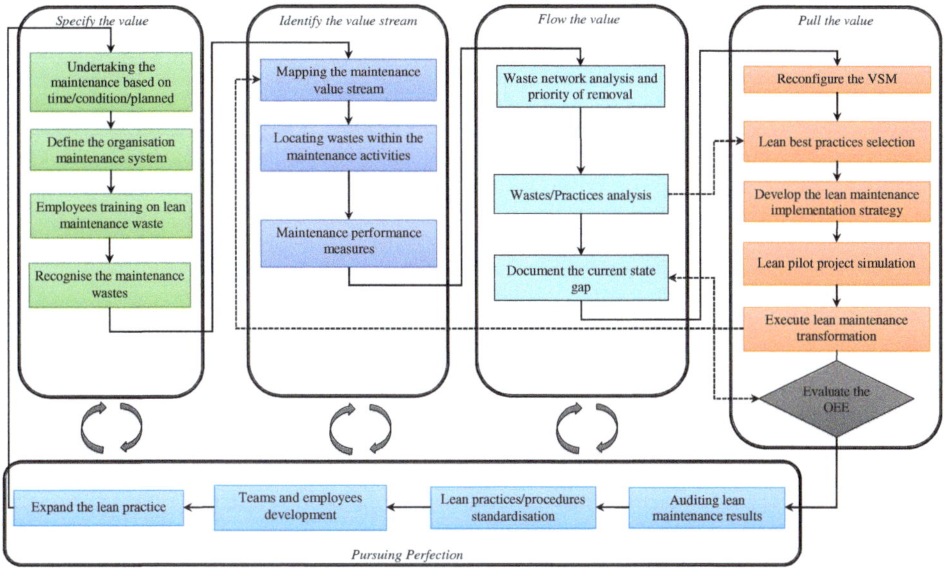

Figure 4. Proposed lean maintenance process.

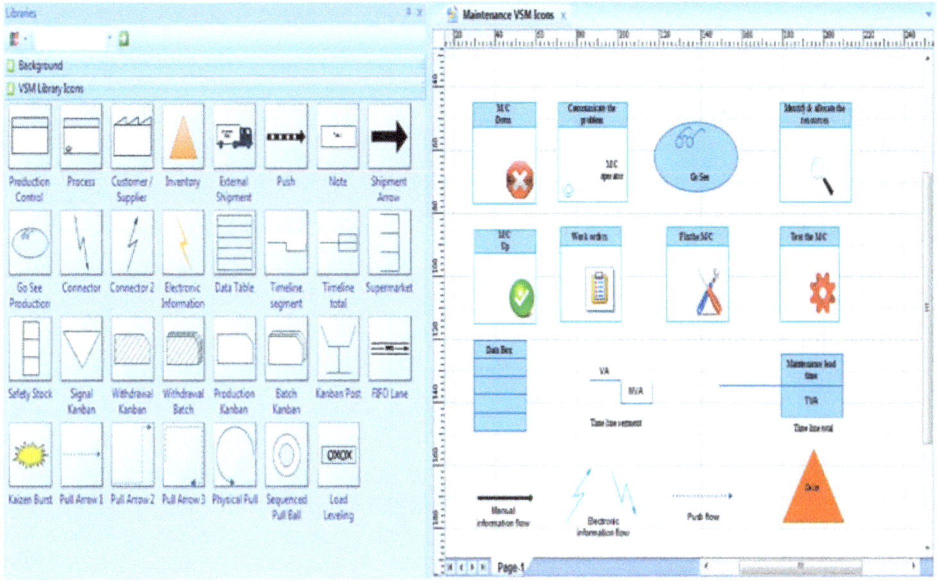

Figure 5. VSM symbols for maintenance activities (developed from Edraw Max software).

professional software package. The discussion of each activity is displayed in Table 4. An example of mapping the maintenance process using these icons for visualising a general current state is displayed in Figure 6.

Table 4. Maintenance VSM symbol description.

	Symbol	Description
(1)	Machine down	Machine need maintenance (e.g. break down, time-based, or condition-based)
(2)	Communicate the problem	Machine operator communicates the maintenance department
(3)	Go see	Maintenance personnel go and check the machine condition and report to the maintenance department
(4)	Identify and allocate resources	Identification and locating of appropriate resources such as tools, spare parts, manpower for fixing an equipment
(5)	Work orders	Generating maintenance work order through the maintenance software
(6)	Machine fix	Steps for fixing the machine
(7)	Test the machine	Testing of machine after repair until first good part is produced
(8)	Delay	Waiting time due to unavailability of resources (e.g. technician, tools, and spare parts)
(9)	Manual information flow	The flow of information from reports
(10)	Electronic information flow	The flow of information from the internet, intranet, local area network, wide area network, and other notes
(11)	Push flow	Represent the physical flow sequence of the maintenance activities
(12)	Time line segment	Represent the VA and NVA time for each activity
(13)	Time line total	Represent the maintenance lead time (down time)
(14)	Data box	Record the information of each maintenance process including cycle time (C/T), changeover time (C/O), and number of employees

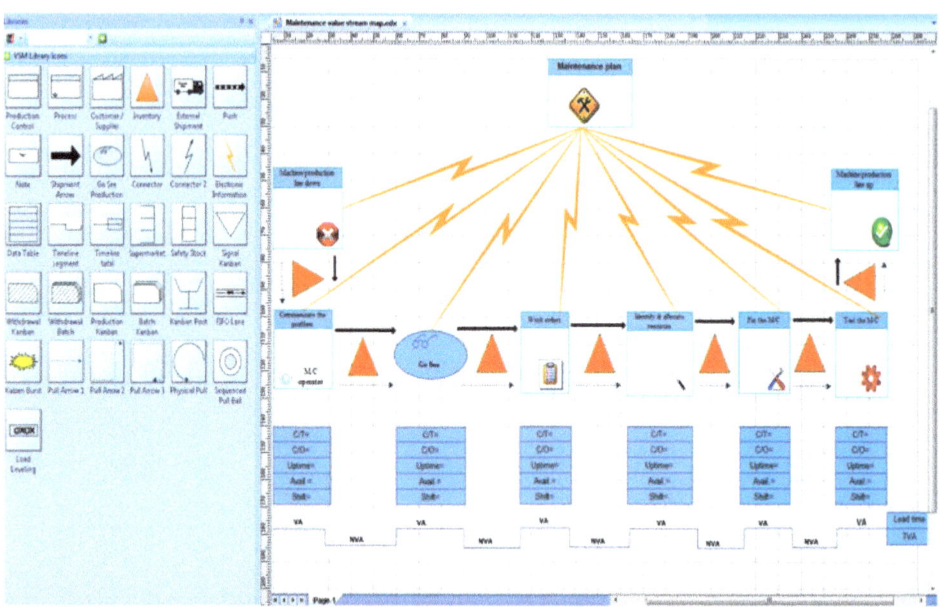

Figure 6. Example of general maintenance state map (developed from Edraw Max software).

6.3. Stage three: flow the value

The process of 'flowing the value' starts from waste network and waste/practices analyses before the results of the analyses are documented at the current state gap of maintenance using the calculation of OEE within an organisation. The documentation of lean practices for maintenance was conducted through the literature presented in Table 3 through marking out the most frequently suggested practices in these studies and then constructed the hierarchical scheme for LM practices as demonstrated in Figure 7. The structure is similar to Shah and Ward (2003), however, for the use of this structure is dedicated to the maintenance process not the manufacturing process. The structure consists of two levels and four bundles: just-in-time (JIT), total quality management (TQM), human resource management (HRM), and TPM and practices assigned under each bundle. The scheme can be used to indicate the association between the eight types of maintenance waste and the LM practices. The success of the LM depends on the application of each bundle. Each practice provides some benefits within the maintenance process. As a result, the performance of a whole maintenance department can be improved. The four LM bundles are briefly explained below.

6.3.1. JIT bundle

The JIT bundle encompasses all practices which are designed to reducing and eliminating unnecessary inventory and waiting in maintenance activities. These practices are Kanban, SMED, work standardisation, Takt Time, visual control, distributed MRO storeroom, and CMMS.

6.3.2. TQM bundle

The TQM bundle aims at continuously improving and sustaining the quality of products and processes through the participation of management, workforce, suppliers, and

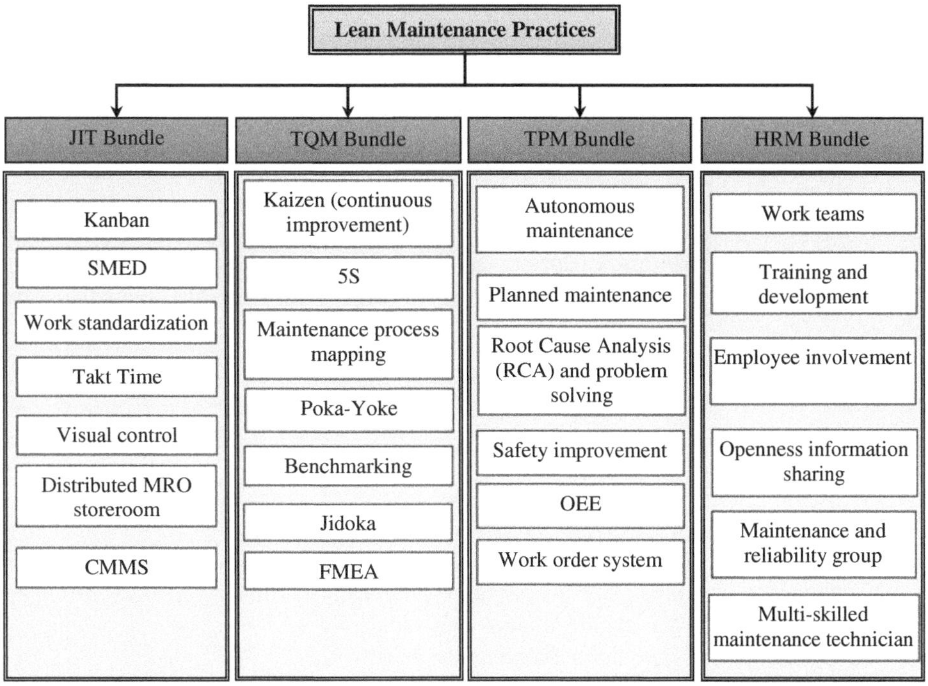

Figure 7. Scheme for lean maintenance practices.

customers, in order to meet or exceed customer expectations (Cua et al., 2001). The practices of TQM are customer focus, leadership, strategic quality planning, use of information and analysis, management of people (participation and partnership), and continuous improvement (Chin, Rao Tummala, & Chan, 2002). This ensures higher customer satisfaction, better quality of products, and higher market share; improving the competitiveness, effectiveness, and flexibility of the whole organisation (Pheng & Chuan, 2001). In this study, the TQM bundle consists of Kaizen, 5S, maintenance process mapping, Poka-Yoke, benchmarking, Jidoka, and FMFA.

6.3.3. TPM bundle

The TPM bundle incorporates all practices that are designed to maximise the equipment effectiveness. The bundle contains autonomous maintenance, planned maintenance, root cause analysis (RCA) and problem solving, safety improvement, OEE, and work order system. The concept of TPM was developed in Japan by Seichii Nakajima aiming at achieving zero losses (e.g. zero breakdowns, zero defects, zero accidents) and attaining the Takt time through improving and maintaining equipment to its highest performance level (Brown et al., 2006). TPM can be integrated with lean in order to distinguishing and attacking six big losses which are reducing the effectiveness of equipment (Rodrigues & Hatakeyama, 2006). These six categories of losses are (1) breakdown losses, (2) set-up losses, (3) minor stoppage/idle losses, (4) reduced speed losses (equipment speed is slow than the designed speed), (5) rework losses, and (6) start-up losses (start-up after periodic repair, start-up after holidays, start-up after lunch breaks, and start-up after suspension) (Waeyenbergh & Pintelon, 2002).

6.3.4. HRM bundle

The HRM bundle encompasses all the practices to ensure that the human resources of an organisation are used in such a way that the employer obtains the greatest possible benefit from their abilities. At the same time, the employees obtain both material and psychological rewards from their work (Hiltrop, 1996). The HRM practices form an organisation performance through increase employees' knowledge, skills, and abilities (KSAs), motivate employees to leverage their KSAs for the firm's benefit and employees empowerment. In addition, work teams, performance appraisal, and information sharing have been recommended to enhance an organisation performance (Huemann, Keegan, & Turner, 2007). In this study, the HRM bundle includes work teams, training and development, employee involvement, openness information sharing, maintenance and reliability group, and multi-skilled maintenance technician.

6.4. Stage four: pulling the value

The fourth stage is to confirm that an equipment (i.e. considered as the customer in the maintenance process) is pulling the value through all maintenance processes. The execution of lean principles takes place at this stage. The stage involves steps in reconfiguring the VSM or designing the future stream map, selecting LM practices, developing the lean transformation strategy, and evaluating the OEE.

6.4.1. Association between maintenance wastes and LM practices

Selecting LM tools is the next step after identifying and discussing the types of waste and lean tools. It assists an organisation to achieve LM targets by associating suitable tools to tackle the spotted waste types. The QFD is suggested in this study to develop the association between these tools and wastes. Mostafa (2011) suggested using QFD to assign the lean practices for each waste. Nonetheless, the study was in a general manufacturing environment rather than a maintenance process. Likewise, Tendayi (2013) applies QFD coupled with AHP to rank the maintenance excellence criteria with some lean tools. The study focused only on a certain case study of rolling stock maintenance. It used limited lean tools and did not include the maintenance wastes. In contrast to Tendayi (2013), this study has identified 26 lean tools for maintenance (see, Figure 7) and eight types of maintenance wastes (discussed in Section 6.1). QFD is a structured tool that identifies important customer expectations and translates them into appropriate technical characteristics which are operational in design, verification, and production. QFD enables resources to be focused on meeting major customer demands. Figure 8 displays the structure of QFD which is frequently called a house of quality (HOQ) because of the shape presented (Bottani & Rizzi, 2006).

The HOQ comprises customer requirements (horizontal axis) and technical characteristics (vertical axis). The customer axis designates what customers require, importance of the requirements, and competitive performance. The customer requirements are often referred as what's. The technical axis describes the technical characteristics that affect customer satisfaction directly for one or more customer expectations. Moreover, on the technical axis are the correlations, importance, and targets of the technical characteristics and technical competitive benchmarking. The technical characteristics are denoted as How's, meaning how to address what's. The technical targets are accordingly called

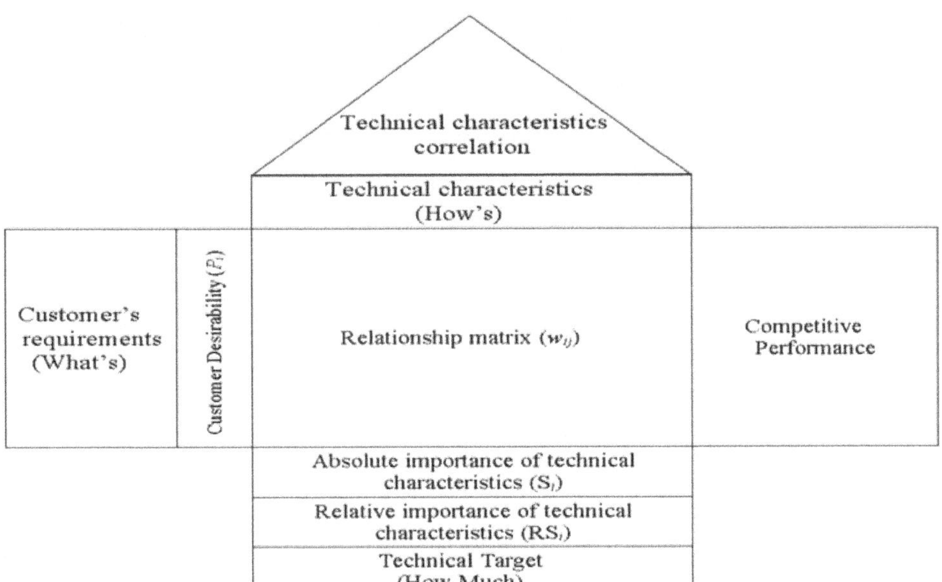

Figure 8. House of quality.

How Much. The interrelationships between customer wants and technical characteristics are evaluated in the relationship matrix.

6.4.2. House of Waste

The HoW is developed in this study to demonstrate the maintenance wastes/lean tools association. HoW identifies the importance of each LM tool in the elimination of each waste type as demonstrated in Table 5. It has two main parts: the horizontal part 'What's' which contains information relevant to the waste types and the vertical part 'How's' which comprises corresponding LM tools. The basic process underlying 'How' resides in the centre of the matrix where intersect of maintenance waste types and lean tools provides an opportunity to examine each waste type versus each tool.

Five steps in developing the wastes/tool association using HoW are briefly described as follows:

(1) List the waste types within the maintenance process in the horizontal axis.
(2) Determine each waste priority (*Pi*), which rates the significance of each waste to the maintenance performance. Otherwise, it can be based on the waste occurrence or cost of removal. A scale of 1 to 9 can be used, where 9 is given to extremely important, 7 strongly important, 5 to very important, 3 to important, and 1 to not important.
(3) List LM tools on the vertical axis. An organisation can state suitable LM tools for their own maintenance process in terms of the application cost of LM tools.
(4) Identify the interrelationships between wastes and tools (w_{ij}). The strength of relationship may be classified into three levels, where a rating of 9 is assigned to a strong relation, 3 to a medium relation, and 1 to a weak relation. Each tool must be interrelated to at least one waste; one waste must also be addressed by at least one tool. This ensures that all wastes are concerned in the maintenance removal planning, and all tools are properly established.
(5) Calculate the score (S_j) and (RS_j) to rank each LM tool. The following formula can be used to calculate the score

$$S_j = \sum_{i=1}^{8} w_{ij} P_i, \ \forall j \tag{1}$$

$$RS_j = \frac{S_j}{\sum_{j=1}^{26} S_j}, \ j = 1, \ldots, 26 \tag{2}$$

where

P_i relative priority of ith waste
i type of maintenance waste, $i = 1, 2, \ldots, 8$
T_j LM tool, $j = 1, 2, 3, \ldots, 26$
w_{ij} numerical VA to position (i, j) of the HoW matrix. This refers to the weight of eliminating waste ith by tool jth. Numerical scale can be (9, 3, 1)
S_j score of tool jth over the total types of waste
RS_j relative score of tool jth over the total score of tools

Table 5. House of Waste.

What's (*maintenance wastes*)	Waste priority	How's (*lean maintenance tools*) Kanban	SMED	Work standardisation	Takt Time	Visual control	Distributed MRO storeroom	CMMS	Kaizen	5S	Maintenance VSM	Poke Yoke	Jidoka	Benchmarking	FMEA	Autonomous maintenance	Planned maintenance	RCA	OEE	Work order system	Work teams	Training and development	Employee involvement	Openness information system	Maintenance & reliability group	Multi-skilled technician	Safety improvement
To eliminate too much maintenance	P_1	w_{11}	w_{12}	w_{13}	w_{14}	…																			…	w_{125}	w_{126}
To eliminate waiting for resources	P_2	w_{21}	w_{22}	w_{23}	w_{24}	…																			…	w_{225}	w_{226}
To eliminate centralised maintenance	P_3	w_{31}	w_{32}	w_{33}	w_{34}	…																			…	w_{325}	w_{326}
To eliminate non-standard maintenance	P_4	w_{41}	w_{42}	w_{43}	w_{44}	…																			…	w_{425}	w_{426}
To eliminate excessive stocks	P_5	w_{51}	w_{52}	w_{53}	w_{54}	…																			…	w_{525}	w_{526}
To eliminate double handling	P_6	w_{61}	w_{62}	w_{63}	w_{64}	…																			…	w_{625}	w_{626}
To eliminate incorrect maintenance	P_7	w_{71}	w_{72}	w_{73}	w_{74}	…																			…	w_{725}	w_{726}
To eliminate underutilisation of maintenance crew	P_8	w_{81}	w_{82}	w_{83}	w_{84}	…																			…	w_{825}	w_{826}
LM score		S_1	S_2	S_3	S_4	S_5	S_6	S_7	S_8	S_9	S_{10}	S_{11}	S_{12}	S_{13}	S_{14}	S_{15}	S_{16}	S_{17}	S_{18}	S_{19}	S_{20}	S_{21}	S_{22}	S_{23}	S_{24}	S_{25}	S_{26}
Relative score		RS_1	RS_2	…																					…	RS_{25}	RS_{26}
Rank																											

The score S_j represents the absolute importance of tool *j*th over all types of wastes. The relative score RS_j refers to the importance of the tool *j*th over all tools. According to Bottani and Rizzi (2006), the technical characteristics are usually ranked based on relative importance rather than absolute importance. Therefore, LM tools should be prioritised based on RS_j. The higher the RS_j, the more important the LM tool that should be incorporated in order to eliminate the waste.

6.5. Stage five: pursue perfection

The last stage is to pursue the complete waste elimination from maintenance processes. This could be achieved through auditing the LM results, standardise the lean tools and procedures, teams and employees developments and expand the lean practice.

7. Implications of proposed LM process

Most of literature works on lean implementations fall into the manufacturing environment albeit LM is essential part for the lean manufacturing system (Soltan & Mostafa, 2014; Tendayi & Fourie, 2013). Plethora of studies included and mentioned lean and maintenance concepts (see Table 2). Nevertheless, the integration of lean principles into the maintenance environment has not evidently received full attention. The common aspects among the previous studies were (1) introducing partial lean tools to be applied in the maintenance process; (2) developing diagrammatic structures (model, pyramid, or scheme) with unspecified sequence of LM implementation; and (3) lack of proper structure. Consequently, some researchers including Davies and Greenough (2010) and Soltan and Mostafa (2014) have proposed more research on incorporating lean principles in the maintenance activities. Therefore, this study proposed a LM process to cover most of the limitations or deficiencies of the existing suggested LM initiatives.

The proposed LM process is a straight forward, comprehensive, and easy to understand for maintenance practitioners and maintenance strategy decision makers. The benefits of the proposed process can be summarised as:

(1) The process is straightforward and easy to comprehend. This is because of the simple structure that is combined in the process.
(2) A systematic procedure of applying the five lean principles into maintenance activities can result in better understanding from an organisation's management view.
(3) All advantages of lean are kept as the process focus. These advantages are integrated and systemised in the steps of the process.
(4) Application of the process establishes pursuing a perfectionist culture in an organisation. This is done through auditing the LM results, standardisation, team's development, and expansion of the practice. Hence, with every iteration of the process, an organisation maintenance activities move a further step towards maintenance excellence.
(5) Aligning the human resource aspects (maintenance employees training and development) with other steps can bring high potential for a comprehensive and sustainable LM implementation process as well as overall process improvement in an organisation.
(6) The process promotes the teamwork and problem solving cultures to ensure high-quality outcomes of the implementation process.

(7) The maintenance process can be converted to be a predictive and proactive system that provides the reliable process using lean principles. As a result, it enhances the maintenance excellence and world-class manufacturing as maintenance and manufacturing are inseparable (i.e. reliability and availability of manufacturing facilities).
(8) The process can be applied to all manufacturing and non-manufacturing organisations that are pursuing the world-class status in their maintenance function. The process suits those organisations either in pursuing to transfer lean thinking to their maintenance departments or starting lean transformation from the maintenance department.
(9) Hitherto, some initiatives have been developed for specific larger size organisations such as shipbuilding and military (see Table 3). However, the proposed process can be accepted within all level of skills and organisational sizes. This is due to the fact that the process is more flexible and can be adjusted according to any maintenance strategies.
(10) The proposed process is precisely documented and discusses the maintenance wastes and LM tools. It is to enhance the understanding the waste elimination and lead to successful LM implementation.
(11) The process presented wastes/tools association to assign LM tools for each waste type. This leads to an effective waste elimination process and sustainable outcomes.

8. Conclusion

Maintenance is a critical contributor to progressing towards world-class manufacturing of an organisation. It has rapidly grown into a very complex undertaking as technologies, competition, and product characteristics have evolved. In order to achieve world-class performance, the maintenance strategies should be linked to manufacturing strategies such as the lean concept. Applying an effective maintenance strategy can ensure a high degree of utilisation, reliability, and availability of manufacturing facilities especially in a continuous production process. This study has introduced a process to adopt lean into maintenance activities. The process highlights types of NVA maintenance activities, a package of VSM symbols to capture the maintenance activities, LM tools scheme, and tools/wastes association. Moreover, it promotes the culture of continuous improvement aiming at maintenance excellence. In general, any improvements attained from implementing the suggested process might take time to transform maintenance activities into LM. Nevertheless, commitment and direct involvement of an organisational management along with employee training and teamwork development can be a catalyst to accelerate the transformation process. The proposed LM process can be applied in real conditions to test its validity and reliability of the process. This can be considered as a further suggestion to the proposed process in this research. Hence, application of the process in different industrial sectors and a wide range of manufacturing companies can contribute to additional empirical evidence.

Acknowledgements

The authors would like to express their gratitude to the anonymous reviewers for their constructive comments and suggestions to improve the quality of the paper.

Disclosure statement

No potential conflict of interest was reported by the authors.

References

Abdulmalek, F. A., Rajgopal, J., & Needy, K. L. (2006). A classification scheme for the process industry to guide the implementation of lean. *Engineering Management Journal, 18*, 15–25.

Ahmed, S., Hassan, M. H., & Taha, Z. (2005). TPM can go beyond maintenance: Excerpt from a case implementation. *Journal of Quality in Maintenance Engineering, 11*, 19–42. doi:10.1108/13552510510589352

Al-Najjar, B., & Alsyouf, I. (2003). Selecting the most efficient maintenance approach using fuzzy multiple criteria decision making. *International Journal of Production Economics, 84*, 85–100. doi:10.1016/S0925-5273(02)00380-8

Ayeni, P., Baines, T. S., Lightfoot, H., & Ball, P. (2011). State-of-the-art of 'Lean' in the aviation maintenance, repair, and overhaul industry. *Proceedings of the Institution of Mechanical Engineers, Part B: Journal of Engineering Manufacture*, 1–16. doi:10.1177/0954405411407122

Baluch, N., Abdullah, C. S., & Mohtar, S. (2012). TPM and lean maintenance – A critical review. *Interdisciplinary Journal of Contemporary Research In Business, 4*, 850–857.

Bevilacqua, M., & Braglia, M. (2000). The analytic hierarchy process applied to maintenance strategy selection. *Reliability Engineering & System Safety, 70*, 71–83. doi:10.1016/S0951-8320(00)00047-8

Bhasin, S. (2015). *Lean management beyond manufacturing*. New York, NY: Springer International Publishing.

Blanchard, B. S. (1997). An enhanced approach for implementing total productive maintenance in the manufacturing environment. *Journal of Quality in Maintenance Engineering, 3*, 69–80. doi:10.1108/13552519710167692

Booth, A., Papaioannou, D., & Sutton, A. (2012). *Systematic approaches to a successful literature review*. Thousand Oaks, CA: Sage.

Bottani, E., & Rizzi, A. (2006). Strategic management of logistics service: A fuzzy QFD approach. *International Journal of Production Economics, 103*, 585–599. doi:10.1016/j.ijpe.2005.11.006

Brown, C. B., Collins, T. R., & McCombs, E. L. (2006). Transformation from batch to lean manufacturing: The performance issues. *Engineering Management Journal, 18*, 3–14.

Bulatovic, M., & Djurovic, D. (2014). Lean concept in the maintenance. *Journal of Trends in the Development of Machinery and Associated Technology, 18*, 135–138. Retrieved from http://tmt.unze.ba/zbornik/TMT2014Journal/30_Journal_TMT_2014.pdf

Burgess, N., & Radnor, Z. (2013). Evaluating lean in healthcare. *International Journal of Health Care Quality Assurance, 26*, 220–235. doi:10.1108/09526861311311418

Carrasqueira, M., & Machado, V. C. (2008). Strategic logistics: Re-designing companies in accordance with Lean Principles. *International Journal of Management Science and Engineering Management, 3*, 294–302. doi:10.1080/17509653.2008.10671056

Chan, F. T. S., Lau, H. C. W., Ip, R. W. L., Chan, H. K., & Kong, S. (2005). Implementation of total productive maintenance: A case study. *International Journal of Production Economics, 95*, 71–94. doi:10.1016/j.ijpe.2003.10.021

Chin, K. S., Rao Tummala, V. M., & Chan, K. M. (2002). Quality management practices based on seven core elements in Hong Kong manufacturing industries. *Technovation, 22*, 213–230. doi:10.1016/S0166-4972(01)00020-7

Chong, M. Y., Chin, J. F., & Hamzah, H. S. (2012). Transfer of total productive maintenance practice to supply chain. *Total Quality Management & Business Excellence, 23*, 467–488. doi:10.1080/14783363.2011.637788

Clarke, G., Mulryan, G., & Liggan, P. (2010). Lean maintenance – A risk-based approach. *Pharmaceutical Engineering, 30*(5), 1–6.
Colbert, F. (2008). The benefits of integrating "lean thinking" concepts into the management of infrared predictive maintenance programs. *Cinde Journal, 29*, 5–12.
Cua, K. O., McKone, K. E., & Schroeder, R. G. (2001). Relationships between implementation of TQM, JIT, and TPM and manufacturing performance. *Journal of Operations Management, 19*, 675–694. doi:10.1016/S0272-6963(01)00066-3
Davies, C. (2003). *The contribution of lean thinking to the maintenance of manufacturing systems* (PhD Thesis). Cranfield University, Cranfield, UK.
Davies, C., & Greenough, R. (2001). *Maintenance survey – Identification of lean thinking within maintenance.* Paper presented at the Proceedings of the 17th National conference on manufacturing research: Advances in manufacturing technology XV, Cardiff, UK.
Davies, C., & Greenough, R. M. (2003). *Testing performance measures within maintenance, through case study research into lean thinking activities.* Retrieved March 24, 2014, from www.plant-maintenance.com/articles/Testing_Performance_Measures.pdf
Davies, C., & Greenough, R. M. (2010). *Measuring the effectiveness of lean thinking activities within maintenance.* Retrieved June 24, 2013, from http://www.plant-maintenance.com/articles/Lean_Maintenance.pdf
De Silva, N., Ranasinghe, M., & De Silva, C. R. (2012). *Maintainability approach for lean maintenance.* Paper presented at the World Construction Conference 2012 – Global Challenges in Construction Industry, Colombo, Sri Lanka. Retrieved from http://www.irbnet.de/daten/iconda/CIB_DC25109.pdf
De Treville, S., & Antonakis, J. (2006). Could lean production job design be intrinsically motivating? Contextual, configurational, and levels-of-analysis issues. *Journal of Operations Management, 24*, 99–123. doi:10.1016/j.jom.2005.04.001
Denyer, D., & Tranfield, D. (2009). Producing a systematic review. In D. A. Buchanan & A. Bryman (Eds.), *The Sage handbook of organizational research methods* (pp. 671–689). Thousand Oaks, CA: Sage.
Djurovic, D., & Bulatovic, M. (2014). *"5 S" as a tool lean concept in the maintenance.* Paper presented at the 8th International Quality Conference, Center for Quality, Faculty of Engineering, University of Kragujevac, Serbia.
Dombrowski, U., & Malorny, C. (2014). Lean after sales service – An opportunity for OEMs to ensure profits. In B. Grabot, B. Vallespir, S. Gomes, A. Bouras, & D. Kiritsis (Eds.), *Advances in production management systems innovative and knowledge-based production management in a global-local world* (Vol. 439, pp. 618–625). Berlin: Springer.
Dunn, S. (2014). 5 keys to lean maintenance and improving maintenance. *Asset Management & Maintenance Journal, 27*, 8–10.
EPA. (2015, 2 February). *Lean thinking and methods: TPM.* Retrieved January 4, 2015, from http://www.epa.gov/lean/environment/methods/tpm.htm
Faccio, M., Persona, A., Sgarbossa, F., & Zanin, G. (2014). Industrial maintenance policy development: A quantitative framework. *International Journal of Production Economics, 147*, 85–93. doi:10.1016/j.ijpe.2012.08.018
Fouladgar, M. M., Yazdani-Chamzini, A., Lashgari, A., Zavadskas, E. K., & Turskis, Z. (2012). Maintenance strategy selection using AHP and COPRAS under fuzzy environment. *International Journal of Strategic Property Management, 16*, 85–104. doi:10.3846/1648715X.2012.666657
Fraser, K. (2014). Facilities management: The strategic selection of a maintenance system. *Journal of Facilities Management, 12*, 18–37. doi:10.1108/JFM-02-2013-0010
George, M. (2003). *Lean six sigma for service: How to use lean speed and six sigma quality to improve services and transactions.* New York, NY: McGraw-Hill.
Georgescu, D. (2010). The maintenance management for lean organization. *Economia. Seria Management, 13*, 437–446.
Georgescu, D., & Militaru, C. (2009, 12–13 November). *7s's System foundation of total productive maintenance house for lean organization.* Paper presented at the The 4th International Conference of the Interdisciplinarity in Engineering, Targu Mures, Romania.
Ghayebloo, S., & Shahanaghi, K. (2010). Determining maintenance system requirements by viewpoint of reliability and lean thinking: A MODM approach. *Journal of Quality in Maintenance Engineering, 16*, 89–106. doi:10.1108/13552511011030345

Gopalakrishnan, P., & banerji, A. (2013). *Maintenance and spare parts management* (2nd ed.). New Delhi: PHI Learning Private Ltd.
Heisler, R. (2003). *Planning and scheduling in a lean maintenance environment*. Retrieved December 20, 2014, from http://www.reliabilityweb.com/index.php/search/results/?q=Randy+Heisler&sa=Go&cx=008700119099951100649%3A5ndeqiorkim&cof=FORID%3A10&ie=UTF-8
Hellingrath, B., & Cordes, A.-K. (2014). Conceptual approach for integrating condition monitoring information and spare parts forecasting methods. *Production & Manufacturing Research, 2*, 725–737. doi:10.1080/21693277.2014.943431
Hiltrop, J.-M. (1996). A framework for diagnosing human resource management practices. *European Management Journal, 14*, 243–254. doi:10.1016/0263-2373(96)00004-7
Holweg, M. (2007). The genealogy of lean production. *Journal of Operations Management, 25*, 420–437. doi:10.1016/j.jom.2006.04.001
Hopp, W. J., & Spearman, M. L. (2004). Commissioned paper to pull or not to pull: What is the question? *Manufacturing & Service Operations Management, 6*, 133–148.
Hou, Y.-H., Zhao, H.-F., Meng, X.-K., & Ma, Y.-X. (2013). Research on lean maintenance system of hydroelectric station equipment. *Industrial Engineering and Management, 3*, 117–121.
Huang, J., Bian, Y., & Cai, W. (2012, 15–18 June). *Weapon equipment lean maintenance strategy research*. Paper presented at the International Conference on Quality, Reliability, Risk, Maintenance, and Safety Engineering (ICQR2MSE), Chengdu, China.
Huemann, M., Keegan, A., & Turner, J. R. (2007). Human resource management in the project-oriented company: A review. *International Journal of Project Management, 25*, 315–323. doi:10.1016/j.ijproman.2006.10.001
Irajpour, A., Fallahian-Najafabadi, A., Mahbod, M. A., & Karimi, M. (2014). A framework to determine the effectiveness of maintenance strategies lean thinking approach. *Mathematical Problems in Engineering, 2014*, 1–11.
Jahanbakhsh, M., Moghaddam, N., & Samaie, H. M. (2013). Lean maintenance (case study: Teen Dairy Industry Co.). *International Research Journal of Applied and Basic Sciences, 4*, 2033–2040.
Jasiulewicz-Kaczmarek, M. (2013). Sustainability: Orientation in maintenance management: Case study. In P. Golinska (Ed.), *EcoProduction and logistics* (pp. 135–154). London: Springer.
Joffe, H. (2011). Thematic analysis. In D. Harper & A. R. Thompson (Eds.), *Qualitative research methods in mental health and psychotherapy* (pp. 209–223). West Sussex: Wiley.
Karim, A., & Arif-Uz-Zaman, K. (2013). A methodology for effective implementation of lean strategies and its performance evaluation in manufacturing organizations. *Business Process Management Journal, 19*, 169–196. doi:10.1108/14637151311294912
Khazraei, K., & Deuse, J. (2011). A strategic standpoint on maintenance taxonomy. *Journal of Facilities Management, 9*, 96–113. doi:10.1108/14725961111128452
Kolanjiappan, S., & Maran, D. K. (2011). Lean practices in aircraft maintenance. *Software Engineering and Technology, 3*, 450–457.
Lee, J., & Wang, H. (2008). New Technologies for Maintenance. In K. A. H. Kobbacy & D. N. P. Murthy (Eds.), *Complex System Maintenance Handbook* (pp. 49–78). London: Springer (eBook).
Levitt, J. (2008). *Lean maintenance*. New York, NY: Industrial Press.
Marks, A. (2008). Effective rotable plant management in a lean maintenance culture. *Asset Management and Maintenance Journal, 21*, 36–43.
Márquez, C. A. (2007). *The maintenance management framework: Models and methods for complex systems maintenance*. London: Springer-Verlag.
McCarthy, D., & Rich, N. (2004). *Lean TPM*. Oxford: Elsevier.
McConnell, K., Lindrooth, R. C., Wholey, D. R., Maddox, T. M., & Bloom, N. (2013). Management practices and the quality of care in cardiac units. *JAMA Internal Medicine, 173*, 684–692. doi:10.1001/jamainternmed.2013.3577
Monden, Y. (1998). *Toyota production system: An integrated approach to just-in-time*. Norcross, GA: Engineering & Management Press.
Mostafa, S. (2011). *A lean framework for maintenance decisions* (Master of Industrial Engineering). Mansoura University, Egypt.
Mostafa, S., Dumrak, J., & Soltan, H. (2013). A framework for lean manufacturing implementation. *Production & Manufacturing Research, 1*, 44–64. doi:10.1080/21693277.2013.862159

Nezami, F. G., & Yildirim, M. B. (2013). A sustainability approach for selecting maintenance strategy. *International Journal of Sustainable Engineering, 6*, 332–343. doi:10.1080/19397038.2013.765928

Nima, R., Hesamodin, N., & Alireza, R. (2014). Presenting an optimal lean maintenance strategy using fuzzy MADM. *International Journal of Management, IT and Engineering, 4*, 263–278.

Okhovat, M. A., Ariffin, M. K. A. M., Nehzati, T., & Hosseini, S. A. (2012). Development of world class manufacturing framework by using six-sigma, total productive maintenance and lean. *Scientific Research and Essays, 7*, 4230–4241. doi:10.5897/SRE11.368

Okoli, C., & Schabram, K. (2010). A guide to conducting a systematic literature review of information systems research. *Sprouts: Working Papers on Information Systems, 10*, 1–50. Retrieved form http://sprouts.aisnet.org/10-26

Önder, M. (2014). *Maintenance as a contributor in green production systems: Interviews with Volvo, Scania, and Dynamate* (Master Thesis). Mälardalen University, Mälardalen, Sweden.

Pedersen, E. R. G., & Huniche, M. (2011). Determinants of lean success and failure in the Danish public sector. *International Journal of Public Sector Management, 24*, 403–420. doi:10.1108/09513551111147141

Pheng, L., & Chuan, C. (2001). Just-in-time management of precast concrete components. *Journal of Construction Engineering and Management, 127*, 494–501. doi:10.1061/(ASCE)0733-9364(2001)127:6(494)

Pinjala, S. K., Pintelon, L., & Vereecke, A. (2006). An empirical investigation on the relationship between business and maintenance strategies. *International Journal of Production Economics, 104*, 214–229. doi:10.1016/j.ijpe.2004.12.024

Potes Ruiz, P., Kamsu Foguem, B., & Grabot, B. (2014). Generating knowledge in maintenance from experience feedback. *Knowledge-Based Systems, 68*, 4–20. doi:10.1016/j.knosys.2014.02.002

Qiang, T., Zhu, B., & Li, L. (2011). *A study on military equipment lean maintenance.* Paper presented at the International Conference on Quality, Reliability, Risk, Maintenance, and Safety Engineering, Xi'an.

Rastegari, A. (2012). *Strategic maintenance management in lean environment* (Master Thesis). Mälardalen University, Västerås, Sweden.

Resta, B., Powell, D., Gaiardelli, P., & Dotti, S. (2015). Towards a framework for lean operations in product-oriented product service systems. *CIRP Journal of Manufacturing Science and Technology, 9*, 12–22. doi:10.1016/j.cirpj.2015.01.008

Rodrigues, M., & Hatakeyama, K. (2006). Analysis of the fall of TPM in companies. *Journal of Materials Processing Technology, 179*, 276–279. doi:10.1016/j.jmatprotec.2006.03.102

Romano, E., Murino, T., Asta, F., & Costagliola, P. (2013). Lean Maintenance model to reduce scraps and WIP in manufacturing system: Case study in power cables factory. *WSEAS Transactions on Systems, 12*, 650–666.

Shah, R., & Ward, P. (2003). Lean manufacturing: Context, practice bundles, and performance. *Journal of Operations Management, 21*, 129–149. doi:10.1016/s0272-6963(02)00108-0

Shah, R., & Ward, P. T. (2007). Defining and developing measures of lean production. *Journal of Operations Management, 25*, 785–805. doi:10.1016/j.jom.2007.01.019

Shahin, A., Shirouyehzad, H., & Pourjavad, E. (2012). Optimum maintenance strategy: A case study in the mining industry. *International Journal of Services and Operations Management, 12*, 368–386. doi:10.1504/IJSOM.2012.047626

Sheng, T. L., & Tofoya, J. (2010, 18–20 October). *The secret of manufacturing excellence: Lean maintenance.* Paper presented at the International Symposium on Semiconductor Manufacturing (ISSM), Tokyo.

Shou, W., Wang, X., Wang, J., Hou, L., & Truijens, M. (2014). Integration of BIM and lean concepts to improve maintenance efficiency: A case study. *Computing in Civil and Building Engineering, 2014*, 373–380.

Smith, R. (2004). What is lean maintenance? Elements that need to be in place for success. *Maintenance Technology, 17*, 15–21.

Smith, R., & Hawkins, B. (2004). *Lean maintenance: Reduce costs, improve quality, and increase market share.* Burlington, MA: Elsevier.

Soltan, H., & Mostafa, S. (2014). Leanness and agility within maintenance process. *International Journal of Engineering Research and Technology, 3*, 553–555.

Stuchly, V., & Jasiulewicz-Kaczmarek, M. (2014). Maintenance in sustainable manufacturing. *Scientific Journal of Logistics, 10*, 273–284. Retrieved from http://www.logforum.net/vol10/issue3/no4

Sullivan, W. G., McDonald, T. N., & Van Aken, E. M. (2002). Equipment replacement decisions and lean manufacturing. *Robotics and Computer-Integrated Manufacturing, 18*, 255–265. doi:10.1016/S0736-5845(02)00016-9

Sunjka, B. P., & Murphy, S. F. (2014). Lean implementation within South African aircraft maintenance organisations. *International Journal of Lean Enterprise Research, 1*, 59–80. doi:10.1504/IJLER.2014.062285

Tendayi, T. G. (2013). *An investigation into the applicability of lean thinking in an operational maintenance environment* (Master of Science in Engineering). Stellenbosch University, South Africa.

Tendayi, T. G., & Fourie, C. J. (2012, July). *A lean maintenance supply chain framework for rolling stock maintenance: A case study.* Paper presented at the The 42nd International Conference on Computers and Industrial Engineering (CIE42), Cape Town, South Africa.

Tendayi, T. G., & Fourie, C. J. (2013). *The combined AHP-QFD approach and its use in lean maintenance.* Paper presented at the 25th Annual Southern African Institute of Industrial Engineering Conference, Stellenbosch.

Thirkell, E., & Ashman, I. (2014). Lean towards learning: Connecting Lean Thinking and human resource management in UK higher education. *The International Journal of Human Resource Management, 25*, 2957–2977. doi:10.1080/09585192.2014.948901

Thiruvengadam, A. (2009). *A practical method for assessing maintenance factors using a value stream maintenance map* (Master of Science). Wichita State University, Wichita, KS.

Tinga, T. (2013). Maintenance concepts. In H. Pham (Ed.), *Principles of loads and failure mechanisms* (pp. 161–186). London: Springer-Verlag.

Tranfield, D., Denyer, D., & Smart, P. (2003). Towards a methodology for developing evidence-informed management knowledge by means of systematic review. *British Journal of Management, 14*, 207–222. doi:10.1111/1467-8551.00375

Van den Heuvel, M. (2008). *Improving maintenance shutdown processes: Reducing delay and increasing work efficiency at Corus's Direct Sheet Plant* (Master Thesis). University of Technology Delft, Delft, The Netherlands.

Verma, A. K., & Ghadmode, A. (2004). An integrated lean implementation model for fleet repair and maintenance. *Naval Engineers Journal, 116*, 79–90. doi:10.1111/j.1559-3584.2004.tb00306.x

Waeyenbergh, G., & Pintelon, L. (2002). A framework for maintenance concept development. *International Journal of Production Economics, 77*, 299–313. doi:10.1016/S0925-5273(01)00156-6

Waeyenbergh, G., & Pintelon, L. (2004). Maintenance concept development: A case study. *International Journal of Production Economics, 89*, 395–405. doi:10.1016/j.ijpe.2003.09.008

Womack, J. P., & Jones, D. T. (2003). *Lean thinking: Banish waste and create wealth in your corporation* (2nd ed.). New York, NY: Free Press, Simon & Schuster.

Yile, L., Hang, X. X., & Lei, Z. (2008). *Lean Maintenance framework and its application in clutch maintenance.* Paper presented at the International Conference on Information Management, Innovation Management and Industrial Engineering, Taipei.

Zarei, M., Fakhrzad, M. B., & Jamali Paghaleh, M. (2011). Food supply chain leanness using a developed QFD model. *Journal of Food Engineering, 102*, 25–33. doi:10.1016/j.jfoodeng.2010.07.026

Zwas, A. (2006). Lean manufacturing techniques in bus and rail maintenance: Study at Chicago Transit Authority in Illinois. *Transportation Research Record: Journal of the Transportation Research Board, 1986*, 54–58. doi:10.3141/1986-09

6

A review of few unconventional machining processes based on the concept of velocity shear instability in plasma

R.K. Tyagi*

Department of Mechanical Engineering, Amity School of Engineering and Technology, Amity University, Noida, India

In this article, published work on unconventional machining process, plasma spray process, plasma etching and abrasive jet machining has been summarized. In unconventional machining process, based on the concept of velocity shear instability in plasma, the material is removed from metallic surface by the impact of ions. Single-crystal silicon surface roughness (up to nanometer sizes) induced by gas plasma in plasma etching based on the concept of velocity shear instability in plasma has been achieved. The material removal from metallic surface can be explained on the basis of erosion phenomena. Momentum transfer from ions to abrasive particles accelerates micron-sized abrasive particles. In this process, material removal from metallic surfaces varies from a few milligrams to a few grams per second. Plasma spray process is based on the concept of velocity shear instability in plasma. In this process, heat generated in plasma and velocity of ions is used to develop high quality surface. Actually, the velocity of ions depend upon different parameters like magnetic field, homogenous D.C electric field, shear scale length, temperature anisotropy, heterogeneity in D.C electric field, density gradient, etc. Therefore, a combination of aforementioned parameters is selected to suit the requirement.

Keywords: velocity shear instability; group velocity of waves; micro-machining process; plasma etching; abrasive jet machining; plasma spray technology

1. Introduction

The purpose of researching the information available in this area is that after analysis of the properties that influence metal removal as well as different processes that are used to serve the purpose, a single machine could be devised for different metal removal requirements, i.e. for range varying from micrometer to nanometer as well as for different applications. The required value of parameters for a particular application can be chosen with the help of theoretical and experimental models suggested earlier. Introduction of this review article has been divided into velocity shear instability, unconventional machining process, plasma spray, etc.

Shanbhag et al. (2003) developed ion-beam micro-contouring process by using ionized argon ions on small optical components. This process is very useful when the dimensions of components are less than mm dimensions. The depth of ion penetration is a function of ion-beam shape and an ion-beam dwell function. Silicon surface can be polished up to nanometer sizes (see Table 1).

*Corresponding author. Email: tyagirk1@gmail.com

Table 1. Characteristic features of unconventional machining methods.

Machining method	Erosion source	Basic character of machining method	Temperature/ pressure	Machining environment
LBM	Photons (10^3–10^6)	Thermal	8000–1200 K	Air
EBM	Electrons (10^3–10^6)	Thermal	3000–8000 K	Vacuum
IBM	Ions (100–500ev)	Physical/Chemical	373–673 K	Vacuum
WJM/FJM	Water Jet	Mechanical	10–700 Mpa	Air
AWJM	Abrasive particles + Water	Mechanical	10–700 Mpa	Air

The machining processes can be classified into three categories on the basis of accuracy achieved: conventional machining, precision machining and ultra-precision machining. Ultra-precision machining is the process by which the highest possible dimensional accuracy is or has been achieved at a given point of time. This is a relative definition which varies with time. It has been predicted that by 2010 AD, machining accuracies in conventional processes would reach 1 μm, while precision and ultra-precision processes would attain accuracies of the order of 0.01 μm (10 nm) and 0.001 μm (1 nm), respectively (Jha & Jain, 2006).

In laboratory plasma, if the number density and velocity of different layers in plasma are different, then in any two adjacent layers, disturbance is created in terms of waves, which is called as velocity shear instability. The instability is generated by plasma species which are in majority, i.e. ions (positive and negative) or electrons.

Ichiki, Kaneko, Hayashi, Tamura, and Hatakeyama (2009) realized velocity shear instability in laboratory by using SF_6 gas and potassium plasma. The electrostatic ion cyclotron instability generated by Ichi et al. is based on non-traditional magnetic field.

However, due to parametric constraints in fusion studies, assumption is made in case of shear-driven phenomena that the shear scale length is much larger than the ion Larmor radius. Although this assumption is in conformity with the experimental observations, it fixes the time scale of variation of the radiant shear-driven phenomenon as much larger than the ion-gyro period (provided that the maximum drift speed is not much larger than the electron thermal speed (Ganguli, Lee, & Palmadesso, 1985).

First time, collisionless drift wave instability has been realized by using segmented tungsten plate and SF_6 gas and potassium plasma. This type of machine is called as Q-machine (Ichiki et al., 2009).

Laboratory study related to energetic ions in ionosphere–magnetosphere system is described with the help of cesium Q machine with a region of non-uniform magnetic field. Electrostatic ion cyclotron waves were generated by above machine (Cartier, Angelo, & Merlino, 2012).

Theoretical value of real frequency and growth in wave has been calculated by taking Marlino et al. data, when considering homogeneous and inhomogeneous DC electric field (Tyagi, Srivastava, & Pandey, 2012).

Machining process can be divided into two parts: one is traditional machining process and second is unconventional machining process. The examples of traditional machining process are turning, boring, milling, shaping, grinding, etc. and unconventional machining processes are abrasive jet machining, ultrasonic machining, water jet machining, electro-discharge machining, electro-chemical machining, etc.

The material removal in different machining process can be performed in different modalities One of the modalities is based upon impact phenomenon, i.e. micro-cracking, micro-cutting, melting and vaporizing of small quantity of metal from workpiece surface layer by ions impact (Slătineanu, Coteaţă, Dodun, Iosub, & Apetrei, 2008).

If the temperature is high enough, a so-called thermal source appears and the workpiece solid material is transformed in molten material or even in vapor (Slătineanu et al., 2009).

In particular, micro-manufacturing refers to the fabrication of products or components where the dimensions of at least one feature are in the micrometer range. Similarly, nano-manufacturing refers to the production of devices where some of the dimensions are in the nanometer range.

Slătineanu et al. 2008 develop some mathematical equation to describe non-traditional machining process based on impact phenomenon for ions/electrons, impact of ions (negative and positive)/electrons on metallic surface with the help of electrical discharge machining process and finding out ion penetration.

Several micro-mechanical manufacturing technologies were studied to characterize geometries such as micro-channel. The technologies used were electro-discharge machining, sand blasting and abrasive jet machining, and performance is compared with lithography capabilities (Jáuregui, Siller, Rodríguez, & Elías-Zúñiga, 2010).

The pulse electro-chemical micro-machining close the gap between the normal ECM with gap width in the middle micrometer range and EC nano-machining (Burkert, Schulze, Gmelin, & Leone, 2009).

Comparison of different micro/nano-manufacturing technologies namely micro-electro-discharge machining, laser ablation, micro-milling, focused ion beam machining, micro-injection molding, hot embossing and electroforming had been studied by Brousseau, Dimov, and Pham (2010).

Garvin, Garmire, Somekh, Stoll, and Yariv (1973) describe sputtering process by plasma ions. The plasma is usually created in an inert gas by DC or RF discharge. Ion beam sputtering is performed by bombarding the target surface with a collimated ion beam in high vacuum.

Plasma etching is a process in which metallic or non-metallic surface roughness can be increased or decreased according to requirement by using plasma species. Plasma etching depends upon various factors, such as true contact area, molecular strength, density, conductivity, etc. Generally surface roughness is calculated in terms of *rms* value (Greenwood & Williamson, 1966).

Plasma etching is a chemical etching process and chemical reaction occurs between a solid atom and a gas atom to form a molecule (Verdonck, 2006).

Actually plasma etching consists of formation of reactant particle, adsorption of the reactant particle, formation of product molecule, etc. Tyagi (2013a) derived an expression and found out the theoretical value of plasma etching when phenomenological power law for the heights of surface roughness was taken into consideration.

The central goal of sputtering by plasma processing is to obtain any high rate, while preserving anisotropy, uniformity and selectivity at the obtained rate. Important advances in basic understanding of plasma-assisted processes have been reported in the recent past, although engineering development of new system continues to rely heavily upon trial and error procedures (Economou & Alkire, 1988).

Using a one-dimensional radial dispersion model, the role of physical factors such as etchant convection and diffusion has been examined, which are determined by the concentration of the etchant. In particular, in (Economou & Alkire, 1988), a detailed study of the etch rate of polymer in an oxygen plasma as a function of pressure, power and flow rate is presented (Economou & Alkire, 1988).

Theoretical value of ion penetration on metallic surface by computerization had been found out when utilizing the parameters of Statineu et al. Surface roughness also has been calculated, when metal is removed by ion beam machining process (Tyagi, 2012a; Tyagi, 2013a; Tyagi, Srivastava, et al., 2012).

Non-traditional machining process by macro-particle impact is known as abrasive jet machining. Generally, macro-abrasive particle are accelerated by compressed air and directed to the contact surface by a nozzle. It has been studied that due to plastic deformation, material removal mechanism causes crack and spalling of ductile material (Bitter, 1963; Finne, 1972).

The material is removed from surface by mechanical erosion process due to impact of high-velocity abrasive particles. The erosion phenomenon in an AJM may be considered to constitute two phases. The first phase consists of transportation phase, i.e. the quantity of abrasive particles flown, and the direction and velocity of impinging particles as determined by the fluid flow condition of solid-gas suspension. The second phase is the determination of the material removal rate or the erosion rate (Roy & Paul, 1987).

The complete mechanism of erosion is complex and involves mechanical, chemical and material properties. The erosion depends upon various factors, such as speed, angle of impact, ductility or brittleness of impinging particles and elasticity of the material. Neema and Pandey proposed an equation for material removal rate by equating the kinetic energy of the impacting particle to the work of deformation during indentation (Neema & Pandey, 1977).

$$Q = kn_a d^3 v^{3/2} (\rho_a / 12\sigma_y)$$

In abrasive jet machining, there is large wear and tear in nozzle. (Tyagi, 2012a) suggested a nozzle-free model of abrasive jet machining, the abrasive particles being accelerated by momentum and high velocity of ions. KSF_6 gas was chosen for the study because it has high electron affinity and higher molecular weight. Wensink and Elwenspoek (Schlautmann, Wensink, Schasfoort, Elwenspoek, & Berg, 2001) studied effect of impact angle and abrasion size. The results obtained reflect that smaller abrasion size and less impact angle improves the machinability. Investigation of the effect of abrasive grit size and mixing ratio on machinability was also done (Balasubramaniam, Krishnan, & Ramakrishnan, 1998).

Plasma spray is a process which includes metallic melting, quenching and solidification in a single operation. The complete plasma spraying consists of injection of metallic powder particles into plasma, melting of powder particles and formation of an adherent coating, i.e. high quality coating. The coating is built up by impact of successive particles by process of flattering, cooling and solidification. The generated surface is fine-grained and homogeneous because cooling rate is high (Herman, 1988).

Plasma spray deposition is one of the most important technologies available for producing the high-performance surfaces required by modern industry. Over the past 25 years, there have been significant advances in the understanding of plasma physics and in the development of spraying equipment and techniques. This has enabled a range of materials including metals, alloys, ceramics and cermets to be plasma-sprayed on to a great variety of substrate types and geometries. During this period, the uniquely aggressive environment within the gas turbine engine has provided not only some of the greatest challenges to plasma spraying technology, but also some of its most successful applications (Gill & Tucker, 1986).

Plasma spraying has undergone a rapid expansion in the past 30 years due to a couple of important advantages. Firstly, it can conveniently treat specimens with a complex geometry. Secondly, the wide spectrum of materials that can be handled by this technique has spurred applications in the area of corrosion-resistant, high-temperature, and ablation-resistant coatings as well as biocompatible films. One of the disadvantages of plasma spraying is the poor adhesion between the substrate and coating, but several measures can be used to improve it. For instance, the thermal gradient at the substrate–coating interface caused by the rapid quenching of the molten particle splats that leads to deposition of an amorphous coating can be reduced. In addition, one can prevent a steep gradient in the coefficients of thermal expansion between the substrate and coating to avoid the formation of strong tensile forces that give rise to crack generation, chipping and/or delamination (Kurzweg, Heimann, Troczynski, & Wayman, 1998).

In cold spray technology, micro-size metallic particles are accelerated with the help of nozzle and air momentum. Tyagi et al. accelerated micro-size metallic particles by momentum transfer of ions to metallic particles. But the main condition of momentum transfer is that energy of plasma ions should not be transferred to micro-size metallic particles. This can be achieved when vacuum is maintained throughout the process. The velocity of micro-sized metallic particles is calculated when density gradient is zero in plasma (Tyagi, 2013b; see Table 2).

The effect of processing parameters for coating was assessed in terms of adhesion, density (micro-hardness), roughness, composition, thickness and continuity in tabular form which presents the range of these characteristics and those of the best coating (Meletis, Nie, Wang, & Jiang, 2002).

Fauchais (2004) wrote a topical review paper, which consists of plasma spraying for metallic and non-metallic particles. Metallic particles size is of micron size. In this work, a model of a plasma gun is also discussed with its dimensions.

Titanium coating on surface is developed by cold spray process with the help of helium and nitrogen gas under different temperature and pressure. The effect of the gas type and temperature on deposition behavior was examined (Cheng & Weng, 2003).

Schmidt, Gaertner, and Kreye (2006) discuss the effect of particle size and temperature of particle on material properties.

Tyagi, Kumar, and Pandey (2012) and Tyagi (2013b) develop a model of machine for plasma coating process, actually this technology is the combination of cold spray and hot spray technology.

Table 2. Surface properties in plasma spray process.

Parameter	Zn coating-I	Zn coating-II	Zn-Al coating
Thickness, μm	12	15	18
Deposition rate (μm/s)	0.37	0.47	0.56
Roughness (Ra), μm	3.5	2.5	2.3
Element contents	Zn	Zn	3.0 at. % Al, Zn bal
Adhesion strength (MPa)	>70	>70	>70
Hardness (Gpa)	0.97	1.06	1.3

2. Decision-making problems

As described in introduction, there are various manufacturing applications based on velocity shear instability in plasma. Some of these applications are non-traditional machining process, surface coating, surface etching, etc. In this review article, all these processes are summarized and suggested that by a single machine, all operation can be performed by using various combination of plasma parameters.

3. Methodology adopted

In mathematical modeling, magnetic and electric field are considered to remain constant w.r.t time though, the value of electric field may or may not be constant w.r.t distance. Vlasov equation is used to find the relation between number of particles which participate in energy transfer from plasma species to waves and those which do not. Larange force equation was applied to calculate displacement, velocity and acceleration of plasma species with the help of electric and magnetic field. This method is known as method of characteristic solution. Number of particles which are present at a particular location and time is described by perturbed distribution function. With the help of perturbed distribution function, dispersion relation and subsequently velocity of ions has been calculated.

Tyagi et al. considered plasma of K^+ and SF_6^- gas. The potassium ions are produced by spraying potassium atoms onto a tungsten plate and SF_6^-. ions are produced from the electrons which are released from another tungsten plate. The generation and control of the parallel velocity shear are accomplished by independently biasing each segment of ion source, i.e. a variable voltage applied between two conjunctive sections. The generation of parallel velocity shear instability canister is guaranteed by the electrostatic energy analyzer and the laser-induced florescence diagnostic technique. The negative ions are produced by introducing sulfur hexafluoride (SF_6) gas into the potassium plasma. An SF_6 molecule has a great electron attachment cross-section for the electron energies less than 1 eV. Due to this production of negative and positive ions, different layers have velocity shear and density gradient in adjacent layer. The detailed description of generation of velocity shear instability is described by Tyagi et al. (2011a, 2011b).

$$\frac{\overline{\omega'}}{\Omega_i} = -\frac{b_1}{2a_1}\left[1 \pm \sqrt{\left(1 - \frac{4a_1c_1}{b_1^2}\right)}\right] \tag{1}$$

where:

$$a_1 = a_2\left(\frac{\Omega_i}{k_{||}\alpha_{||}}\right)^2, \quad b_1 = \frac{\Omega_i}{k_{||}\alpha_{||i}}b_2 - \frac{2k_\perp\Delta'}{k_{||}^2\alpha_{||i}^2}a_2\Omega_i,$$

$$a_2 = \frac{\eta_e}{\eta_i}\frac{T_{\perp i}}{T_{||i}} + \frac{T_{\perp i}}{T_{||i}} - \Gamma_n(\mu_i)\frac{T_{\perp i}}{T_{||i}}, \quad b_2 = \frac{\Gamma_n(\mu_i)k_\perp}{2k_{||}}\varepsilon_n\rho_i\frac{\alpha_{\perp i}}{\alpha_{||i}} - \frac{\Gamma_n(\mu_i)k_\perp}{2k_{||}} - \frac{\Gamma_n(\mu_i)k_\perp n\Omega_i}{2k_{||}^2\alpha_{||i}},$$

$$c_1 = \frac{\Gamma_n(\mu_i)T_{\perp i}}{2T_{||i}}\left(1 - \frac{k_\perp}{k_{||}}A_i\right) - \frac{b_2k_\perp\Delta'}{k_{||}\alpha_{||i}} + \frac{k_\perp^2\Delta'^2}{k_{||}^2\alpha_{||i}^2},$$

$$\eta_i = 1 - \frac{\overline{E_i'}(x)}{4\Omega_i^2}, \quad \eta_e = 1 - \frac{\overline{E_e'}(x)}{4\Omega_e^2}, \quad \overline{\omega'} = \overline{\omega} - n\Omega_i,$$

$$E(x) = E_0\left(1 - \frac{x^2}{a^2}\right), \quad \bar{E}(x) = \frac{e_s E(x)}{m_s}$$

$$\Omega_s = \frac{e_s B_0}{m_s}, \quad \alpha_{\perp s} = \left(\frac{2k_B T_{\perp s}}{m_s}\right)^{1/2}, \quad \alpha_{||s} = \left(\frac{2k_B T_{||s}}{m_s}\right)^{1/2},$$

$$\xi = \frac{\overline{\omega} - (n+p)\Omega_s - k_\perp \Delta'}{k_{||}\alpha_{||s}}, \quad \Delta' = \frac{\partial \Delta}{\partial t}, \quad \Delta = \frac{\bar{E}(x)t}{\Omega_s}\left[1 + \frac{E''(x)}{E(x)} \cdot \frac{1}{4}\left(\frac{v_\perp}{\Omega_s}\right)^2\right],$$

$$A_s = \frac{1}{\Omega_s}\frac{\delta v_{oz}(x)}{\delta x}, \quad \varepsilon_n = \frac{\delta \ln n_0(x)}{\delta x}, \quad A_T = \frac{\alpha_{\perp s}^2}{\alpha_{||s}^2} - 1, \quad \overline{\omega} = \omega - k_{||} v_{oz}(x),$$

$$\Gamma_n(\mu_s) = \exp(-\mu_s) I_n(\mu_s), \quad \mu_s = \frac{k_\perp^2 \rho_i^2}{2}, \quad (s = i, e).$$

where, $\mathbf{E}(x)$ is the inhomogeneous DC electric field, perpendicular to external magnetic field $\mathbf{B}_0$ and parallel to ion flow. The detailed description of all variables, which used in the given expression, is well thought-out in (Tyagi et al., 2011b).

The dimensionless real frequency and ion's velocity have been calculated by mathematical model and computerization technique for Equation (1) for heterogeneous DC electric field. For heterogeneous DC electric field, the condition $\frac{x}{a} \leq 1$ has been taken.

The kinetic energy of the ions under the action of equivalent voltage U is determined by Tyagi (2012b).

$$E = eU = \frac{mv^2}{2} \tag{2}$$

Clearly, the metal removal rate is dependent on generated voltage U, which defines the depth of ion penetration in the workpiece material. The thickness of this surface layer for the free penetration of the electrons has been distinct by the Shenland's relation:

$$\delta = 2.2 \cdot 10^{-12} \cdot \frac{U^2}{\rho} \ [\text{cm}](3) \tag{3}$$

where, ρ is the workpiece material density (in g/cm^3) and U is the acceleration voltage (in V).

The complete depiction of ion penetration with effect of plasma factors on ions penetration is presented by (Tyagi, Srivatava, et al., 2012).

By using an atomic force microscope, the topographical charts of the surface have been documented. The root mean square value of the roughness altitude was calculated from power law. The pragmatic law for surface roughness (*rms*) was found to obey the following phenomenological power law, in which $\beta = 1$ and $\eta = 0.45$ (Pétri et al., 1994):

$$rms \propto \frac{1}{\sqrt{E}} (J^+/J_F)^n t^\beta \tag{4}$$

where, J^+ is the ion flux impinging on the substrate, E is the kinetic energy of the ion, J_F is the SF_6 atom flux and t is the exposure time.

$$n_0 = n_p \times v/4$$

n_0 = No of ions striking the workpiece, n_p = plasma density and v = velocity of ions which is found out by Equation (1).

Plasma density is $10^{22-23}/m^3$ (Tyagi, 2012a).

$$Q = kn_a d^3 v^{3/2}(\rho_a / 12\sigma_y)$$

where, d is the size or diameter of an abrasive particle; ρ_a is the density of the abrasive material; v is the velocity of the abrasive particle and σ_y is the yield stress of the work material.

4. Results and discussions

4.1. Result and discussion of unconventional machining process

Tyagi, Srivastava, et al. (2012) found out ion penetration vs. $k_\perp \rho_i$ for different combinations of plasma factors, such as magnetic field, magnetic field and heterogeneity in DC electric field. The value of ion penetration varies from 0.137 to 2.0 μm when plasma parameters are temperature anisotropy 1.5, temperature ratio 2; DC electric field varies from 12 to 20 V/m; and shear scale length from $A_i = 0.5$ to $A_i = 0.55$.

Shanland et al. in Romania experimentally penetrated the ions on metallic surface when value of applied voltage varies from 5000 to 200,000 V. The value of ions penetration varies from $\delta = 7.05 \times 10^{-8}$ to 1.128×10^{-2} cm. When ions penetrate the metallic surface, due to momentum of ions, adjacent particles oscillate. Due to oscillations, upper surface of metal melts and is removed from surface by circulation of fluid.

In electron-beam machining, material from surface is removed by the impact of high velocity electrons. The electrons are generated in electron gun. Jets of electrons are accelerated up to speed of light by applying voltage up to 50 kV between anode and cathode. Magnetic lens are used to focus the accelerated electron beam on workpiece. The energy density of electron beam is of the order of 108 W/cm. The kinetic energy of the electrons is converted into heat which is sufficient to cause rapid melting and vaporization of the workpiece material. To eliminate scattering of the beam of electrons, the work is done in a high vacuum chamber. Electron-beam machining can be applied to a wide range of materials. As a result of extremely high energy density of the beam and the short duration of beam–workpiece interaction, thermal effects on the workpiece material are limited to a heat affected zone that seldom exceeds 0.025 mm in depth. The high beam power density also enables high aspect ratio holes to be drilled, often as large as 15–1 (Velayudham, 2007).

4.2. Results and discussion of plasma etching process

Tyagi (2013a) theoretically found out surface roughness for different combinations of plasma parameters. These plasma parameters are ratio of ion flux to the neutral reactant flux (J^+/J_F), heterogeneous DC electric field, exposure time, magnetic field, heterogeneity (x/a) in DC electric field etc. The maximum value of surface roughness was 3.24 nm and minimum value of surface roughness was 1.32 nm.

Pétri et al. (1994) in calculated experimental value of surface roughness on silicon surface by considering following parameters: ratio of ion flux to the neutral reactant flux (J^+/J_F) varies from 20 to 80, the exposure time varies from 2 to 12 min, the magnetic field is 0.10 T, the homogeneous DC electric field is 15 V and other parameters also described in [5]. The value of silicon surface roughness varies from 2 to 20 nm.

4.3. Results and discussion of abrasive jet machining process

Matthew, Chastagner, and Albert (2007) found out the effect of three parameters on edge erosion. These two parameters are distance between nozzle and edge, and angle of impingement varies from 0 to 30°. Input air pressure was fixed at 552 kPa. In the first experiment, the author maintained l = 0 mm and $\alpha = 0°$, while t was varied from 0 to 15 in 1 s increments. Another test was conducted under the same conditions with $t = 30$ s. This experiment studies the effect of time of blasting. In yet another experiment, the effect of distance between nozzle and edge tip was investigated. The value of l was varied from 0 to 5 in 1 mm steps, while $t = 7$ s and $\alpha = 0°$. Also, the effect of angle α, which was varied from 0° to 30° in 5° increments (with l = 0 mm and $t = 7$ s) was studied. Three repeated tests were conducted for every test condition. All tests eroded a spot on the edge. In production, the nozzle is expected to move along the edge to generate the desired edge radius. The spot test in this study is the first step in understanding the magnitude of the edge radius and collateral damage in AJM.

Tyagi (2012a) presented a new technique of abrasive jet machining based on the concept of velocity shear instability in plasma. The effect of different plasma parameters on metal removal from metallic surface was studied. The plasma parameters were different values of electric field (V_0), heterogeneity in D.C electric field, magnetic field, heterogeneity in electric field, shear scale length, etc. The maximum value of metal removal is 2.2 gm/s and minimum value is 0.2 gm/s.

Figure 1 shows the variation of metal removal (gm/s) vs. magnetic field (B_0) for different values of electric field (V_0) by fixing other parameters. The value of metal removal (gm/s) decreases by increasing the value of magnetic field (Tyagi, 2012a). The maximum value of metal removal is 0.2 gm/s when the value of magnetic field is 0.2 T and the value of homogeneous DC electric field is 24 V/m with other fixed parameters listed in the figure caption. Heterogeneity in D. C. electric field is a useful parameter to control ion penetration into metallic surfaces. It has been shown that under the parameters considered, the maximum value of ion penetration is 41 μm (at value of the magnetic field 0.11 T, voltage 16 V and inhomogeneity 0.2). Moreover, the theoretical

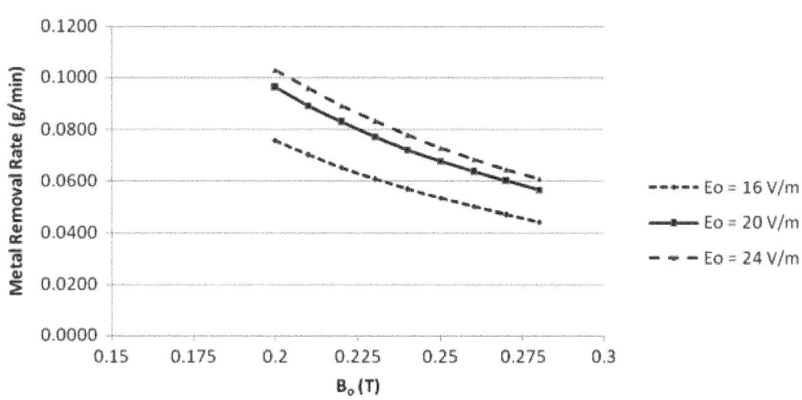

Figure 1. Variation of metal removal rate vs. magnetic field for different values of E_0 and other parameters are $A_i = 0.5$, $T_e/T_i = 2$, $\theta_1 = 88.5^0$, $A_T = 1.5$, $\varepsilon_n\rho_i = 0.2$, $X/a = 0$, plasma density = $10^{22}/m^3$, abrasive particle = 10^4/min, size of abrasive particles = 0.45 mm, density of abrasive = 2.3 gm/cm^3 and density of workpiece = 2.8gm/cm^3.

results show that the ion penetration increases with corresponding decrease of the magnetic field value and inhomogeneity in the DC electric field and by increasing of the DC electric field value (Tyagi, Srivastava, et al., 2012).

5. Conclusion

In this review article, theory and model of non-tradition machining process, plasma spray technology, Plasma etching and abrasive jet machining have been discussed. The outline of this review article is divided into some industrial applications based on velocity shear instability in plasma. The computational/experimental results of this review article will be useful to fabricate a new plasma spray machine, unconventional machining process, plasma etching and abrasive jet machining or to increase the effectiveness of existing plasma spray machine and non-tradition machining process. This article shows the flexibility of using the plasma parameters for controlling/achieving some unconventional machining operations.

Acknowledgment

I would like to express my sincere gratitude to Mrs Shraddha Sharma for the non-stop prop up of writing this review article, for her patience, motivation, enthusiasm and immense knowledge. Her support helped me in all the time of writing of this article. I could not have imagined having a better advisor for this review article.

References

Balasubramaniam, R., Krishnan, J., & Ramakrishnan, N. (1998). Investigation of AJM for deburring. *Journal of Materials Processing Technology, 79*, 52–58. Retrieved from http://dspace.library.iitb.ac.in

Bifano, T. G. (2000). Ion-beam machining of millimeter scale optics. *Applied Optics, 39*, 599–611. Retrieved from people.bu.edu/tgb/PDF_files/21_Ion.pdf

Bitter, J. B. A. (1963). A study of erosion phenomena part I. *Wear, 6*, 5–21. Retrieved from www.ewp.rpi.edu/hartford/~ernesto/F2007/FWM/Papers/Bitter2.pdf

Brousseau, E. B., Dimov, S. S., & Pham, D. T. (2010). Some recent advances in multi-material micro- and nano-manufacturing. *The International Journal of Advanced Manufacturing Technology, 47*, 161–180. Retrieved from link.springer.com/article/10.1007%2Fs00170-009-2214-5

Burkert, S., Schulze, H. P., Gmelin, T., & Leone, M. (2009). The pulse electrochemical micromachining (PECMM). *International Journal of Material Forming, 2*, 645–648. Retrieved from link.springer.com/article/10.1007%2Fs12289-009-0464-2

Cartier, S. L., Angelo, N. D., & Merlino, R. L. (2012). A laboratory study of ion energization by eic waves and subsequent upstreaming along diverging magnetic field lines. *Journal of Geophysical Research Space Physics, 91*, 8025–8033. Retrieved from onlinelibrary.wiley.com

Cheng, J. L., & Weng, Y. L. (2003). Deposition characteristics of titanium coating in cold spraying. *Surface and Coatings Technology, 167*, 278–283. Retrieved from www.sciencedirect.com/science/article/pii/S0257897202009192

Economou, D. J., & Alkire, R. C. (1988). A mathematical model for a parallel plate plasma etching reactor. *Journal of The Electrochemical Society, 135*, 2786–2794. Retrieved from jes.ecsdl.org/content/135/11/2786

Fauchais, P. (2004). Understanding plasma spraying. *Journal of Physics D: Applied Physics, 37*, R86–R108. Retrieved from www.engr.sjsu.edu/sgleixner/PRIME/.../Fauchais_plasma%20spraying.pd

Finne, I. (1972). Some observations on the erosion of ductile materials. *Wear, 19*, 81–90. Retrieved from www.sciencedirect.com/science/article/pii/0043164872904449

Ganguli, G., Lee, Y. C., & Palmadesso, P. (1985). Electrostatic ion-cyclotron instability caused by a nonuniform electric field perpendicular to the external magnetic field. *Physics of Fluids, 28*, 761–763. Retrieved from http://scitation.aip.org/content/aip/journal/pof1/28/3/10.1063/1.865096

Garvin, H. L., Garmire, E., Somekh, S., Stoll, H., & Yariv, A. (1973). Ion beam micromachining of integrated optics components. *Applied Optics, 12*, 455–459. Retrieved from www.ncbi.nlm.nih.gov/pubmed/20125327

Gill, B. J., & Tucker, R. C. (1986). Plasma spray coating processes. *Materials Science and Technology, 2*, 207–213. Retrieved from http://www.ingentaconnect.com/content/maney/mst/1986

Greenwood, J. A., & Williamson, J. B. P. (1966). Contact of nominally flat surfaces. *Proceedings of the Royal Society A: Mathematical, Physical and Engineering Sciences, 295*, 300–319. Retrieved from rspa.royalsocietypublishing.org/content/295/1442/300

Herman, H. (1988). Plasma-sprayed Coatings. *Scientific American, 259*, 112–117. Retrieved from http://www.osti.gov/scitech/biblio/6966003

Ichiki, R., Kaneko, T., Hayashi, K., Tamura, S., & Hatakeyama, H. (2009). Parallel-velocity-shear-modified drift wave in negative ion plasmas. *Plasma Physics and Controlled fusion, 51*, 1–11. Retrieved from http://iopscience.iop.org/0741-3335/51/3/035011

Jáuregui, A. L., Siller, H. R., Rodríguez, C. A., & Elías-Zúñiga, A. (2010). Evaluation of micro-mechanical manufacturing processes for microfluidic devices. *The International Journal of Advanced Manufacturing Technology, 48*, 963–972. Retrieved from http://link.springer.com/search

Jha, S., & Jain, V. K. (2006). Nanofinishing techniques. In *Micromanufacturing and nanotechnology* (pp. 171–195). Berlin: Springer. Retrieved from http://link.springer.com/chapter/10.1007%2F3-540-29339-6_8

Kurzweg, H., Heimann, R. B., Troczynski, T., & Wayman, M. L. (1998). Development of plasma-sprayed bioceramic coatings with bond coats based on titania and zirconia. *Biomaterials, 19*, 1507–1511. Retrieved from www.ncbi.nlm.nih.gov/pubmed/9794527

Matthew, W. Chastagner, & Albert, J. S. (2007). Abrasive jet machining for edge generation. *Transactions of NAMRI/SME, 35*, 359–366. Retrieved from wumrc.engin.umich.edu/.../07NAMRCChastagnerABRASIVEJETMACH

Meletis, E. I., Nie, X., Wang, F. L., & Jiang, J. C. (2002). Electrolytic plasma processing for cleaning and metal-coating of steel surfaces. *Surface and Coatings Technology, 150*, 246–256. Retrieved from www.bennati.it/.../Electrolytic%20plasma%20processing%20for%20clea

Neema, M. L., & Pandey, P. C. (1977). Erosion of Glass when act upon by an abrasive jet. In *Proceedings of International Conference on Wear of Materials*. India.

Pétri, R., Brault, P., Vatel, O., Henry, D., André, E., Dumas, P., & Salvan, F. (1994). Silicon roughness induced by plasma etching. *Journal of Applied Physics, 75*, 7498–7506. Retrieved from springer.com/article/10.3103%2FS1068375513010158

Roy, P. K., & Paul, A. K. (1987). Some aspect of abrasive jet machining. *Journal of Institution of Engineers (India), 68*, 27–30. Retrieved from www.academia.edu/.../ANALYSIS_OF_ABRASIVE_JET_MACHINING

Schlautmann, S., Wensink, H., Schasfoort, R., Elwenspoek, M., & Berg, A. (2001). Powder-blasting technology as an alternative tool for microfabrication of capillary electrophoresis chips with integrated conductivity sensors. *Journal of Micromechanics and Microengineering, 11*, 386–389. Retrieved from iopscience.iop.org/0960-1317/11/4/318

Schmidt, T., Gaertner, F., & Kreye, H. (2006). New developments in cold spray based on higher gas and particle temperatures. *Journal of Thermal Spray Technology, 15*, 488–494. Retrieved from http://link.springer.com/article/10.1361%2F105996306X147144

Shanbhag, P. M., Feinberg, M. R., Sandr, G., Horenstein, M. N., Tseng, T. G., & Chen, A. (2003). Electron beam lithography in nanoscale fabrication: Recent development. *Electronics Packaging Manufacturing, IEEE Transactions, 26*, 141–149. Retrieved from ieeexplore.ieee.org/iel5/6104/27729/01236879.pdf

Slătineanu, L., Coteaţă, M., Dodun, O., Iosub, A., & Apetrei, L. (2008). Impact phenomena in the case of some non-traditional machining processes. *International Journal of Material Forming, 1*, 1391–1394. Retrieved from link.springer.com/content/pdf/10.1007/s12289-008-0124-y.pdf

Slătineanu, L., Dodun, O., Coteaţă, M., Gonçalves-Coelho, A. M., Beşliu, I., & Pop, N. (2009). Machining methods based on the impact effects. *International Journal of Modern Manufacturing Technologies, 1*, 83–88. Retrieved from esaform2008.insa-lyon.fr/proceedings/MS20/p_Sl_351.pdf

Tyagi, R. K. (2012a). Abrasive jet machining by means of velocity shear instability in plasma. *Journal of Manufacturing Science and Engineering, 14*, 323–327. Retrieved from www.sciencedirect.com/science/article/pii/S1526612512000400

Tyagi, R. K. (2012b). Effect of different plasma parameters on plasma energy by means of velocity shear instability. *European Journal of Applied Engineering and Scientific Research, 1*, 57–61. Retrieved from www.scholarsresearchlibrary.com

Tyagi, R. K. (2013a). Theoretical analysis of silicon surface roughness induced by plasma etching. *Surface Engineering and Applied Electrochemistry, 49*, 78–82. Retrieved from link.springer.com/article/10.3103%2FS1068375513010158

Tyagi, R. K. (2013b). An innovative approach for cold spray technology. *Asian Journal of Current Engineering and Maths, 2*, 342–344. Retrieved from www.innovativejournal.in

Tyagi, R. K., Srivastava, K. K., & Pandey, R. S. (2011a). Analysis of electrostatic ion-cyclotron instability driven by parallel flow velocity shear. *Surface Engineering and Applied Electrochemistry, 47*, 370–377. Retrieved from link.springer.com/article/10.3103%2FS1068375511040144

Tyagi, R. K., Srivatava, K. K., & Pandey, R. S. (2011b). Analytical and experimental study of velocity Shear Instability in the presence of inhomogeneous perpendicular D.C electric field. *Journal of Engineering and Technological Research, 3*, 118–126. Retrieved from academic journals.org/article/article1380115988_Tyagil%20et%20al.pdf

Tyagi, R. K., Kumar, A., & Pandey, R. S. (2012). Surface coating by velocity shear stability in plasma. *Theoretical Foundation of Chemical Engineering, 46*, 594–600. Retrieved from link.springer.com/article/10.1134%2FS0040579512050193

Tyagi, R. K., Srivastava, K. K., & Pandey, R. S. (2012). Non-traditional machining processes by means of velocity shear instability in plasma. *Surface Engineering and Applied Electrochemistry, 48*, 64–68. Retrieved from link.springer.com/article/10.3103%2FS1068375512010152

Velayudham, A. (2007). Modern manufacturing processes: A review. *Journal on Design and Manufacturing Technologies, 1*, 30–40. Retrieved from journals-sathyabama.com/archives/dmt/DMT-Int-Journal-Oct-2007-6.pdf

Verdonck, P. (2006). *Plasma etching*. Brazil: UNICAMP. Chapter 1. 11 p.

7

Production scheduling for continuous manufacturing systems with quality constraints

Jalel Ben Hmida[a]*, Jim Lee[a], Xinchun Wang[a] and Fathi Boukadi[b]

[a]Systems Engineering, University of Louisiana at Lafayette, P.O. Box 44170, Rougeou, Hall Room # 244, Lafayette, LA 70504, USA; [b]Department of Petroleum Engineering, University of Louisiana at Lafayette, P.O. Box 44408, Madison Hall, Room # 134, Lafayette, LA 70504, USA

This research is motivated by a real world production scheduling problem in a continuous manufacturing system involving multiple objectives, multiple products and multiple processing lines with various inventory, production and quality constraints. Because of the conflicting objectives, a global optimization approach is considered as not feasible by the plant management. Given a customer demand forecast, two practical heuristic or sequential optimization algorithms are developed to generate daily production schedules for two primary objectives: minimize shipment delays (pull-backward procedure) and minimize average inventory levels (push-forward procedure). A third heuristic algorithm (reduce switch-over procedure) which is based on the current management practice is also developed to serve as a benchmark. A factorial experiment was performed to evaluate the performance of the heuristic procedures and to identify factors that might affect the performance differences among heuristics. Since each heuristic is designed to give priority to one of the three conflicting objectives, none of them is absolutely superior to the other algorithms in all aspects. However, the first two heuristic procedures performed better than the current management practice in shipment delays and average daily inventory. The production schedules generated by the two procedures also satisfy the quality constraints. The experimental results also showed that the performance of the algorithms is significantly affected by product mix, inventory levels, and demand pattern.

Keywords: production scheduling; continuous manufacturing system; product quality; multiple objectives; heuristic algorithms

1. Introduction

Scheduling is concerned with efforts to see that the resources of a manufacturing system are well utilized so that the products are produced within reasonable accord with the customer demand. Numerous planning and scheduling methods have been developed for discrete part manufacturing systems, and published research has shown that the approach for scheduling production depends on the type of manufacturing system involved. Reviews of the characteristics of continuous manufacturing systems in Vollmann, Berry, and Whybark (1992), Askin and Standridge (1993) suggest that the continuous manufacturing systems require different modeling approaches.

*Corresponding author. Email: jalel.benhmida@louisiana.edu

This research is motivated by an actual production scheduling problem in a continuous manufacturing system of a large producer of chemical products. On average, the chemical plant receives more than 200 customer orders in a month, totaling about four million pounds of chemical products. The production system is capable of producing a large number of different types of products through multiple processing lines. A large number of process constraints exist due to varied capabilities of the production lines and the processing requirements. Most of the products can be produced on more than one line and some of the processes require the sharing of special tools. Some products have precedence constraints, and some products have similar production conditions that should be scheduled for production consecutively.

Currently, scheduling daily production in the system is subjectively based on the management's experience. With an increasing emphasis on the multiple objectives of on-time shipment, low inventory, and production quality; the management of the plant needs a scheduling tool to improve the production scheduling for better system performance. The three objectives identified by the plant management are described as below:

(1) *On-time shipments*. The current overdue shipment rate (OSR) is unacceptable by the top management. The lack of on-time delivery may prevent the plant from being considered as a major supply candidate by the downstream industries. A measure of on-time shipment is commonly considered as the most important performance indicator of a continuous manufacturing system.
(2) *Low inventory*. To reduce overdue shipments and if processing lines are available, finished products are normally produced before the due dates. This results in high level of inventory. While "zero inventory" may not be practically feasible, efforts must be made to reduce the overdue shipments and the inventory level at the same time.
(3) *Product quality*. Production run length is defined as the number of days a processing line is scheduled to produce the same product type. Longer runs can produce finished products with consistent quality. Frequent product switchovers in the processing line can result in quality problems. However, a run length larger than necessary can increase the inventory level.

The management of continuous manufacturing systems of this type faces the dilemma of meeting customer delivery dates while operating the system efficiently. This involves conflicting objectives. The conflict arises because improvement in one objective can be made to the detriment of one or more of the other objectives. Moreover, various production and quality constraints need to be satisfied. All of these concerns are addressed in this research. Currently the approach used by the plant management gives priority to product quality where the product switchover is minimized. A scheduling tool, which can generate several daily production schedules based on algorithms with priorities given to different objectives, would be helpful to the plant management.

2. Literature review

Production scheduling, perhaps because of its considerable quantifiable structure and richness of virtually endless variations, continues to attract many researchers for scholarly work. Patterson, Slowinski, Talbit, and Weglarz (1989) stated that most of the realistic scheduling problems could not be solved in a computationally efficient manner. Linear programming (LP) is commonly used to solve production planning and scheduling problems optimally, but the problem formulation is normally over-simplified and the solution is used only at the strategic level. Heuristic rules have been developed to deal with practical scheduling problems on the shop floor.

Deepark and Grossmann (1990) proposed a multi-period linear programming model for the simultaneous production planning and scheduling of multi-product batch manufacturing systems that may consist of several non-identical parallel lines. The model involves a simplistic representation of the scheduling problem where many possible constraints are not considered. Gupta, Bector, and Gupta (1990) provided an approach to determine the optimal sequence of products with due date in a single machine system. Linear programming models were developed for three different due date assignment methods. Minton, Johnston, Phillips, and Laird (1992) described a simple heuristic approach to solve large-scale constraint satisfaction and scheduling problems. Anwar and Nagi (1997) formulated an integrated scheduling and lot-sizing problem. An efficient heuristic is developed that schedules operations by exploiting the critical path of a network and iteratively groups in order to determine the lot-size that minimizes the make span as well as set-up and holding costs. Certain aspects of scheduling problems such as operational pre-conditions (e.g. product sequencing and cleaning requirements) are important to avoid compromising final product quality, safety and feasibility of operations. (Amaro & Barbosa-Póvoa, 1999) developed a Mixed Integer Linear Programming (MILP) formulation so as to account for operational conditions. The formulation is solved using a Branch and Bound (B&B) procedure. In some production systems, early shipments are forbidden. Scudder, Hoffmann, and Rohleder (1993) explored various policies on shipments using a simulation model and also came up with best policies.

Many scheduling problems are multi-objective that can be efficiently solved by a heuristic approach. Daniels (1994) described a multi-objective scheduling situation where information concerning the relative importance of the criteria is available through interaction with a given decision maker. Yim and Lee (1996) used Petri nets and heuristic search for multiple objectives scheduling in a flexible manufacturing systems (FMSs). Tung, Lin, and Nagi (1999) presented a hierarchical approach to scheduling in a FMS that pursues multiple performance objectives. Viana and de Sousa (2000) used heuristics to solve multi-objective resource constrained project scheduling problem. Lou, Liu, Zhou, Wang, and Sun (2012) present a method for developing a pro-active scheduling model that anticipates known uncertainties and a reactive model that accounts for and handles unknown uncertainties.

One approach to avoid frequent product switchover in production is to assign a long setup time between product switchovers. He and Kusiak (1992) considered a mixed-integer formulation of the single-machine scheduling program with sequence-dependent setup times and precedence constraints. Seki and Kogure (1996) developed a stochastic-demand model to deal with lot scheduling problem for multi-item continuous demand utilizing a single machine with setup times. A recent study by Wagner and Smits (2004) used local search heuristic for stochastic economic lot scheduling problem that considered multiple product in a single facility with limited capacity and significant setup times and costs.

Recent research in scheduling has also incorporated various management concepts. Modarress, Ansari, and Willis (2000) presented a technique using Just-In-Time (JIT) and Statistical Process Control (SPC) methods to minimize the short run production in an environment with demand uncertainties. Bose and Pekny (2000) presented Model Predictive Control (MPC) concept for planning and scheduling problem. A forecasting model was developed to calculate target inventory. The importance of Supply Chain (SC) production planning was studied in Caramanis, Anli, and Paschalidis (2003). Their model produces weekly production schedule that minimizes inventory and backlog costs subjected to non-linear constraints on production imposed by weekly varying dynamic lead times and inventory hedging policies.

Giannelos and Georgiadis (2003) investigated scheduling in a continuous manufacturing process. They developed a model for scheduling fast consumer good manufacturing process using MILP techniques combined with continuous representation of time so as to alleviate some of the computational problems incurred by discrete time-index formulations. Kogan, Leu, and Perkins (2002) suggested polynomial-time algorithm for the continuous-time scheduling problem in a manufacturing system where multiple product types are produced on a set of parallel machines. The main objective of the study was to schedule the production so that inventory, production, and planning horizon costs are minimized.

Review of scheduling literature indicates that most of the realistic scheduling problems are solved by efficient heuristic methods. These methods are developed primarily for discrete manufacturing systems and do not consider the effects of scheduling on product quality. In this paper, mathematical models which formulate the problem in continuous manufacturing systems with multiple objectives are presented first. Several heuristic algorithms are then proposed to solve the scheduling problem with priority given to different objectives. The management can determine an acceptable production schedule based on the proposed decision support procedures. The performance of the heuristic procedures is evaluated based on a simulation experiment.

3. Problem description

The scheduling problem addressed in this research can be stated as follows: *A* continuous manufacturing system can produce *M* types of products. There are *N* processing lines, each being capable of producing a subset of *M* products. If finished products are produced earlier than the due date, they are stocked in the warehouse where each product has a limited storage area. The production rate varies by products on a line, and only one product can be produced by one line at any time.

Table 1 shows a sample of the data of maximum inventory level allowed and production rate of each processing line by product. For the purpose of modeling, the

Table 1. Sample data for inventory levels and processing line capabilities.

		Production rate by line (×1000 lb/day)				
Product	Maximum inventory (×1000 lb)	*A*	*B*	*C*	*D*	*E*
1SB	150	25	25	0	0	0
1SC	262	25	25	0	0	0
2SB	80	24	0	24	0	0
2SC	126	24	0	24	0	0
3SB	150	23	0	0	23	0
3SC	189	23	0	0	23	0
4B	150	19	0	0	0	19
4C	112	19	0	0	0	19
5	180	34	0	0	0	0
6B	120	0	29	0	0	0
6C	180	0	29	0	0	0

setup time is included in our estimation of production rate by processing line. For example, if line A is used to produce product 1SB and the minimum run length is 4 days. The total production in this run of 4 days is estimated to be 100 (thousand pounds), and the daily production is considered to be $100/4 = 25$ (thousand pounds). The real production time of the last day in a run is normally shorter than the other days, as the line must be cleaned and adjusted (i.e. setup time) for another run with a different product. As a result, the total estimation is normally slight lower than the actual production to cover the loss of setup time. When the run length is long, it is possible that the production run can be completed 20–30 min early. The effect of the difference does not significantly affect the estimation on the production rate.

For example, if line A is used to process product 1SB for one day, then 25 thousand pounds of 1SB can be produced. While most of the products can be produced on more than one line, some of the products can be produced only on one line (e.g. product 5 can be produced on Line A only). A sample of customer orders by days is shown in Table 2. Based on the customer orders, the scheduling problem is to determine the daily production schedule for each processing line with production, inventory and quality constraints.

To formulate the problem we require the following notation:

T planning horizon
t day in planning horizon ($t = 1, \ldots, T$)
M number of product types
N number of processing lines
i,j product index ($i, j = 1, \ldots, M$)
k processing line index ($k = 1, \ldots, N$)
L_i waiting period for product i in days. Waiting period is the number of days that a product needs to stay in warehouse before shipment for quality assurance
U_i actual due date of product i
E_i adjusted due date of product i based on waiting period
HL_i minimum run length (in days) of product i
IH_i maximum inventory allowed for product i
I_{it} inventory of product i in day t
D_{it} demand of product i in day t
Q_{ikt} production quantity of product i on line k in day t
QO_i quantity of overdue product i
W_{kt} times of product switchovers on line k in day t
C_{ik} daily production rate for product i on line k
P set of jobs $[i, j]$, where product i precedes product j
G_g set of product group which should be produced sequentially
g represents the type of groups
R_r set of products that require the same special tool (resource) r for processing
r represents the type of resource
RA_r number of units of resources available for products in set R_r
H_{ik} production run length (in days) of product i on line k
B_{ik} production starting day of product i on line k
O_{it} demand of product i that cannot be produced on time in day t
X_{ikt} a zero-one variable. X_{ikt} is 1 if product i is scheduled on line k in day t

Table 2. Sample monthly customer order (in thousand units).

	Product type												
Due day	1SB	1SC	2SB	2SC	3SB	3SC	4B	4C	5	6B	6C	7	8B
1	0	15	20	0	0	0	0	0	35	0	0	0	0
2	0	0	0	0	0	0	0	10	0	0	35	0	0
3	20	10	0	20	0	30	0	0	0	20	0	0	0
4	0	30	0	0	0	0	0	0	0	0	0	0	0
5	5	0	0	0	0	0	0	0	0	0	0	0	0
6	10	0	15	0	0	0	0	0	0	0	0	0	0
7	0	0	0	0	0	0	0	0	5	0	0	0	0
8	0	30	0	0	0	0	0	0	0	0	0	0	0
9	20	0	0	0	0	0	0	0	0	10	35	0	40
10	0	0	30	0	0	0	0	0	0	0	0	0	0

The three objectives to be considered in the production scheduling are: minimizing overdue shipment, minimizing inventory, and minimizing product switchovers. They are discussed below:

(1) OSR measures the amount of customer demands that cannot be met by a production schedule. In this research, it is expressed as ratio of overdue shipment amount to the total demands in the planning horizon. The overdue shipment of production day t (O_{it}) is demand (D_{it}) minus the inventory $I_{i,t1}$ and minus the production of product i on all lines (Q_{ikt}).

$$\text{OSR} = \sum_i QO_i / \sum_t \sum_i D_{it} \tag{1}$$

$$\text{where } QO_i = \sum_t O_{it} \text{ and } O_{it} = D_{it} - I_{i,t-1} - \sum_k Q_{ikt}$$

$$\text{If } O_{it} < 0, \text{ let } O_{it} = 0$$

(2) Average daily inventory (ADI) reflects the average daily amount of finished products produced before the due date. The inventory of product i on day t (I_{it}) is the initial inventory plus production on day t and minus the demand on day t. The ADI is determined by the total inventory divided by the planning horizon T. It is shown as follow:

$$\text{ADI} = \sum_t \sum_i I_{it} / T \tag{2}$$

$$\text{where } I_{it'} = I_{io} + \sum_{t=1}^{t} \sum_k Q_{ikt} - \sum_{t=1}^{t} D_{it}$$

$$\text{If } I_{it'} < 0, \text{ let } I_{it'} = 0$$

(3) Total Switch Times (TST) of the product is used to check the quality of a product as frequent product switchovers could lead to poor product quality. For everyday t and every line k, product switchover count W_{kt} is 1 if the product scheduled on day t is different from $t-1$. The TST can be defined as:

$$\text{TST} = \sum_k \sum_t W_{kt} \tag{3}$$

where $W_{kt} = 1$ if the product scheduled on day t is different from $t-1$.

Several types of production, inventory and quality constraints are considered in scheduling the daily produbelow.

(1) *Production capacity and capability constraints*. The quantity of product i that can be produced on line k in day t, Q_{ikt} should be less than the capacity C_{ik} of line k. The capacity of line k is a function of product type i. For every i and every k,

$$Q_{ikt} \leq C_{ik} \tag{4}$$

where C_{ik} is 0 if line k is not capable of producing product i.

(2) *Inventory constraints*. Each product is allocated limited storage space in the warehouse. A maximum inventory level IH_i is imposed on each product i.

$$I_{io} + \sum_t \sum_k Q_{ikt} - \sum_t D_{it} \leq \text{IH}_i \tag{5}$$

(3) *Minimum run length constraints*. To achieve the desired quality level, some products have minimum run length requirements as defined by HL_i. For every i, H_{ik} represents the actual run length in days of product i on line k:

$$H_{ik} \geq HL_i \tag{6}$$

(4) *Precedence constraints*. If product i must be produced before product j, then the production starting date of product i on line k (B_{ik}) should be completed before the line starts to process product j.

$$B_{ik} + H_{ik} \leq B_{jk} \tag{7}$$

In our example, product 2SB must be produced prior to product 4B and 4C.

(5) *Group constraints*. Products with similar production conditions should be produced sequentially to ensure product quality. If product i and j are in the same group, it is desired to produce these product back to back on line k.

$$B_{ik} + H_{ik} = B_{jk}, \text{ or } B_{jk} + H_{jk} = B_{ik} \tag{8}$$

For example, Product 17B, 18B, 20 require similar production conditions. Then they should be produced sequentially in the production schedule.

(6) *Resource sharing constraints*. The processing operations of some products require the use of special equipment with capacity constraints. Let RA_r represent the number of tools available and R_r is a set of products required in the production. At any time, no more than RA_r products in R_r can be produced.

$$\sum_{i \in R_r} \sum_k X_{ikt} \leq RA_r \tag{9}$$

where $X_{ikt} = 1$ if i is scheduled on k.

For example, no more than two S type products can be produced at one time on all lines as the processing of a S type product requires the use of a special tool 1, and $RA_1 = 2$.

(7) *Preferable product-line assignment*. Since most of the products can be produced on more than one processing line, priorities on the product-line assignment can be established by the management. The assignment may affect the production efficiency and

the quality of the product in some systems. For example, products 2SB and 3SB are preferably produced on line B though they can also be produced on line A.

The number of constraints for this multi-line, multi-product facility is more than one hundred, and the very challenging task is how to satisfy these constraints in the scheduling process while considering the conflicting objectives. This research proposes the development of practical heuristic scheduling algorithms to generate various production schedules based on the demand data and the constraints. With the scheduling tool, the management will be able to generate several feasible production schedules with estimated overdue shipment, daily inventory levels and the number of product switchovers. This information will allow the management to determine the actual daily production schedule that balances the three objectives.

4. Heuristic scheduling algorithms

The research problem is to construct production schedules that minimize (1) OSR, (2) ADI and (3) TST while satisfying the various production, inventory and quality constraints. With three conflicting objective functions, it is infeasible to develop a global optimization model which optimizes all three functions. Therefore, three heuristic or sequential optimization algorithms are developed in this research to optimize each of the three objectives. The Pull-Backward Heuristic Procedure (PB) first produces an initial production schedule to minimize the OSR. The initial schedule is then adjusted to reduce ADI without affecting the OSR. This procedure is basically a sequential optimization method which optimizes the OSR and ADI in two steps. The Push-Forward Heuristic Procedure (PF) first produces an initial schedule which minimizes the ADI. The initial schedule is then adjusted to reduce OSR without affecting the ADI. Similar to the PB schedule, the PF is a sequential optimization method which optimizes the ADI first then OSR in two steps. Finally, the Reduce Switchover Heuristic Procedure (RS) is used to produce an optimal solution on TST first. Then the OSR is considered next, followed by ADI.

The methodology underlying the three procedures is similar and follows the three basic steps: problem decomposition, product sequencing, and demand shifting. First, the planning horizon is broken into several smaller periods. If the planning horizon is one month, the length of a period can be about one week. All orders of the same product in a period are lumped together to form one product demand. The production for each of the periods is scheduled according to the lumped demand. After all the period schedules are obtained, they are merged together to form the schedule for the whole planning horizon.

In creating a period schedule, two sequencing rules are used to create priorities. The Earliest Due Date (EDD) rule is chosen as the primary rule, where the choice is made based on the product that has the earliest due date. The Shortest Processing Time (SPT) rule, where the choice is made based on the processing operation with the shortest duration, is used as a secondary rule. For example, if 1SB, 1SC and 2SB are produced in line A (see Table 2) in the first 7-day period, 1SB will be produced last as it has an order of 20 due on day 3 compared to the due date of 1 for 1SC (with order quantity 15) and 2SB (with order quantity 20). Since 1SC and 2SB have the same due date, they are sequenced based on SPT rule. The processing priority is given to 2SB because the demand of 35 thousand pounds of 2SB can be produced with a shorter processing time than the time required to produce 55 thousand pounds of 1SC.

When the production cannot be scheduled for on-time delivery due to capacity constraints, the demand can be shifted to another processing line according to line capability. If other lines are also occupied, the demand can be shifted to the next period. To satisfy precedence constraints, group constraints, and resource constraints; the product demand can be shifted to other dates based on the three objectives of the scheduling. Different approaches of shifting are employed for different heuristic procedures. However, the demand of a product is always shifted forward to satisfy the maximum inventory constraints. The procedures are summarized below.

4.1. Pull-Backward Heuristic Procedure (PB)

The PB procedure is basically a sequential optimization method which optimizes the OSR and ADI in two steps. If the demand is greater than the production capacity in a given period, then not all production runs can be scheduled before the due dates. In this case, an attempt is made to schedule the production in the previous period to avoid shipment delay. However, the ADI may go up.

Step 0. Initialization

Adjust the due date of customer order U according to waiting period L to form an adjusted demand date E in planning horizon. Given a product i, let $E_i = U_i - L_i$ for all products with a non-zero waiting period. Break planning horizon T into PN scheduling periods; let v be the length of the period and n be the nth period ($n = 1, \dots, \mathrm{PN}$) in the planning horizon, Then $\mathrm{PN} = \frac{T}{v}$. If PN is non-integer, it is rounded down to the nearest integer.

Step 1. Product-processing line assignment

Assign products to processing lines based on preferable product-line assignment, subject to capability constraints. One product is assigned to one line.

Step 2. Period schedule generation

Construct a trial production schedule based on demand for period n subject to production capacity constraints and minimum production run length.

(2.1.) Compute the total demand due in period n by product and by line. Let DD_{ikn} represents this demand for product i assigned to processing line k. For each k, sort due day E_i for all products in ascending order, resulting in k sorted lists. If a product has several orders in period n, E_i is computed based on the order with the earliest due date. Products with smaller E_i values have higher priorities to be scheduled for production. If two or more products have the same due day, the product with the smallest processing time (SPT) has a higher priority to be scheduled than other products.

(2.2.) For each processing line k, sequence products for the line based on the priority established in the sorted list from step 2.1.

(2.3.) Determine production run length H for each of product scheduled in the period. Run length of a product in a period is determined according to the total quantity of demand, capability and minimum run length constraints:

$$H_{ik} = \begin{cases} DD_{ikn}/C_{ik} & \text{for} \quad H_{ik} > HL_i \\ HL_i & \text{for} \quad H_{ik} \leq HL_i \end{cases} \tag{10}$$

where DD_{ikn} is the demand of product i on line k in period n. H_{ik} is rounded up to the next integer if DD_{ink}/C_{ik} is not an integer. Repeat this step for all k. (For example, product 1SC in the first 5-day period in Table 2 has $DD_{ik1} = 55{,}000$ and $C_{ik} = 25{,}000$. DD_{ink}/C_{ik} is 55,000/25,000, or 2.2. Therefore, H_{ik} would be rounded up to 3.)

(2.4.) Shift the unfilled demand of a product to another processing line in accordance with capability constraints. If H_{ik} cannot be totally assigned to line k in the period, shift the unscheduled demand to another line which is not fully utilized. If product i has an unfilled demand in period n and cannot be assigned to any other processing line, shift the unfilled demand to period $n+1$.

Step 3. Period schedule revision

(3.1.) Modify the schedule to meet the precedence constraints. If product j has been scheduled at time t_1 before product i at time t_2 where $[i, j]$ has precedence constraint P, then re-schedule i at t, where $t<t_1$. Adjust production schedule on the processing line accordingly. Repeat this step for all processing lines.
(3.2.) Modify the schedule to meet the group constraints. If production of a set of products $[i, j]$ in G_g is separated in the schedule and product i is in the earliest position, shift all other products in the same group next to and immediately following the product i. Shift the production of products without group constraints to a later day in the schedule if needed.

Step 4. Period iteration
Repeat step 1 to step 3 until PN schedules $S_1, S_2, \ldots, S_{PN}$ are obtained for each of the PN periods.
Step 5. Schedule Combination
Merge period schedules and check constraints.

(5.1.) Merge subsets of schedules $S_1, S_2, \ldots, S_{PN}$ to form a complete schedule S.
(5.2.) Exchange the position of products scheduled at the same line if precedence constraints are violated.
(5.3.) Check resource constraints for the production schedule across all processing lines. If a constraint is violated in day t, shift the demand of the product with lower DD_{ikn} to an earlier day if possible. If the product cannot be produced on an earlier day, shift the demand to day $t+1$. Adjust the production schedule on the same processing line correspondingly.
(5.4.) Satisfy the maximum inventory constraints. Assume the production on product i in line k can be shipped in the same day to meet the demand. Determine the inventory level for all products. If the inventory of product is greater than the maximum level allowed, shift that demand to day $t+1$ and then check the resource constraints. The production schedule will be shifted to the future if the new schedule violates the resource sharing constraints.

4.2. Push-forward heuristic procedure (PF)

The PF is a sequential optimization method which optimizes the ADI first then OSR in two steps. If the demand is greater than the production capacity in a given period, however, the demand is shifted to the next period to avoid inventory. Based on the previous description, only two steps need to be replaced. They are described below.

(3.1.) Modify the schedule to meet the precedence constraints. If product j has been scheduled at time t_1 before product i at time t_2 where $[i, j]$ has precedence constraint

P, then re-schedule j at t, where $t > t_2$. Adjust production schedule on the processing line accordingly. Repeat this step for all processing lines.
(3.2.) Modify the schedule to meet the group constraints. If production of a set of products $[i, j]$ in G_g is separated in the schedule and product i is in the earliest position, shift all other products in the same group to be produced immediately before the product i. Shift the production of products without group constraints to a later day in the schedule if needed.
(5.3.) Check resource constraints for the production schedule across all processing lines. If a constraint is violated in day t, shift the demand of the product with lower DD_{ikn} to day $t+1$. Adjust the production schedule on the same processing line correspondingly.

4.3. Reduce Switchover Heuristic Procedure (RS)

The RS procedure focuses on minimizing product switchovers on processing lines which is similar to the current practice of the plant management. It is used to produce an optimal solution on TST first. Then the OSR is considered next, followed by ADI. Based on the current practice, once a processing line is assigned, the production run is continued until the total demand for the planning horizon (normally one month) is satisfied. The advantage of the current practice is that the method is easy to implement. The disadvantage is that both the OSR and ADI are not acceptable.

To implement the RS algorithm, the number of days in a period used in the previous procedures must be increased to allow a longer run. A test on various period lengths shows that the length of the period should be doubled or the number of periods (PN) should be reduced by about 50% in this procedure to observe a statistically significant different number of product switchovers. With RS, all the steps remain unchanged as procedure PB. Only the PN is changed. Other steps remain unchanged as procedure PB.

From the above discussions, it is noted that different production schedules can be obtained by using different heuristic procedures. A computer program written in C language is then implemented to include the three heuristic procedures. Many variables, such as the initial assignment of products to the processing lines and the number of periods in the planning horizon can be changed in the program. Consequently, alternative schedules can be produced for the management to make scheduling decisions.

5. Simulation experiments

Simulation experiments were carried out to evaluate the performance of the three heuristic procedures. The purpose of the experiment was to compare the performance of the procedures by answering the following questions:

(1) What is the magnitude of the relative performance of one procedure over another?
(2) If the performances of the two procedures are statistically significant different, then what are the factors related to the problem attributes affecting their differences?

To answer these questions, four factors that may have influence on the performance measures are identified. They are as follows:

(1) *Group ratio (G).* The group ratio is expressed as a ratio of the number of

products with group constraints to the total number of products produced in the system. For example, if the total number of products is 50, and five of them are involved in group constraints, then the group ratio is 5/50 or 10%. Three levels are tested: 10, 25, and 50%.

(2) *Precedence strictness ratio (P).* The precedence constraints specify the products that must be produced before others. The precedence strictness ratio is defined as the number of precedence requirements among the products over the number of feasible pairs of products. For example, in a system producing five products, the maximum number of precedence constraints is 5!/2!3! or 10. If only one constraint exists for a pair of products, then the precedence strictness ratio is 1/10 or 0.1. Two levels are tested in the experiment: 0.1 and 0.2. These two ratios are relatively low because the type of products in the continuous manufacturing system we studied does not have too many precedence constraints.

(3) *Maximum inventory (I).* The maximum finished-goods inventory allowed in the system is basically limited by the space in the warehouse. To reduce the inventory cost, it is desired to reduce the maximum inventory level. Two levels of maximum inventory are tested. A level of 100% represents the actual maximum inventory allowed in a chemical plant, which is about 4500 thousand pounds of products in total. The level of 67% means that the maximum inventory allowed is reduced to 4,500 × 67% or 3000 thousand pounds in total.

(4) *Demand (D).* Three sets of real demand data were collected from a chemical plant. They are referred to as D1, D2, and D3. Each data set contains the daily demand for over 60 product types over a month period. The average total daily demands for D1, D2 and D3 are 96.6, 108.1 and 109.4 thousand pounds respectively. The standard deviations of the total daily demand of the three demand patterns are 93.90 for D1, 74.63 for D2, and 75.24 for D3.

Since the production scheduling issues are unique, a well-known data is not available in the literature. Using the three data sets from industry, a factorial experiment was designed and performed to evaluate the performance of the three heuristic procedures and to identify the factors affecting the three objectives scheduling.

Table 3 summarizes the factors and associated levels tested in the experiment. A full factorial experimental design is used, so the number of basic experiments designed for each heuristic procedure is $3 \times 2 \times 2 \times 3 = 36$. In order to account for variations in the experiments, five replications are created for each of the basic experiments. Considering three heuristic procedures and five replications per experiment, the total number of experiments would be $36 \times 5 \times 3 = 540$.

The method to replicate the experiment is based on a 10% random perturbation on demands. For every customer order in the planning horizon, a uniformly distributed random number between (0, 1] is generated to determine the direction of the perturbation. If the random number is between (0, 0.25], then the value is decreased by 10%. If the random number is between (0.25, 0.5], the order is decreased by 5%. If the random number is between (0.5, 0.75], the corresponding value will be increased by 5%. Finally, the value is increased by 10% if the random number is between (0.75, 1). This allows the demand data of replications to be varied within a 20% range. For example, if the original demand for a product is 25, it may be changed to 22.5 (i.e. 25 × .90), 23.75 (i.e. 25 × 0.95), 26.25 (i.e. 25 × 1.05) or 27.5 (i.e. 25 × 1.10) in the replications.

Three performance measures corresponding to each of the three scheduling objectives are used in the experiments: (1) OSR, (2) ADI and (3) Total switch times (TST).

Table 3. Experiment factors and levels.

Factor	Level
Group (G)	10%; 25%; 50%
Precedence (P)	10%; 20%
Maximum Inventory (I)	100%; 67%
Demand (D)	D1; D2; D3

The three performance measures have been defined in Equations (1)–(3). Based on OSR, ADI and TST, a total cost index (TCI) can be estimated to represent an overall economical performance for a schedule where:

$$\mathrm{TCI} = (400\ \mathrm{QO} + \mathrm{ADI} + 2500\ \mathrm{TST})/100 \tag{11}$$

The quantity QO is the total amount (in thousand pounds) of overdue shipment products, ADI is the ADI in thousand pounds, and TST is the number of times the switchover occurs in the production schedule. Note that the coefficients of the three parameters are estimated to present a certain relation between parameters, and the values vary in accordance with different production systems.

6. Analysis of experimental results

Based on the experimental design described in the previous section, a total of 540 experiments were performed. The performance measures of the 180 experiments for each of the heuristic procedures were averaged and presented in Table 4.

Among the three objectives of the scheduling, OSR and ADI can be considered as the primary or key performance measures for comparing different production schedules. As the product quality is critical to the success of the business, every production schedule developed by management must satisfy the run length constraints to insure that the quality of products is acceptable. Therefore, TST is simply used as a benchmark for schedule comparison purpose. In the real world a smaller TST is easier to implement and control, so it can be also used by management to evaluate the feasibility of schedule implementation. With the consideration of the minimum run length constraints in our scheduling problem, all three algorithms proposed can schedule production with acceptable product quality.

Table 4 shows that each heuristic procedure performs best for one of the three objectives as expected. Procedure RS results in an average of 12% OSR for the test problems. If the reduction of OSR is critical, then procedure PB can reduce the rate from 12% to 9%, and procedure PF is also slightly better than RS by 1%. For the ADI, PB is better than RS by 1%, and PF can reduce the inventory by more than 5%.

Table 4. Summary performance by Heuristic.

	Heuristic procedure			% Difference	
Performance measure	PB	PF	RS	PB-RS	PF-RS
OSR	0.09[a]	0.11	0.12	3	1
ADI	73,164	69,883[a]	73,852	1	5
TST	42.40	40.33	36.51[a]	−14	−9
TCI	3032[a]	3233	3211	6	1

[a]Indicates the best performer by performance measure.

However, the product switchover (TST) for PB and PF will go up by 14 and 9%. Based on the TCI with the coefficients defined in equation (11), procedure PB is the most economical procedure among the three. However, this superiority can be diminished if the coefficients in equation (11) are changed.

The summary results shown in Table 4 reveal that the performance measures of the three heuristic procedures are different. To determine whether the differences are significantly away from 0, t tests were performed. Under the null hypothesis, there should be no difference between the performance measures obtained from each procedure. If the null hypothesis is rejected at a selected level of significance, it is concluded that the difference exists. The assumption of normality on the data was first tested and verified. The test was set at 0.01 level of significance.

Four types of variables were used to measure the relative difference of the heuristic procedures. Variables DOSR(i, j) measures the difference in OSR between two heuristic procedures i and j:

$$\mathrm{DOSR}(i,j) = \mathrm{OSR}(i) - \mathrm{OSR}(j) \tag{12}$$

Variables for measures of ADI, TST and TCI are expressed as percentage difference between two procedures i and j:

$$\mathrm{PADI}(i,j) = [\mathrm{ADI}(i) - \mathrm{ADI}(j)]/\mathrm{ADI}(j) \tag{13}$$

$$\mathrm{PTST}(i,j) = [\mathrm{TST}(i) - \mathrm{TST}(j)]/\mathrm{TST}(j) \tag{14}$$

$$\mathrm{PTCI}(i,j) = [\mathrm{TCI}(i) - \mathrm{TCI}(j)]/\mathrm{TCI}(j) \tag{15}$$

The reason of using only difference in OSR is that the OSR is already a percentage value. Moreover, the value for measure of percentage difference between two procedures sometimes might be infinite since OSR(j) could be near zero.

The difference of the four statistics (OSR, ADI, TST and TCI) and the results of the t tests appear in Table 5. A '*' is marked in the "Significance" column if the hypothesis that the difference is 0 cannot be rejected at the 0.01 level of significance.

Table 5. t-test results by performance measure.

Variable	No. Obs	M (%)	SD	Significance
DOSR(PB,RS)	180	−2.24	0.036	0.0001
DOSR(PB,PF)	180	−1.96	0.052	0.0001
DOSR(PF,RS)	180	−0.28	0.064	0.554*
PADI(PF,RS)	180	−5.43	0.017	0.0001
PADI(PF,PB)	180	−4.77	0.019	0.0001
PADI(PB,RS)	180	−0.94	0.015	0.0001
PTST(RS,PB)	180	−14.80	0.081	0.0001
PTST(RS,PF)	180	−9.71	0.032	0.0001
PTST(PF,PB)	180	−7.69	0.192	0.0001
PTCI(PB,RS)	180	−5.17	0.214	0.0002
PTCI(PB,PF)	180	−4.70	0.214	0.0037
PTCI(RS,PF)	180	−4.43	0.428	0.071*

*Indicates that the difference is insignificant at level 0.01.

Table 6. ANOVA results by performance measure.

Variable	Significant effects
DOSR(PB,PF)	G, I, D, G*D, I*D, G*I*D
DOSR(PB,RS)	G, I, D, G*D, I*D
PADI(PB, PF)	G, I, D, G*D, G*I*D
PADI(PB, RS)	G, I, D, G*D
PADI(PF, RS)	G, I, D, G*D, G*I*D
PTST(PB, PF)	G, D, G*D
PTST(PB, RS)	G, D, G*D, I*D
PTST(PF, RS)	G, D, G*D
PTCI(PB, PF)	G, I, D, G*D, I*D, G*I*D
PTCI(PB, RS)	G, G*D, I*D

The *t* tests show that procedure PB significantly outperforms PF and RS, but the difference between PF and RS is not significant. The OSR generated by procedure PB is about 2% less than that of procedure PF and is 2.24% less than that of procedure RS. Procedure PF can generate the lowest level of ADI, and procedure PB ranks second in ADI. The *t* tests indicate that all the differences are significant, although the percent deviations are not very large. Procedure RS is the best on TST as the approach gives priority to reduce the number of product switchovers. The differences among the three procedures are statistically significant different. The statistics of the TCI show that procedure PB is significantly superior to the other two procedures for a system with cost coefficients similar to Equation (11). The difference between PF and RS is not significant. The results show that no procedure is absolutely superior to other procedures in every aspect. On the other hand, the experiment shows that the three procedures perform as desired since each of them is given a different priority.

The preceding observation suggests that further statistic analyses can be done to evaluate the effects of various system factors and their interaction effects on the difference between the procedures. To identify important factors affecting the performance of the heuristic procedures, an analysis of variance (ANOVA) was done using the SAS software system to evaluate the effects of the four factors and all their possible interactions. The ANOVA results on the four performance measures are summarized in Tables 6.

The variable column identifies the data set tested in the ANOVA model. For example, DOSR(PB, PF) represents the set of experimental data showing DOSR between procedures PB and PF. The level of significance is set at 0.05. All significant effects are listed in the tables.

The observations from ANOVA results are summarized as below:

(1) Group ratios (G) and demand patterns (D) significantly affect the performance differences in most of the cases.
(2) The interaction effects of G*D, I*D and G*I*D can also significantly affect the difference in performance measures.
(3) Three of the four factors (G, I, D) affect the performance. It seems that the performance of the heuristic strongly depends on the operating conditions. Only the precedence constraints (P) do not significantly affect the difference in performance measures.

7. Conclusion

This research presents a multiple objective formulation of production scheduling problem in continuous manufacturing systems and the development of three heuristic procedures to schedule daily production. Each heuristic procedure gives priority to one of three objectives. A computer program was developed to implement the heuristics. Many variables such as assignment of product to processing lines, coefficients of the cost function and the number of periods in the planning horizon can be changed in the program. Consequently, these heuristic procedures can be considered as "decision support" rules because the management will still need to make decisions themselves based on the various schedules produced.

A factorial experiment was designed and performed to evaluate the performance of the three heuristic procedures and to identify the factors that affect the performance differences. The analyses of the results show that no procedure is absolutely superior to the other procedures in all aspects. This is not a surprise because each of the three algorithms produces a local optimal solution. However, both the PB and PF algorithms can generate schedules with better OSR and ADI than the current management approach modeled by the RS algorithm. To provide a composite performance measure, a TCI was suggested. It is noted that the rank of the heuristics can be changed as a function of the cost coefficients in TCI.

Based on the ANOVA results, the differences between procedures are significantly affected by group ratio, maximum inventory and demand pattern. This information is of important value because it suggests that if one of the above factor changes in the manufacturing system, the method to schedule production needs to be re-evaluated. Further research is currently underway to study the sensitivity of the resource availability to the performance measures, and the development of a decision support system based on the heuristics proposed in this research.

References

Amaro, A. C. S., & Barbosa-Póvoa, A. P. F. D. (1999). Scheduling of industrial distribution manifolds with pre-conditions. *European Journal of Operational Research, 119*, 461–478.

Anwar, M. F., & Nagi, R. (1997). Integrated lot-sizing and scheduling for just-in-time production of complex assemblies with finite set-ups. *International Journal of Production Research, 35*, 1447–1470.

Askin, R. G., & Standridge, C. R. (1993). *Modeling and analysis of manufacturing systems*. New York, NY: John Wiley.

Bose, S., & Pekny, J. F. (2000). A model predictive framework for planning and scheduling problems: A case study of consumer goods supply chain. *Computers & Chemical Engineering, 24*, 329–335.

Caramanis, M. C., Anli, O. M., & Paschalidis, I. C. (2003). Supply chain (SC) production planning with dynamic lead time and quality of service constraints. *Proceedings of the IEEE Conference on Decision and Control, 5*, 5478–5485.

Daniels, R. (1994). Incorporating preference information into multi-objective scheduling. *European Journal of Operational Research, 77*, 272–286.

Deepark, B. B., & Grossmann, I. E. (1990). Simultaneous production planning and scheduling in multiproduct batch plants. *Industrial & Engineering Chemistry Research, 29*, 570–580.

Giannelos, N., & Georgiadis, M. (2003). Efficient scheduling of consumer goods manufacturing processes in the continuous time domain. *Computers & Operations Research, 30*, 1367–1381.

Gupta, Y. P., Bector, C. R., & Gupta, M. C. (1990). Optimal schedule on a single machine using various due date determination methods. *Computers in Industry, 15*, 245–254.

He, W., & Kusiak, A. (1992). Scheduling manufacturing systems. *Computers in Industry, 20*, 163–175.

Kogan, K., Leu, Y., & Perkins, J. (2002). Parallel-machine, multiple-product-type, continuous-time scheduling: Decomposable cases. *IIE Transactions, 34*, 11–22.

Lou, P., Liu, Q., Zhou, Z., Wang, H., & Sun, S. X. (2012). Multi-agent-based proactive–reactive scheduling for a job shop. *The International Journal of Advanced Manufacturing Technology, 59*, 311–324.

Minton, S., Johnston, M. D., Phillips, A. B., & Laird, P. (1992). Minimizing constraint satisfaction and scheduling problems. *Artificial Intelligence, 58*, 61–205.

Modarress, B., Ansari, A., & Willis, G. (2000). Controlled production planning for just-in-time short-run suppliers. *International Journal of Production Research, 38*, 1163–1182.

Patterson, J. H., Slowinski, R., Talbit, F. B., & Weglarz, J. (1989). *An algorithm for a general class of research and resource constrained scheduling problems. Advances in Project Scheduling*. Amsterdam: Elsevler Science.

Scudder, G. D., Hoffmann, T. R., & Rohleder, T. R. (1993). Scheduling with forbidden early shipments: Alternative performance criteria and conditions. *International Journal of Production Research, 31*, 2287–2305.

Seki, Y., & Kogure, K. (1996). Lot scheduling problem for continuous demand. *International Journal of Production Economics, 44*, 7–15.

Tung, L., Lin, L., & Nagi, R. (1999). Multiple-objective scheduling for the hierarchical control of flexible manufacturing systems. *International Journal of Flexible Manufacturing Systems, 11*, 379–409.

Viana, A., & de Sousa, J. P. (2000). Using metaheuristics in multiobjective resource constrained project scheduling. *European Journal of Operational Research, 120*, 359–374.

Vollmann, T. E., Berry, W. L., & Whybark, D. C. (1992). *Manufacturing planning and control systems*. Boston, MA: Irwin.

Wagner, M., & Smits, S. (2004). A local search algorithm for the optimization of the stochastic economic lot scheduling problem. *International Journal of Production Economics, 90*, 391–402.

Yim, S., & Lee, D. Y. (1996). Multiple objective scheduling for flexible manufacturing systems using Petri nets and heuristic search. *Proceedings of the IEEE International Conference on Systems, Man and Cybernetics, 4*, 2984–2989.

8

Influence of surface pretreatment in resistance spot welding of aluminum AA1050

Ihsan K. Al Naimi[a]*, Moneer H. Al Saadi[b], Kasim M. Daws[a] and Niels Bay[c]

[a]*Mechanical Department, College of Engineering, Baghdad University, Baghdad, Iraq;* [b]*Welding Department, University of Kerbala, Baghdad, Iraq;* [c]*Department of Mechanical Engineering, Technical University of Denmark, Lyngby, Denmark*

Resistance spot welding (RSW) of aluminum alloys implies a major problem of inconsistent quality from weld to weld due to problems of varying thickness of the oxide layer. The high resistivity of oxide layer causes strong heat development, which has significant influence on electrode life and weld quality. An experimental study of the influence of pretreatment on weld quality in RSW of AA1050 sheets with three thicknesses, comparing welding of as-received sheet with pretreated sheet by either pickling in NaOH or glass-blasting were investigated. Different weld settings were applied with low-, medium-, and high-energy inputs. The as-received sheet showed higher electrical contact resistance because of thicker oxide layer. Lower values were noticed with pickled surfaces, whereas the lowest electrical contact resistance was obtained when glass blasting, resulting in the roughest surface topography, which facilitated breakdown the oxide layer. Highest strength and smaller scatter in strength were obtained by pickling in NaOH.

Keywords: resistance spot welding; contact resistance; oxide film; surface pretreatment

1. Introduction

The increasing demands for energy savings have implied special focus in automotive industry on lighter vehicles production, among other solutions by introducing skin panels of aluminum instead of steel. Weight savings up to 46% are reported (Wheeler, Sheasby, & Kewley, 1987), and furthermore resistance to corrosion is improved.

Resistance spot welding (RSW) is the most dominant process in sheet metal joining, particularly in automotive industry due to low cost, high productivity, flexibility, easy automation and maintenance, and minimum requirements for skilled labor (Brown, Newton, & Boomer, 1995; Cho, Li, & Hu, 2006; Spinella, Brockenbrough, & Fridy, 2005). The process is also widely applied in other industries of sheet product manufacturing, e.g. other transportation industries and in production of kitchen utensils.

It should, however, be emphasized that RSW of aluminum is more problematic than steel due to higher electrical and thermal conductivity, higher coefficient of expansion, lower melting temperature, and an oxide layer, which has high electrical resistance and high melting temperature (2050 °C). The latter together with the fact that the effective contact resistance grows considerably as the oxide film grows implies large scatter in

*Corresponding author. Email: ihsan_kad@yahoo.com

quality of RSW aluminum sheets, which therefore require close production control (Boomer, Hunter, & Castle, 2003; Kim, Park, Hwang, & Kang, 2009; Mathers, 2002).

The high thermal and electrical conductivity of aluminum require 2–3 times higher current and shorter weld time, typically 25% of that used to spot weld steel. Accurate control and synchronization of current and electrode force is required due to the narrow weld temperature range (*Resistance Welding Manual*, 2003). The problems are especially pronounced when welding unalloyed aluminum AA 1XXX.

Aluminum is highly reactive to oxygen, and after removal of the oxide layer by mechanical or chemical means, a new oxide layer will immediately form on its surface in normal atmosphere. The layer is beneficial as it protects the base metal from corrosion, but creates problems in connection with resistance welding, where it causes severe electrode degradation and scatter in weld quality due to (Han, Thornton, Boomer, & Shergold, 2010; Patrick, Auhl, & Sun, 1984; Sun, 1982). Spot welding in aluminum AA1050 is less stable than in alloyed aluminum, and the electrodes will stick to the sheet after 50 welds because of the oxide film problems (Pedersen, 2010). Contributions to the contact resistance comes from the oxide layer as well as the contaminant film, which may include dirt, lubricant, chemicals, and water vapor (Rashid, 2011; Sun, 2003). Even though maintenance of uniform surface conditions is taken carefully into consideration, the electrical contact resistance will vary from weld to weld implying inconsistent weld quality.

Contact resistance was experimentally investigated by Song, Zhang, and Bay (2005) using a Gleeble system. They showed that interface normal pressure had significant but decreasing influence on the contact resistance with increasing normal pressure, whereas the influence of temperature was less pronounced as pressure increased. Surface roughness together with elastic–plastic properties of the materials also influences the electrical contact resistance (Dzekster & Ismailov, 1990; Zheng, Shan, Hu, & Luo, 2006). Crinon and Evans (1998) showed that the effect of the oxide film is greatest for specimens with the smoothest surface.

Expulsion during RSW may either occur at the faying surfaces or the electrode/workpiece interfaces. The latter may severely affect surface quality and electrode life. The risk of expulsion is especially high in spot welding of aluminum due to the very dynamic and unstable character of the process, related to the requirements of high current and short welding time as compared to welding steels (Mathers, 2002; Senkara, Zhang, & Hu, 2004; Zhang & Senkara, 2012).

A large number of investigations in the literature deal with the influence of surface preparation on RSW of aluminum sheets. Rönnhult, Rilby, and Olefjord (1980) studied the weldability of aluminum alloy AA5252, both as-received and after etching in NaOH and oxalic acid, and proved that removing the oxide layer led to a significant improvement in weld quality. Etched, mill finish, and pretreated/lubricated AA5754 surfaces were examined by Thornton, Newton, Keay, Sheasby, and Evans (1997), they suggested that etched surfaces that had a thin and uniform oxide layer, gave the most consistent surface resistance and weld strength, which was in good agreement with other studies by Pickering and Hart (1994) and Li, Hao, Zhang, and Zhang (2007). Rashid, Fukumoto, Medley, Villafuerte, and Zhou (2007) suggested a special lubricant to lower the tendency to sticking of the workpieces to the electrodes thereby increasing the electrode life. The company Fronius International (2014) developed a technique called DeltaSpot using continuous tape running between the electrodes and the sheets to be joined thereby increasing the electrode life remarkably. Han (2010) reported that frequent, light dressing of the electrodes removing pickup of aluminum increases the electrode life significantly.

Although different techniques and materials were investigated by these researchers, the influence of surface pretreatment before RSW of aluminum AA1050 and the use of mechanical pretreatment by glass blasting have not been clearly identified in any previous work. Blasting with glass beads is easy to automatize and may not cause higher costs than other surface pretreatments in mass production. Furthermore, the treatment is more environmentally friendly than chemical treatment.

The present study focuses on the influence of surface pretreatment prior to RSW of aluminum AA1050, aiming at minimizing variations in weld strength caused by the oxide layer. Mechanical pretreatment by glass blasting and chemical treatment by pickling with NaOH are compared with welding of as-received material.

2. Experimental procedure

2.1. Workpiece material, electrodes, and welding equipment

The experiments were conducted at the Technical University of Denmark (DTU) welding .6, 1.0, and 1.5 mm sheets of unalloyed aluminum AA1050 on a TECNA AC welding machine with specifications, as listed in Table 1. The properties and composition of the workpiece sheets determined by spectrum analysis are shown in Table 2. Samples of dimensions 16 × 115 mm were cut from delivered sheets, with the longitudinal dimension in the rolling direction. They were welded as a lap joint ready for subsequent shear-tensile testing. A hard wood fixture was used to mount the samples in good alignment with the electrodes (see Figure 1).

The electrode tips (Female Cap) were of type A according to ISO 5821-2009 and the American standard RWMA No. FF-25. They were made of zirconium–copper alloy (CuCrZr) with the following chemical composition; Cr: .7–1.2%, Zr: .06–.15%, bal. Cu. The electrodes were of dome configuration, Ø16 mm in diameter with a spherical end

Table 1. Machine specifications, TECNA AC welder.

Specifications	Values	Specifications	Values
Controller	TE-180, 16 functions	Nominal power at 50%	250 kVA
Supply voltage	380 V	Phases	1
Frequency	50 Hz	Supply pressure	6.5 bar
Max. welding current	68 kA	Electrode force per 1 bar	3.14 kN
Max. welding force	18.85 kN	Throat depth	250 mm
Max. welding power	810 kVA	Water cooling	12 l/min

Table 2. Strip material specifications.

Trade name	Thickness (mm)	Tensile (MPa)	Hardness (HV)	Nominal composition (wt%)				
				Fe	Si	Mn	Others	Al
1050	.6	105	30	.255	.173	.021	.051	99.5
1050	1.0	105	30	.378	.100	.018	.004	99.5
1050	1.5	127	45	.350	.070	.010	.070	99.5

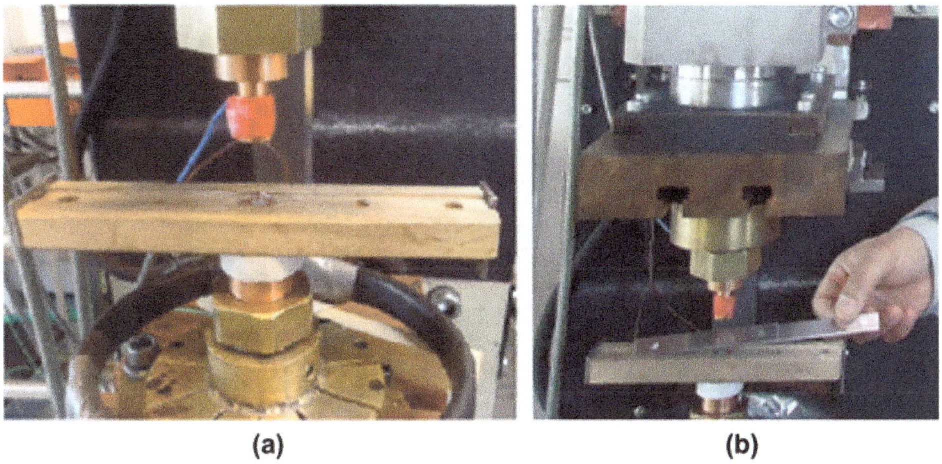

Figure 1. Setup for welding, (a) hardwood-fixture for mounting strips, (b) removing welded strips. Pretreatment of the workpiece surfaces was done by two methods, mechanical and chemical.

surface of radius 40 mm. Close to the tip of the two electrodes, an Ø1.5-mm hole was drilled and copper wires were inserted to measure the secondary voltage over the weld.

The welding current was measured by a Rogowski coil together with a precalibrated TECNA-1430 conditioner, and a Kistler piezoelectric force transducer was used to measure the electrode load. The acquired data were treated on a PC by specially developed software in LabVIEW.

The following parameters of the RSW process were calculated for each experiment, RMS current I (A), RMS voltage U (V), welding time C (s), and the electrode force P (kN). Tensile-shear testing was carried out using a universal testing machine at a deformation rate of 2 mm/min to determine the weld strength S(N). Vickers microhardness measurements were performed with a load of 50-g testing on weld cross-sections in longitudinal direction through the diameter of the nugget at intervals of .5–1.0 mm. Macrographs and micrographs of the welds were made in light optical microscope. Moreover, high-resolution images were made in SEM, using EDS for quantitative chemical analysis.

2.2. Shot blasting

The mechanical treatment was done by shot blasting with 100-μm glass beads. Each strip was subjected to 30 s blasting at an air pressure of 200 kPa (see Figure 2). In the main series of experiments, surface treatment was done on both sides of the strips, whereas additional experiments were performed dressing only the outer side of the strips contacting the electrodes.

2.3. Pickling

Pickling was performed in an aqueous solution of sodium hydroxide (NaOH) (see Figure 3). During pickling of aluminum alloy strips, the surface oxides or hydroxides

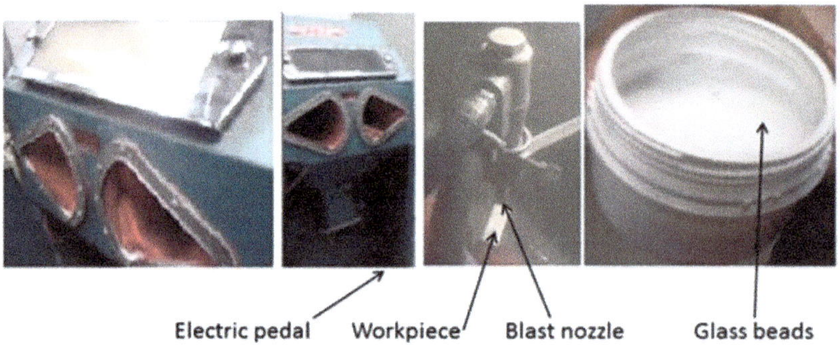

Figure 2. Glass blast surface treatment.

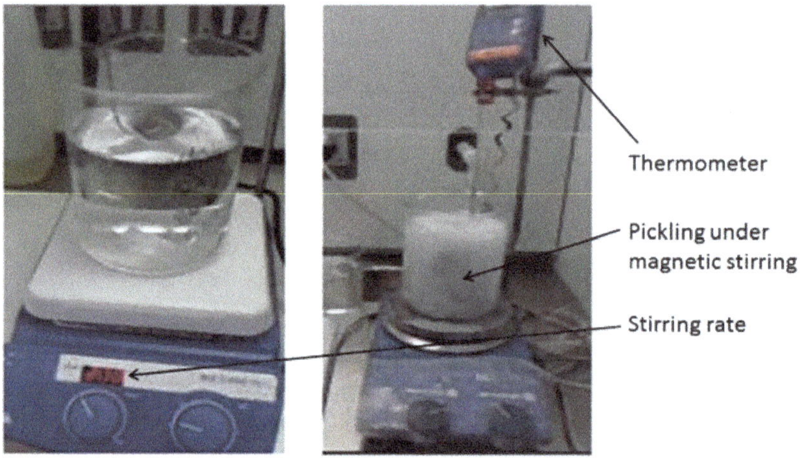

Figure 3. Pickling setup.

are dissolved. The pickling rate is dependent on several variables including; (1) concentration, (2) solution temperature, (3) composition of the surface sheet, and (4) stirring rate of solution.

An aqueous solution of 60 g of NaOH per liter of water was chosen with a bath temperature of 60 °C. Magnetic stirring was done in a fume cupboard at a rate of 200 rpm. The strips were subsequently rinsed in hot water and ethanol. Preliminary experiments with 5-min pickling time resulted in severe sticking of the workpieces to the electrodes as well as expulsions between the electrodes and the strips. Due to this, pickling time was reduced to 2 min thereby minimizing these problems.

As earlier mentioned, an oxide layer is re-established on the aluminum specimens immediately after surface dressing. In order to ensure similar conditions for all experiments with surface dressing, this was carried out within 3 h before spot welding. In this way, a substantially smaller oxide layer than that of the as-received aluminum can be expected.

2.4. *Plan of experiments*

The experiments were planned as a general factorial with three replicates per condition (Cho et al., 2006), and the results were recorded as the average value of each condition. The welding parameters and their associated values are given in Table 3. For each of the three sheet thicknesses, three different weld currents were chosen, representing low, medium, and high welding energy. During welding, possible expulsion and sticking to the electrodes were observed and recorded.

3. Results and discussion

3.1. *Surface roughness*

The two surface treatments were each characterized by five surface roughness measurements on three different strips. Table 4 shows the resulting R_a values. Pretreatment by glass blasting show high values of roughness, in the order of 10 times the value of the as-received surfaces. Chemical surface treatment by pickling gives a bit higher R_a values than the as-received, but the standard deviation (SD) is lower.

Table 3. Experimental setup (welding parameters and values).

	Welding energy					
	Low		Medium		High	
Sheet (mm)	Times (cycles)	Current (kA)	Times (cycles)	Current kA	Times (cycles)	Current (kA)
.6	.04 (2)	20	.1 (5)	23	.18 (9)	26
1.0	.04 (2)	23	.1 (5)	26	.18 (9)	29
1.5	.04 (2)	26	.1 (5)	29	.18 (9)	32

Note: Electrode force 1.85–2.45 kN.

Table 4. Surface roughness values.

Sheet (mm)	As-received	Pickling in NaOH	Glass-blast
.6	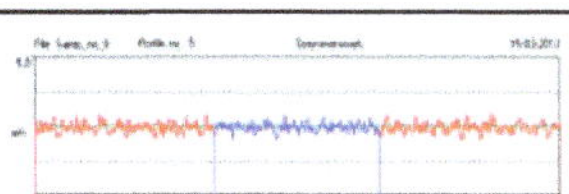 R_a = .287 μm, SD = .046	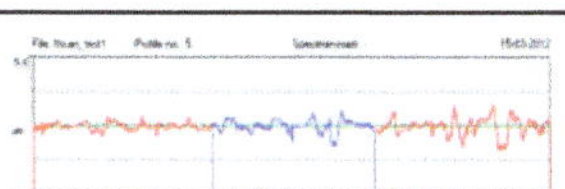 R_a = .339 μm, SD = .011	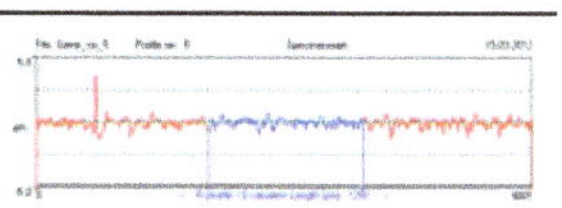 R_a = 2.650 μm, SD = .115
1.0	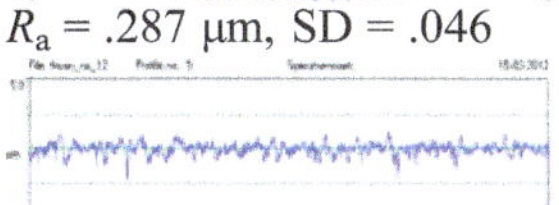 R_a = .300 μm, SD = .051	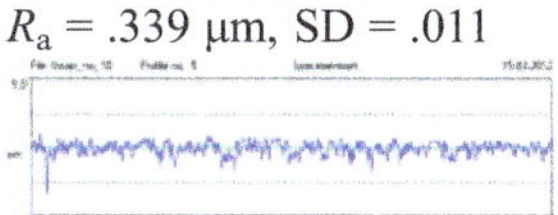 R_a = .385μm, SD = .037	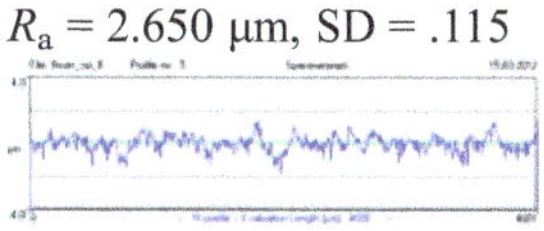 R_a = 4.703 μm, SD = .987
1.5	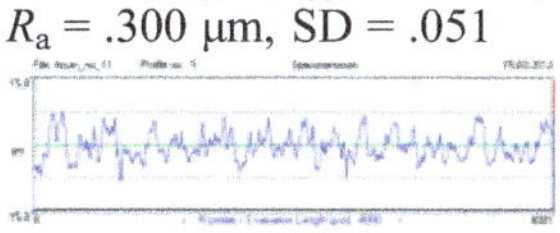 R_a = .238 μm, SD = .039	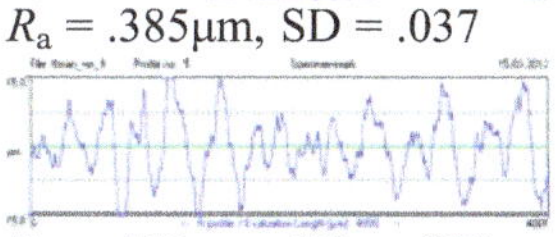 R_a = .411 μm, SD = .028	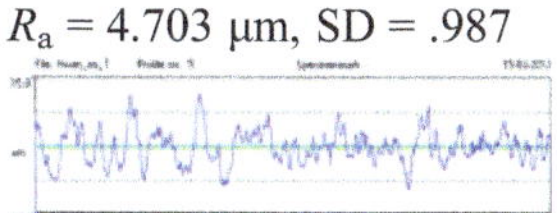 R_a = 3.230 μm, SD = .453

3.2. Electrical resistance as a function of surface condition

In order to study the influence of surface preparation on electrical contact resistance, the total electrical resistance between the two electrodes was determined for each surface dressing condition and each sheet thickness. The resistance was calculated by dividing

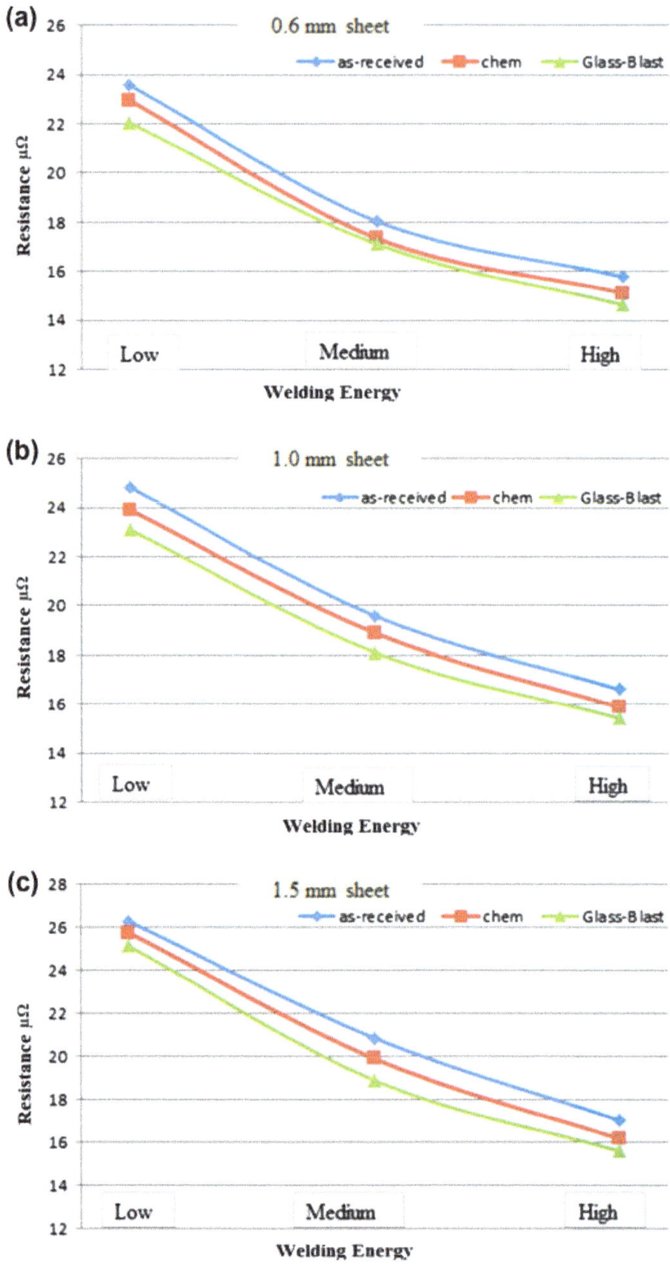

Figure 4. Electrical resistance as a function of energy input and surface dressing for (a) .6 mm; (b) 1.0 mm; (c) 1.5 mm sheets.

the RMS voltage by the RMS current determined by sampled current and voltage data using the LabVIEW software program. The error due to induction of emf was minimized by twisting the wires, when measuring the voltage between the electrodes. Although it is the entire electric resistance between the electrodes, which is determined in this way, a comparison between the individual surface treatments gives a good idea of their influence on the accumulated contact resistance between workpieces and between workpieces and electrodes.

Figure 4 shows the total electrical resistance between the two electrodes as a function of the energy input for each surface condition and each sheet thickness. It is noticed that the as-received sheet has the highest electrical resistance for all sheet thicknesses due to the thicker oxide layer increasing the contact resistance. Medium contact resistance is obtained with chemical pickling, which lowers the contact resistance by decreasing the oxide layer thickness. The lowest electrical resistance is obtained with glass-blasted sheets, which besides removing the thick oxide film creates a rough surface causing easy breakup of the newly formed oxide layer, when the applied electrode load deforms the contacting surface asperities. It is, furthermore, noticed that the electrical resistance decreases with increasing welding energy (increased current and welding time). This might be explained by easier breakdown of the layer of oxides and contaminant film at high temperature as also observed by others (Crinon & Evans, 1998; Han et al., 2010; Song et al., 2005). Finally, a small increase in the electrical resistance is observed with increasing sheet thickness, as to be expected.

3.3. Microhardness tests and microstructure examinations

Figure 5 shows the microhardness profiles for RSW of 1.0-mm sheets with the three different surface treatments. The hardness of the fusion zone FZ and the heat-affected zone (HAZ) is noted to be lower than that of the base metal for all cases due to annealing of the work-hardened base metal during welding. A few points indicate relatively low microhardness due to presence of void defects nearby, e.g. points 5 and 7 in as-received sheets and point 4 in pickled sheets.

The comparison of hardness profiles in Figure 5(d) shows the hardness in the FZ and the HAZ to be highest for welds with pretreatment by glass blasting and lowest for the case of as-received sheets. This is explained by the influence of electrical contact resistance on heat generation discussed earlier. Hardness profiles for welded 1.5-mm sheets showed the same trend.

Macrographs also showed the nugget size when welding as-received sheets to be larger than pretreated sheet surfaces. This is due to the smaller electrical contact resistance in the faying surfaces generating less heat in the latter cases. For example, the nugget sizes were 5.1, 5.0, and 4.7 mm in 1.0-mm sheets, and 7.0, 6.54, and 5.35 mm in 1.5-mm sheets for as-received, pickled, and glass-blasted sheets, respectively. The minimum nugget size was obtained by glass-blasted sheets, due to minimum electrical contact resistance.

Figure 6 shows micrographs of the 1.0-mm glass-blasted strips welded with 29-kA welding current and .1-s (five cycles) welding time. In the oval nugget region, recrystallized, small, equiaxial grains, and insoluble particles of $FeAl_3$ (black) are noticed as well as a narrow zone of columnar grains along the nugget edge nugget. Next to this zone is the HAZ, which consist of dendritic grains. Moreover, some porosities (large, black areas) are evident. No significant difference in the microstructures was found in welds between the as-received and the pretreated sheets.

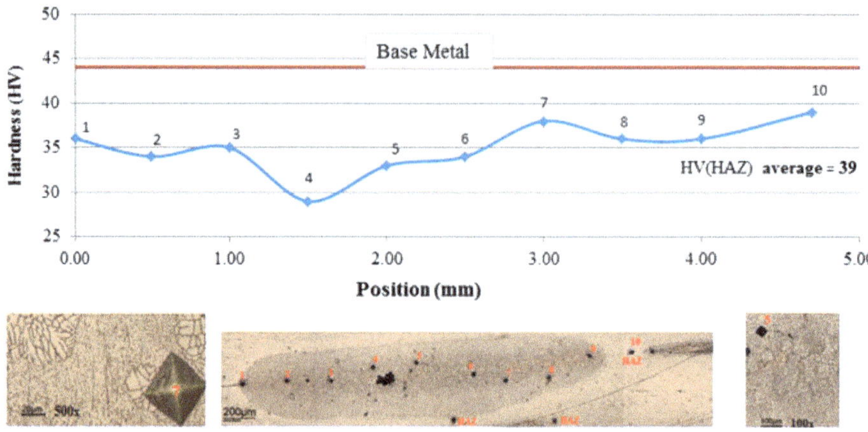

(a) Glass-blasted pretreatment surfaces

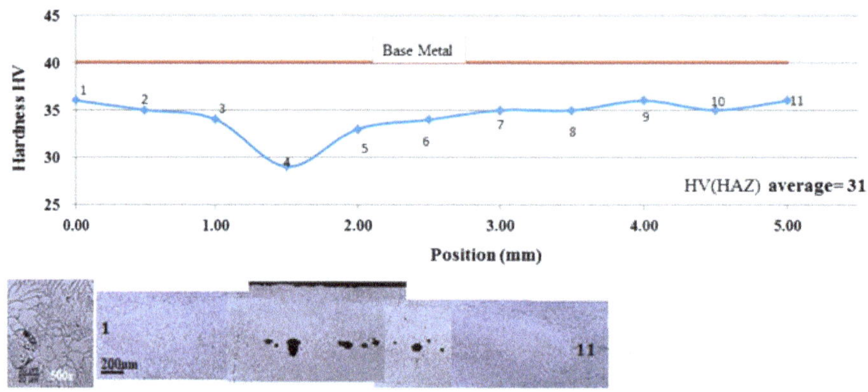

(b) Pickling in NaOH pretreatment surfaces

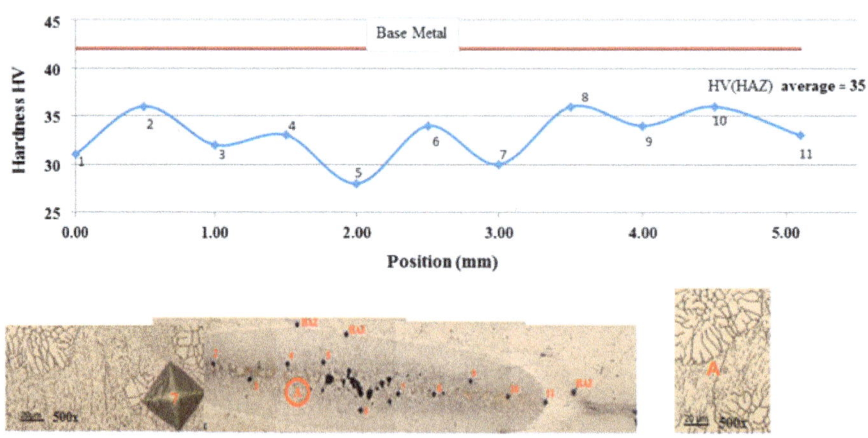

(c) As-Received sheet

Figure 5. Microhardness profiles of 1.0-mm sheet, welding parameters 29 kA, .1 s (five cycles).

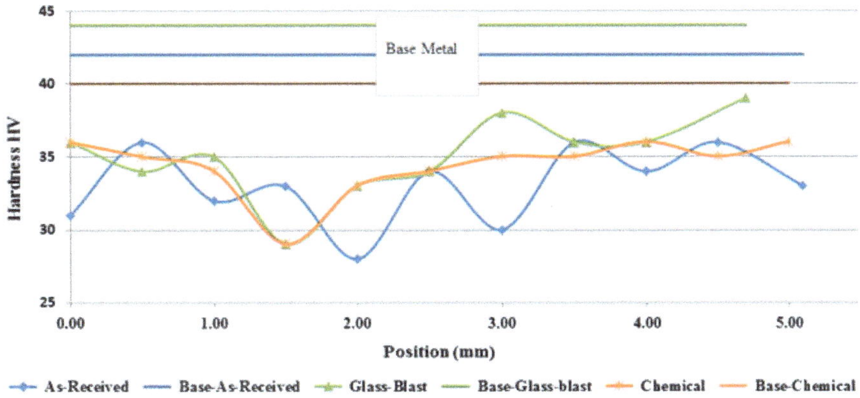

(d) All surface dressing conditions

Figure 5. (*Continued*).

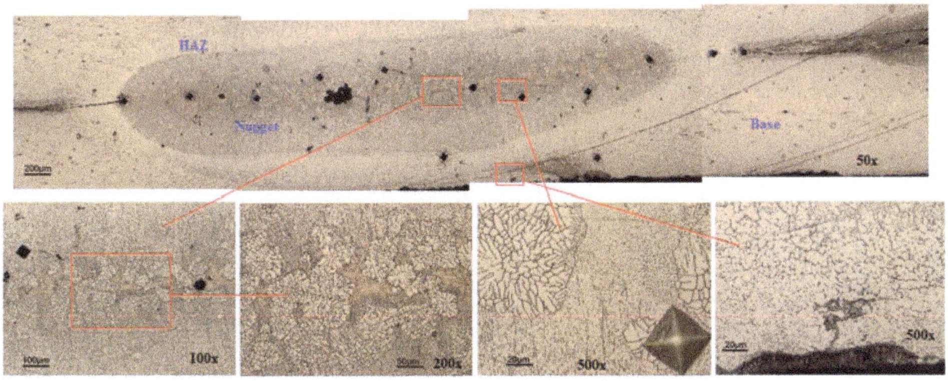

Figure 6. Micrographs of RSW of 1.0-mm sheet, welding parameters 29 kA, .1 s (five cycles).

3.4. Tensile-shear tests

Tensile-shear tests of the welded joints indicate their strength and the failure mode. Figure 7 shows the strength as function of the welding energy input specified in Table 3 for the three different surface treatments (as-received, pickled, and glass blasted). The surface treatment is generally done on both sides of the strips, but for the case of glass blasting and medium energy input, one-sided treatment of the outer sides contacting the electrodes is also investigated. The data represented by an average value and a scatter bar are based on an average of three identical tests. It is seen that an increase in energy input increases the strength. The mode of failure was recorded in each test and classified as one of the following three types: (1) interfacial failure (nugget fracture in shear), (2) plug failure (nugget pull-out), and (3) failure in the HAZ. Failure mode 1 typically occurred at low input energy indicating insufficient heating, mode 2 at medium input energy indicating satisfactory heating, and mode 3 at high input energy indicating over-heating and softening of the region near the nugget. A few abnormal test results were observed to fall outside these types of failure. In line with other studies (Mathers, 2002;

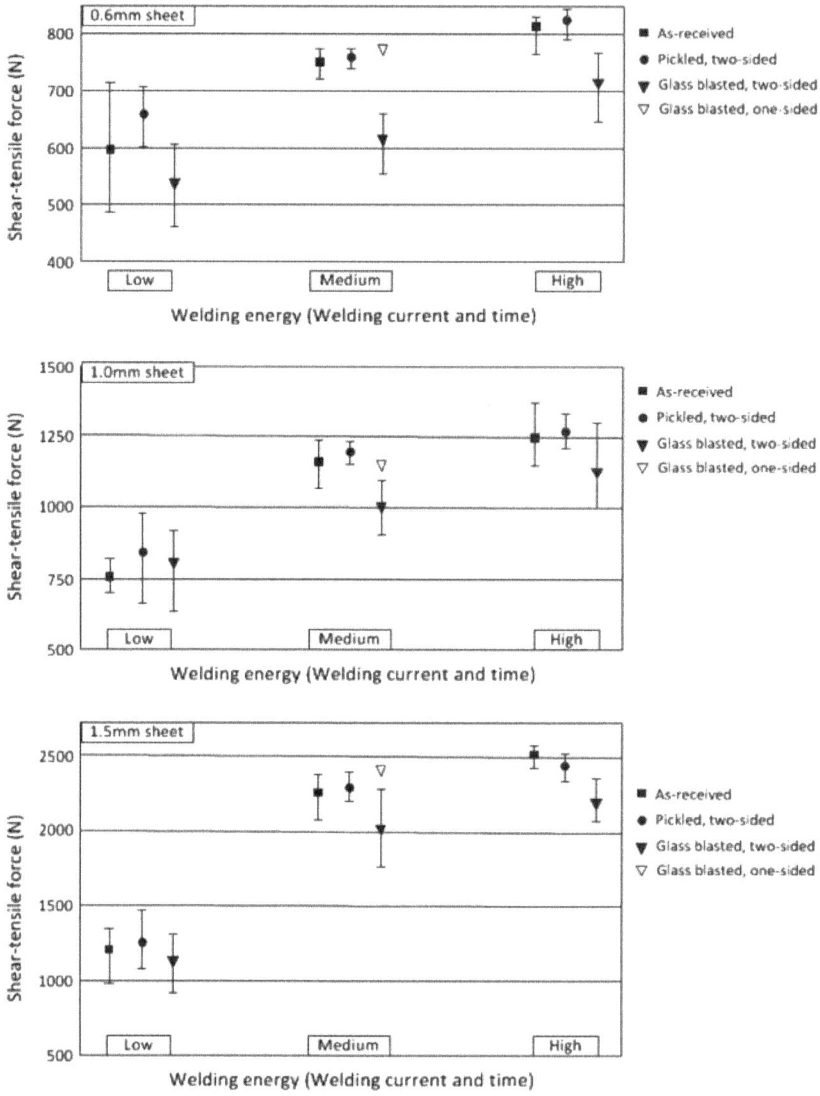

Figure 7. Shear-tensile force vs. welding energy input for different surface treatments and sheet thicknesses.

Newton, Browne, Thornton, Boober, & Keay, 1994; Rashid, 2011; Senkara et al., 2004; Shi & Guo, 2013), expulsion and severe sticking of the workpiece to the electrodes occurred at high-energy input. Furthermore, fracture in mode 3 was usually in the form of ductile tearing around the nugget.

The scatter in weld strength is less when surface treating by pickling than for the as-received strips. Largest scatter is generally noticed for the two-sided, glass-blasted sheets, which may be attributed to the large surface roughness, low interface resistance, and difficulties in controlling this manually operated surface dressing. The treatment by pickling is also best regarding average weld strength independent of the sheet thickness.

The values for medium energy input are 760, 1193, and 2283 N for .6, 1.0, and 1.5-mm sheets, respectively. The lowest strength is obtained for the glass-blasted sheets due to the lowest values of electric contact resistance; see Figure 4, which leads to small nugget size. The values for medium heat input are 616, 1008, and 2020 N for .6, 1.0, and 1.5-mm, respectively. The low electrical contact resistance of the glass-blasted, rough surfaces may be attributed to easier breaking up of the oxide layer as also observed by other authors (Dzekster & Ismailov, 1990; Han et al., 2010; Song et al., 2005).

If only the surfaces contacting the electrodes are glass blasted leaving the faying surfaces untreated, a significant increase in strength occurs, i.e. 775, 1147, and 2408 N for .6, 1.0, and 1.5-mm sheet thickness, respectively, for medium-energy input. The reason for this is higher heat generation due to higher electrical contact resistance at the faying surfaces. This treatment is recommendable, since it also ensures less sticking to the electrodes, and thus increased electrode life. The fact that pickling improves weld quality as regards level and consistency compared to the as-received surface condition even if the surface roughness is similar indicates that the latter is not of overriding importance controlling the electric contact resistance although increasing roughness may facilitate breakdown of the oxide layer. The main importance is the thickness of the oxide layer as also stated by other researchers (Patrick et al., 1984; Song et al., 2005; Sun, 1982).

3.5. SEM and EDS examinations

Micrographs were also made in SEM to study microstructure and surface profile and carry out high-resolution mapping determining chemical composition of areas of special interest by X-ray spectroscopy (EDS). When welding as-received sheet (with thick oxide layers), considerable indentation into the aluminum sheets was observed, as well as rapid degradation of the electrodes. This was due to pitting of the electrode surfaces and dissolving or diffusion of copper into the aluminum workpiece material and vice versa (see Figure 8(a)).

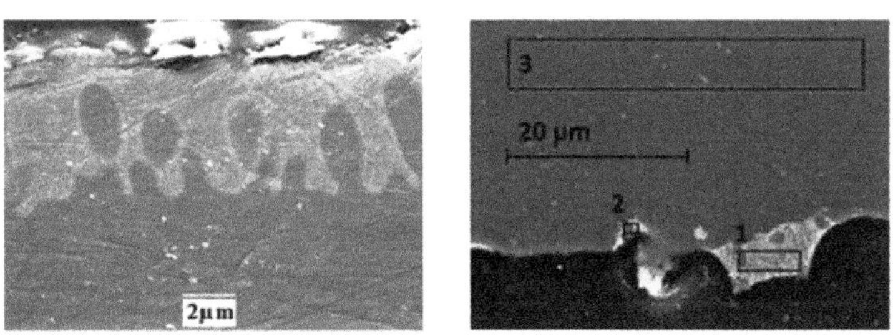

(**a**) 1.0 mm as-received sheet (**b**) 1.5 mm chemical pretreatment

	Al	O	Cu	Total
1	69.4	1.0		100
2	66.5	2.2		100
3	100			100
Max.	100	2.2	31.3	
Min.	66.5	1.0	29.6	

(c) EDS analysis

Figure 8. SEM images RSW and measured composition referring to Figure 5(b).

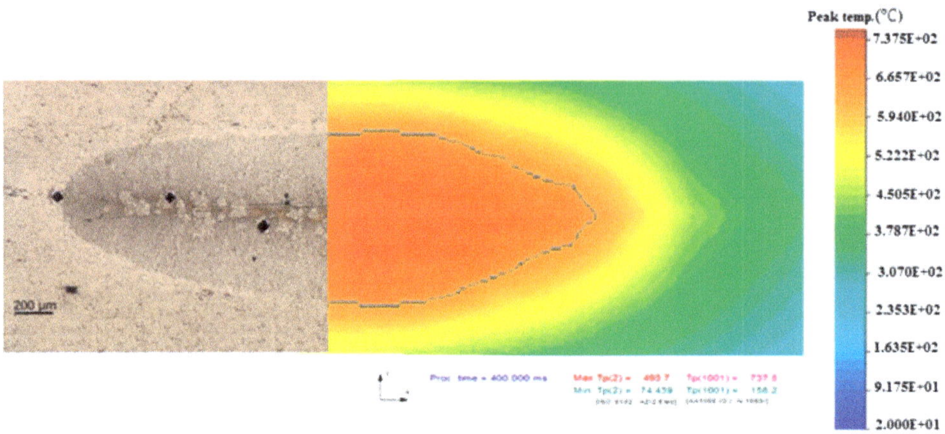

Figure 9. Comparison of experimental and predicted nugget shape.

Welding of pretreated sheets showed less indentation and dissolving of electrode material in the workpiece due to a thinner oxide layer causing less heating (Figure 8(b)). The reason for a few defects in this case may be pitting of the electrode surface. The white colored areas in the SEM images indicate the presence of oxide as well as dissolving of copper into the workpiece, as shown in the EDS analysis table in Figure 8(c).

3.6. Numerical simulation

The commercial finite element program SORPAS® developed especially for resistance welding (SWANTEC Software & Engineering ApS. SORPAS®) was used in the present work. It combines mechanical, electrical, thermal, and metallurgical models analyzing the development of heat and distributions of current, voltage, temperature, stresses, and strains (Bay & Zhang, 2007; Zhang, 2003). Using a new feature called Weld Schedule Specifications optimized weld current, electrode force, welding time, and holding time were calculated, see Table 5, for welding of as-received as well as surface-dressed sheet adjusting the input value of the electrical resistance at 20 °C in the materials database of the program to lower value in case of surface dressing according to Figure 4. It is clearly noted that especially the proposed welding current for the pretreated surfaces is lower than that for as-received sheets.

Figure 9 shows a comparison between predicted and actual weld nugget with as-received strip surface, in 1.0-mm sheet for 27-kA welding current and .1-s five cycles welding time. It is noticed that the nugget shape is rather well predicted.

Table 5. WSS of weld planning optimization SORPAS.

Sheet	As-Received (mm)			Pretreatment (mm)		
	0.6	1.0	1.5	0.6	1.0	1.5
Welding Current (kA)	23.9	26.27	31.64	22.87	24.58	27.26
Welding Time (cycles)	2	4	5	2	4	4
Electrode Force (kN)	1.16	1.44	1.78	1.04	1.44	1.49

4. Conclusions

The experiments showed that consistency in weld quality is improved when the oxide layer of the as-received aluminum surfaces is properly removed. However, the resistance of an oxide layer is beneficial at the interface between the faying surfaces in order to increase local heating to form the weld nugget, whereas the resistance at the contact between electrodes and workpiece should be as low as possible to avoid sticking and poor electrode life. The most significant conclusions drawn from this experimental work are as follows:

(1) The surface condition of the aluminum sheets has significant influence on the weldability and the electrode lifetime.
(2) The highest value of electrical contact resistance is obtained with as-received sheet due to a thicker oxide layer.
(3) Pretreatment by pickling in 60 °C NaOH gives highest weld strength and lowest scatter. Optimum pickling time is 2 min.
(4) Lowest contact resistance is obtained by glass blasting, which besides removing the original oxide film provides a rough surface decreasing the contact resistance due to breakdown of the oxide layer by asperity deformation.
(5) One-sided surface treatment by glass blasting of the surfaces contacting the electrodes is favorable leading to less electrode sticking, and thus increased electrode life and furthermore better weld strength than two-sided glass blasting.
(6) The microhardness in the FZ and HAZ is lower than in the base metal for all cases due to annealing of the as-received, workhardened material.
(7) SEM and EDS examinations confirm the presence of copper (electrode material) in the welded aluminum strips especially when welding as-received sheet indicating degradation of the electrodes due to dissolving and diffusion, which may furthermore cause pitting and accelerated electrode wear due to the pitting on the surface of the electrodes.
(8) Numerical simulation with SORPAS® gives good prediction of appropriate welding parameters and nugget size.

Acknowledgments

This work is part of cooperation on independent research between Baghdad University and Technical University of Denmark (DTU). The authors would like to thank co-workers at the Department of Mechanical Engineering, DTU, especially Peter Sanderhoff, Steffen Munch, Rolf Jensen, and Chris V. Nielsen for helping with the experimental work. First author, Ihsan Al Naimi, is grateful for the financial support by Iraqi Ministry of Higher Education and Scientific Research.

Disclosure statement

No potential conflict of interest was reported by the authors.

Funding

The work of Ihsan Al Naimi was supported by Iraqi Ministry of Higher Education and Scientific Research.

References

Bay, N., & Zhang, W. (2007, May). Thermomechanical modelling of resistance welding. *Proceedings of International Conference on Computational Methods for Coupled Problems in Science and Engineering* (162–165). Ibiza, Spain.

Boomer, D. R., Hunter, J. A., & Castle, D. R. (2003). *A new approach for robust high-productivity resistance spot welding of aluminum* (SAE Technical Paper -01-0575).

Brown, D. J., Newton, C. J., & Boomer, D. (1995). Optimization and validation of a model to predict the spot weldability parameter lobes for aluminum automotive body sheet. Advanced Technologies & Processes. In *Proceedings of International Body Engineering Conference, IBEC* (pp. 100–106).

Cho, Y., Li, W., & Hu, S. J. (2006). Design of experiment analysis and weld lobe estimation for aluminum resistance spot welding. *AWS Welding Journal, 85*, 45s–51s.

Crinon, E., & Evans, J. T. (1998). The effect of surface roughness, oxide film thickness and interfacial sliding on the electrical contact resistance of aluminum. *Material Science and Engineering Journal, A242*, 121–128.

Dzekster, N. N., & Ismailov, V. V. (1990). Some methods for Improving Aluminum Contacts. In *Proceeings of 36th Annual Holm Conference on Electrical Contacts* (pp. 518–520). Chicago, IL, Illinois Institute of Technology.

Fronius International. (2014, September). Retrieved from http://www.fronius.com/cps/rde/xchg/SID-3AC27A9B-F495FC44/fronius_international/hs.xsl/79_33390_ENG_HTML.htm#.VBmeZaPU-70

Han, L. (2010, September 22–24). Development of RSW Al for LWV manufacture. In *Proceedings of the 6th International Seminar on Advances in Resistance Welding* (pp. 55–70). Hamburg.

Han, L., Thornton, M., Boomer, D., & Shergold, M. (2010). Effect of aluminium sheet surface conditions on feasibility and quality of resistance spot welding. *Journal of Materials Processing Technology, 210*, 1076–1082.

Kim, D. C., Park, H. J., Hwang, I. S., & Kang, M. J. (2009). RSW of aluminum alloy sheet 5J32 Using SCR type and inverter type power supplies. *International Scientific Journal, 38*, 55–60.

Li, Z., Hao, C., Zhang, J., & Zhang, H. (2007, April). Effects of sheet surface conditions on electrode life in resistance welding aluminum. *Welding Journal*, 81–89.

Mathers, G. (2002). *The welding of aluminum and its alloys*. Cambridge: Woodhead.

Newton, C. J., Browne, D. J., Thornton, M. C., Boober, D. R., & Keay, B. F. (1994). *The fundamental of resistance spot welding aluminum*. Sheet Metal Conference VI; Paper E2.

Patrick, E. P., Auhl, J. R., & Sun, T. S. (1984). *Understanding the process mechanism is key to reliable resistance spot welding aluminum auto body components* (SAE Technical Paper 840291).

Pedersen, K. R. (2010). *Resistance welding of aluminum alloys* (Master thesis). Department of Mechanical Engineering, Technical University of Denmark.

Pickering, E. R., & Hart, C. J. (1994). *Optimizing resistance spot welding on aluminium alloy 6111 auto body sheet* (SAE Technical Paper, 940662).

Rashid, M. (2011). Some tribological influences on the electrode-worksheet interface during resistance spot welding of aluminum alloys. *Journal of Materials Engineering and Performance, 20*, 456–462.

Rashid, B. M., Fukumoto, S., Medley, J. B., Villafuerte, J., & Zhou, Y. (2007, March). Influence of lubricants on electrode life in resistance spot welding of aluminum alloys. *Welding Research,* 62–70.

Resistance Welding Manual. (2003). Bridgeport, NJ: RWMA Resistance Welding Manufacturers Association.

Rönnhult, T., Rilby, U., & Olefjord, I. (1980). The surface state and weldability of aluminium alloys. *Materials Science and Engineering, 42*, 329–336.

Senkara, J., Zhang, H., & Hu, S. J. (2004). Expulsion prediction in resistance spot welding. *AWS, Welding Journal, 83*, 123s–132s.

Shi, Y., & Guo, H. (2013). Fatigue performance and fatigue damage parameter estimation of spot welded joints of aluminium alloys 6111-T4 and 5754. *Fatigue & Fracture of Engineering Materials & Structures, 36*, 1081–1090.

Song, Q., Zhang, W., & Bay, N. (2005). An experimental study determines the electrical contact resistance in resistance welding. *AWS Welding Journal, 84*, 73s–76s.

Spinella, D. J., Brockenbrough, J. R., & Fridy, J. M. (2005). Trends in aluminum resistance spot welding for the auto industry. *AWS Welding Journal, 84*, 34–41.
Sun, M. A. (2003). *Fundamental study of contact resistance behavior in RSW aluminum* (PhD dissertation). School of The Ohio State University, Ohio State University.
Sun, T. S. (1982). *Electrode deterioration mechanisms in resistance spot welding of aluminum* (AlCOA Internal Report No. 53-82-3).
SWANTEC Software and Engineering ApS. SORPAS® (Version 10.6). Retrieved from http://www.swantec.com
Thornton, M. C., Newton, C. J., Keay, B. F. P., Sheasby, P. G., & Evans, J. T. (1997). Some surface factors that affect the spot welding of aluminium. *Transactions of Institute of Metal Finishing, 75*, 165–170.
Wheeler, M. J., Sheasby, P. G., & Kewley, D. (1987). *Aluminum structured vehicle technology. A comprehensive approach to vehicle design and manufacturing aluminum* (SAE Technical Paper 870146).
Zhang, H., & Senkara, J. (2012). *Resistance welding – Fundamentals and applications* (2nd ed.). Boca Raton, FL: Taylor & Francis Group, LLC.
Zhang, W. (2003). Design and implementation of software for resistance welding process simulations. Welding & Joining, SP-1766, SAE. In *World Congress* (pp. 105–113). Detroit, MI, USA.
Zheng, Z. T., Shan, P., Hu, S. S., & Luo, Z. (2006). Resistance spot welding of aluminum alloy with great resistance medium. *Transactions of the China Welding Institution, 27*, 5–8.

9

Sequencing jobs in an engineer-to-order engineering environment

Douglas H. Grabenstetter and John M. Usher*

Department of Industrial and Systems Engineering, Mississippi State University, MS 39762, USA

Engineer–to-order (ETO) firms produce complex – one of a kind – products and desire shorter lead times as a key component to cost competitiveness. In ETO firms, the engineering process is the largest controllable consumer of lead time. Given that lead time is a function of completion rate and scheduling policy, one critical process is to accurately sequence jobs in front of the engineering function. However, unlike other manufacturing models, such as make–to-stock or make-to-order models, the design for an ETO product is not realized until after the engineering process has been completed. Hence, the only information available does not include data normally required by most sequencing algorithms. Therefore, the problem becomes the determination of an accurate schedule within a complex transactional process for jobs which have not even been designed yet. This paper investigates this topic in the context of the engineering process within the ETO model. Based on research conducted in conjunction with multiple firms, common factors are identified which drive complexity, and a new framework and algorithm are presented for using these factors to sequence jobs. Using discrete event simulation, the performance of this new algorithm is found to be a significant improvement over current industry and published methods.

Keywords: engineer-to-order; sequencing; design

1. Introduction

The engineer-to-order (ETO) model is used by a significant number of manufacturers across multiple sectors. Indeed, ETO firms comprise approximately one-fourth of all North American manufacturing and are growing at a rate of 20% (Cutler, 2005). One reason for this level of growth is the increased demand for customized products which, in turn, has caused markets to become more and more fragmented. Use of the ETO model provides firms with increased agility and flexibility thereby, allowing them to be ideally positioned to respond to rapid market shifts through production of one of kind products. Despite these advantages, the very nature of the ETO environment drives complexity. For instance, ETO products are manufactured and assembled in low quantities to individual customer specifications. In addition, the goods manufactured using ETO models tend to be complex in nature, resulting in deep and complicated multilevel product structures and routings which often contain job-specific and engineered-to-order components (Hicks & Braiden, 2000; Hicks, Song, & Earl, 2007). In the ETO environment, the engineering process is the largest controllable consumer of lead time taking up to one half of the total lead time (Little, Rollins, Peck, & Porter, 2000; Pandit & Zhu, 2009). Since ETO firms produce products which are often used in large projects, it

*Corresponding author. Email: usher@ise.msstate.edu

is not unusual for their customers to impose large cost penalties for lateness. Further, given that lead time is a function of completion rate and scheduling policy, one critical process is to accurately sequence jobs in front of the engineering function and therefore is an important topic for these companies.

Yet, despite the aforementioned need, there has not been much sequencing research focused on the ETO engineering environment. Therefore, this paper investigates sequencing approaches used in the literature in order to generate good (not necessarily optimal) job sequences. A new complexity sequencing approach is introduced that has been tested using real-world data from a typical ETO industrial firm and found to meet the needs of ETO firms in that it is quick, easy to use, accessible, and accurate. Indeed, the approach has been found to significantly outperform traditional sequencing methods used in the ETO environment.

2. Background

When discussing ETO firms, it is important to consider a couple of important constraints. The first of these is that the engineering process is a bottleneck resource per Gosling and Naim (2009), Cutler (2005, 2009), and Pandit and Zhu (2009). Indeed, it is a core process and is never outsourced (Hicks, McGovern, & Earl, 2000). Further, competition from other firms has caused some firms to reduce their costs by 50% (Hicks, McGovern, & Earl, 2001). Since resources for the engineering processes are costly (Qi & Tu, 1998), these staffs in turn have been scaled down (Hicks et al., 2001). As a result, engineering staffs are now often resource constrained. Secondly, most ETO firms use enterprise resource planning (ERP) systems, per Stevenson, Hendry, and Kingsman (2005), which are substantial investments per Worthen (2002). Hence, trying to convince these firms to throw out their ERP systems likely will be a hard sell! Unfortunately, despite the need for research on the engineering process, the vast majority of research performed for ETO firms by such authors as Hicks and Pongcharoen (2006, 2009), Hicks and Braiden (2000), Hicks et al. (2000, 2001, 2007), Pongcharoen, Hicks, and Braiden (2004), Pongcharoen, Hicks, Braiden, and Stewardson (2002), and Song, Hicks, and Earl (2001, 2002, 2006) recommends complex global shop floor scheduling approaches which seek to replace ERP systems. Coates, Ritchey, Duffy, Hills, and Whitfield (2000) provide the only discussion found which focuses on the ETO engineering process. They present an agent-based design decision evaluation tool for large make-to-order products. This tool chooses the design concept which takes the least amount of projected time with minimum resource use. The purpose of the agents is to perform this task while dealing with a diverse set of design tools using data from heterogeneous distributed computing systems. Seven agents are identified with each of the agents fulfilling a specific role performing several activities. A resource manager agent optimizes the utilization of the resources. If the utilization of a particular resource falls below a given target, or if an alternative resource's efficiency level is better than the current resource, the resource manager triggers a scheduling agent to create a task matrix which contains task dependencies and task durations. The scheduling agent determines in which order certain design candidates are evaluated. The scheduling problem is defined by the authors as a job shop problem (as opposed to a single-machine setting) which has been proven to be NP-Hard (Brucker, Drexl, Moehring, Neumann, & Pesch, 1999). Therefore, the authors make use of a multicriteria genetic algorithm (GA) which "optimizes" multiple objective functions simultaneously. The GA produces a pareto-optimal set of schedules using the task matrix and chooses the schedule which maximizes the utilization of the resources, minimizes the number of resources, and minimizes the time to complete design concept evaluations. The main problem with the GA approach lies with its complexity as driven by the scheduling problem definition. Since the scheduling problem is defined as a job shop problem, computation time will likely be an issue due to problem logic complexity.

In contrast to what has been proposed by others, this paper offers a more robust approach which uses the output of ERP and focuses on scheduling just the engineering process not each process of the entire system. The new methodology addresses the inherent scheduling complexity of the ETO environment through the use of dispatching rules (DR). Compared to the aforementioned methods, DR are low in computational complexity, having only to compute which job to process next as opposed to having to determine the entire sequence of all the jobs in the queue. While use of DR will not guarantee optimal solutions, good solutions are often achieved for either static or dynamic, single or multiple machine environments. As well, DR are very easy to implement. Therefore, it is no surprise to find widespread use of DR by ETO firms.

There are three basic types of sequencing approaches found in the literature which use due dates as an input: allowance-based, slack-based, and ratio-based. In his study, Baker (1984) investigates approaches from each of these three categories and finds that the performance of slack-based approaches is equivalent to the allowance-based approaches. Referring to Table 1, Baker and Trietsch (2009) find that shortest processing time (SPT) performs quite robustly for the proportion tardy (PT) and mean tardy (MT) objective functions. Additionally, Baker reports that modified due date (MDD) performs well for the single-machine dynamic maximum lateness (*T*) problem and that early due

Table 1. Brief summary of dispatching rule metrics (adapted from Baker & Trietsch, 2009).

Metric	Best rule(s)	Other rules compared	Authors
PT	S/OPN	EDD/MST	Conway (1965a), Conway, Maxwell, and Miller (1967)
	SPT	MST,EDD,S/OPN	Elvers (1973)
	SPT	EDD, ODD, MST, SCR, SCR,OST	Muhlemann, Lockett, and Farn (1982)
	SCR, A/OPN	S/OPN	Miyazaki (1981)
	SPT	MST, OST, EDD, ODD, SCR, OCR	Kanet and Hayya (1982)
	SPT, S/OPN, MST	EDD	Elvers and Taube (1983)
MT	MST	S/OPN, SPT	Gere (1966)
	SPT	S/OPN, EDD	Conway (1965a,1965b), Conway and Maxwell (1962), Carroll (1965)
	S/OPN	SPT	Weeks (1979)
	SPT, ODD	MST, OST, SCR, OCR	Kanet and Hayya (1982)
	A/OPN	SCR, S/OPN	Miyazaki (1981)
	SPT, EDD	MST, SCR	Baker and Bertrand (1981)
	MST, SPT, EDD	OST, S/OPN, ODD, SCR	Muhlemann et al. (1982)
CMT	S/OPN	SPT, EDD	Conway (1965a,1965b), Conway and Maxwell (1962)
	SCR	SPT, EDD, ODD, MST, S/OPN, OST	Muhlemann, Lockett, and Farn (1982)
	OCR	SPT, EDD, ODD, MST, OST, SCR	Kanet and Hayya (1982)

date (EDD) maximizes the minimum service level in the stochastic single machine environment. Further, EDD has been found by this author to be the method of choice among ETO firms.

3. Sequencing research in the ETO engineering process

As previously mentioned, the available research in the literature is not directly applicable to the ETO environment. In addition, very little information exists about the ETO engineering process and how engineering due dates are calculated. Therefore, subject-matter experts (SMEs) from five ETO firms of various industrials (listed in Table 2) were consulted remotely via telephone, or, where feasible, face-to-face. With each of the ETO firms, research was conducted regarding the front-end processes and flows they employed including due-date setting methods.

The engineering process is distinct from traditional shop floor processes as its inputs and outputs are purely intellectual and not physical in nature. Hence, in order to gain an idea of how sequencing is performed in ETO firms, SMEs at each site were interviewed about their individual process flows including the inputs and outputs, as well as the details of the tasks at each step. Not surprisingly, the challenges of the ETO environment have forced each of the firms to adopt similar process steps. The data gathered are summarized in the form of a generic IDEF0 diagram (see Figure 1).

In step A1, an initial job due date is set. The lead times are set intentionally wide enough to cover the amount of uncertainty present. Once the order is obtained, the firm must provide a committed job due date in step A2. The job information is forwarded on to the engineering step (A3). Research found that several constraints exist in regard to sequencing for the ETO design environment. First, the ETO environment is complex, and each job is unique and often complicated. In addition, there is a limited amount of information available. Referring to Figure 2, the first step in the engineering process (A3.1) is a review and sequencing process where the job information is scanned for long lead time non-inventoried components. In addition, the job data are assessed to ensure that vital data required by the engineering discipline in play (e.g. wiring diagrams, ladder diagrams, and other details) do not contain clean-order defects. These clean-order defects encompass missing information not provided by the sales force, incorrect information in that it is wrong, infeasible or non-sensical, and conflicting information between two or more pieces of data. For instance, a ladder diagram may call for logic which does not translate properly into the expected components.

Clean-order defects are a major source of variability in the ETO environment and cause rework loops in the engineering process (steps A3.2 and A3.4). These issues in

Table 2. Description of the ETO industrial firms studied.

Firm	Market	Product	Annual sales	Employees
A	Industrial	Motor control center	\$60MM	256
B	Construction	Switchboards	\$25MM	20
C	Industrial	Process instrumentation	\$38MM	80
D	Construction	Switchgear	\$81MM	317
E	Industrial	Busway	\$63MM	232

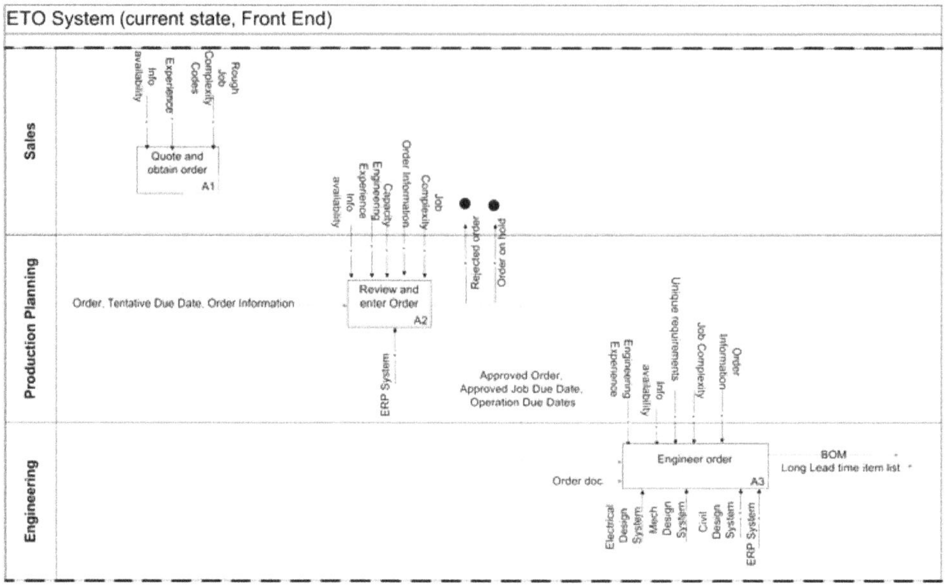

Figure 1. ETO engineering flow of front-end process.

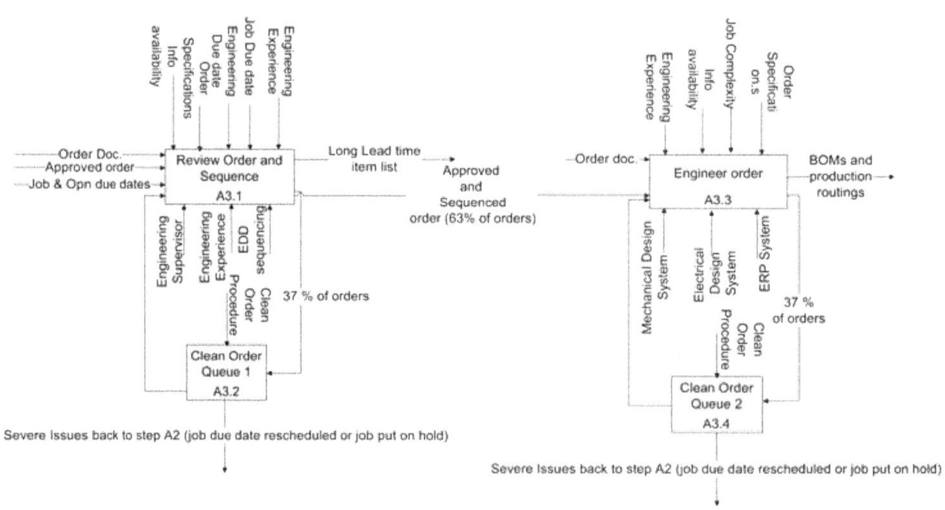

Figure 2. ETO engineering process.

ETO firms are problematic. Indeed, a four-year study conducted by Firm A found that 37% of orders contain some sort of clean-order error. Normally, the cycle time to solve clean-order defects may take up to 2 weeks. However, if the customer is found to be the source of the defect, the ETO sales force is contacted and negotiations then occur, which can drive up the resolution cycle time. According to Firm A, such a defect occurrence can take 60 days or more to resolve. Additionally, if the customer is at fault or if a change is initiated by the customer, the due date could be moved out. If the clean-order negotiations become intractable, it is also possible that the order will be put on hold or even canceled.

Once the clean-order review has been completed, the job is then sequenced in step A3.1. As previously mentioned, research with the five partner firms has revealed that the EDD method of sequencing is used as a general practice. The main reason for using EDD is its simplicity. Each of the firms stated that they prefer simple, easy to use methods that can be run using a spreadsheet. This follows the findings of Stoop and Wiers (1996) that simple and robust approaches are the most broadly used in practice. Otherwise, the firms will likely not use the methodology (O'Neil, 1991). The scheduling process is typically conducted on a weekly basis for several reasons. First, Pandit and Zhu (2009) and Little et al. (2000) point out that the ETO engineering process can consume up to one-half of the production lead time. Hence, the engineering task is considered to be a long duration process. Second, the engineering step is a complex, intellectual process and interruptions to the process can lead to an inflated error rate. Thus, when an engineer begins work on a job, if at all possible, it is desired that they be allowed to complete the engineering task without interruption (non-preemption). Otherwise, constant rescheduling is known to cause schedule nervousness.

Upon the job being sequenced and assigned to an engineer, the next step in the process (A3.3) is the actual engineering task where the job inputs are transformed into a tangible product design. Given the high complexity of ETO products, the engineers themselves typically possess a high degree of experience and are difficult to replace. In addition, due to the high product complexity present, each of the partner ETO firms makes use of a proprietary computer-aided job engineering configuration software system (CAJECSS). When engineering the job, the engineer chooses predefined job characteristics found in the system. The CAJECSS then uses logic tables to generate the appropriate engineering output (e.g. wiring diagrams and assembly drawings). In addition, it also generates a top-level bill of material which is used in the subsequent production planning step to order materials. Only when the engineering process is finished, can the remainder of lower level discrete components be procured. Once the components are on-hand, the manufacturing process itself can begin.

These findings lead to an overriding question: How can scheduling capabilities which co-exist with existing ERP systems be developed for the highly complex ETO engineering environment while, at the same time, remaining realistic, robust, acceptable, and easily implemented? One might conjecture that the answer simply lies in applying what has been researched already in the job shop literature. However, much of this research presupposes that the process time of each job is known or can be easily estimated. However, because each job is unique, in the ETO environment, this simply is not the case. Restated, the literature survey and ETO firm research have shown that what is needed is a sequencing policy which:

- robust and relatively simple to implement and use (spreadsheet driven),
- accurate (an improvement over EDD), and
- complementary to an ERP system (Recall that many firms have invested heavily in their ERP systems and are in no hurry to replace them).

4. Regression-driven complexity-based flow time prediction

Based upon the previous discussions, the engineering process can be modeled as a dynamic job shop with the following assumptions:

- arrivals are dynamic,
- machines process one job at a time,
- setup times are independent of job sequence,
- machines are continuously available, and
- neither idle time nor preemption is allowed.

SPT is well known to be a robust sequencing algorithm. However, given the nature of the ETO engineering environment, nothing is known ahead of time in regards to job processing time. To overcome this problem, Grabenstetter and Usher (2013) identified seven engineering factors that can be used in this context to predict job processing times using a regression-based approach. These factors are:

- x_1 – number of functional requirements
- x_2 – number of basic components
- x_3 – number of design interdependencies
- x_4 – number of technologies
- x_5 – number of regulations and standards
- x_6 – number of subsystems
- x_7 – presence of a reference job

Although multiple regression has been used by others such as Salman and Dogru (2004), no such procedure has ever been employed which uses ETO-specific complexity factors (CFs). Once the regression procedure generates a predicted processing time for each job *j*, they can then be used as an input to SPT. The framework for this approach (referred to as RegSPT) is comprised of three parts that are complementary in nature. The first component employs multiple linear regression using a basic CF equation composed of the seven CFs.

$$\mathrm{CF}_j = \beta_1 x_1 + \beta_2 x_2 + \beta_3 x_3 + \beta_4 x_4 + \beta_5 x_5 + \beta_6 x_6 + \beta_7 x_7 \quad (1)$$

where $\mathrm{CF}j > 0$, x_1, x_2, x_3, x_4, x_5, x_6, $x_7 \geq 0$ and initially x_1 through x_6 are defined as discrete integer variables, and initially x_7 is defined as a binary ordinal categorical variable (1, 0).

The second element of the framework is an *estimation algorithm* (EA) which is used to determine the values of the coefficients of the CF equation. This step is necessary for each different ETO firm given that the exact continuous distribution function of the dependent variable data and the exact probability distribution function of the dependent variable data will vary somewhat with each firm. As well, the EA will need to be rerun periodically because it is likely that business conditions will change over time and process improvements will occur. These changes may affect the efficacy of the input data used to build the actual regression equation currently in use. This means every p periods, the EA will need to be run using up-to-date historical data for the last p periods for both the response variable and the independent variables x_1, x_2, x_3, x_4, x_5, x_6, and x_7. Because not all ETO firms are exactly alike, the length of the update cycle, p, and the sample size, z, will need to be determined by each firm. The EA is defined as:

Step 1: Gather dependent variable data for the last p periods.
Step 2: Gather independent variable data for the last p periods.
Step 3: Generate updated slope coefficients (β): using Equation (1) as a starting point, generate a statistically valid regression model using the seven aforementioned factors.

The final component of the framework is the RegSPT algorithm (RSA) which uses the resulting CF equation as a part of the sequencing process to generate the complexity-based processing times for each job order. Operationally, when a job is received, the production planning step assembles data that are found in the specific job's packet. The independent variable data (x_i's) are then entered at the engineering step into the regression equation for the job at hand to determine the job unique complexity-driven processing time. Once processing times have been predicted for each job to be sequenced, the SPT sequencing algorithm is then utilized. RSA is defined as:

Let n = the number of jobs.
Step 1: Initialization: Upon arrival of job j, Set $j = 1$.
Step 2: For Equation (1), determine x_1, x_2, x_3, x_4, x_5, x_6, and x_7 using the job information provided by the sales department.
Step 3: Evaluate Equation (1) to generate the predicted process time.
Step 4: Let $j = j + 1$.
Step 5: If $j = n$, Go to step 6. Else, go to Step 1.
Step 6: Sequence job j … n in ascending order of predicted processing times. If ties result, schedule by earliest due date.
Step 7: Stop.

5. Example implementation of the RegSPT method

To illustrate the application of the framework for generating job sequences, actual data from Firm A was chosen because it closely matches the criterion listed in the literature for ETO firms (Hicks et al., 2007). When performing regression analysis, it is important to follow a strategy of solid statistical practice. Therefore, the strategy listed in Kutner, Nachtsheim, and Neter (2004) was followed.

Step 1 – The first step of the EA process is to gather data from the last p periods. At this point, it should be stated that due to differences in software and operational practices, the execution of this task will vary with each firm. Some firms may make use of MS Excel® tables to store these data, while others may store them in their ERP system. At Firm A, response variable data were gathered which represent actual historical engineering labor flow time data in days. These data were resident in their ERP system and consisted of two years' worth (1276 distinct jobs) of data which were exported to a spreadsheet program. Again, for our tests, in order to rule out the impact of sampling bias, the time period from which the data were gathered came from the most recent period of economic stability and represents steady state/normal business conditions. The response variable statistics were mined directly from the data-set. Outliers were identified, and, if determined by Firm A to be inappropriate, were removed.

Step 2 – The independent variable data to be gathered must represent actual historical engineering data used to engineer each of the corresponding jobs identified in Step 1. Once the raw data are obtained, the data need to be separated into the aforementioned seven factors. Again, due to differences in the format in which this data is stored and operational practices, the difficulty of this task will vary with each firm. In this example, similar to some other ETO companies, Firm A makes use of a sophisticated proprietary CAJECSS. With this system, the engineer chooses predefined job characteristics found in the system. Then using logic tables, the CAJECSS generates appropriate wiring diagrams and assembly drawings. In addition, it also generates a top-level Bill of Material which is used in the subsequent production planning step to order materials. For two of the seven factors (number of subsystems and presence of a reference job), the data could easily be extracted directly. However, data categorization for the other five factors was not as simple. These data were discovered to be embedded within multiple levels of tables found within the CAJECSS. Further, given the multilevel nature of the tables, and that factor x_3 is built upon interdependence of design characteristics, the data for each factor could not be extracted using simple sorting techniques. Therefore, in order to accomplish separation of these data into their appropriate factors, a binary mapping table using MS Excel® was developed which took advantage of the logic statements found within the configuration system. Upon reviewing the raw data from the CAJECSS, the amount of raw data was found to be a staggering amount. As a result, the mapping table was then utilized to develop a computerized algorithm to separate the data into the five aforementioned factors. The algorithm was written in the Visual Basic® for Applications programming language. The run time was very quick, requiring less than 30 s. These tools are reusable and once developed, can be used when future updates of the system are required.

Once the data are gathered, the next step is to validate the data and gain an understanding of the data quality. Direct diagnostic plots of the response variable are normally not very important in regression analysis as the values of the observations on the response variable are functions of the levels of the predictor variables. However, in order to ensure that the study is of the highest quality, a cursory look at the data should be performed. The data for each dependent and independent variable were plotted using box plots and reviewed with the SMEs in order to identify and remove gross data errors and outliers.

Step 3 – Next, the regression model is generated using Equation (1) and a standard residual analysis performed to ensure that the model meets the statistical criterion for multiple regressions. Since the distribution of residuals was initially found to be non-normal, a logarithmic transformation was performed on the response, which is referred to in our example as Log PT. Referring to Tables 3 and 4, the model is found to be appropriate. The model itself has a p-value of zero, which is well below the normal .050 alpha threshold. Further, all of the factors except for x_2 and x_5 similarly are well

Table 3. Model with factor significance.

$$\text{Log PT} = .635 + .00148\,x_1 + .0422\,x_2 + .143\,x_3 - .0427\,x_4 - .00118\,x_5 + .264\,x_6 - .590\,x_7$$

Predictor	Coef	SE Coef	T	p
Constant	.63492	.03167	20.05	.000
x_1 – # Func. req.	.00148	.00025	6.01	.000
x_2 – # Basic comp.	.04222	.01216	3.47	.001
x_3 – # Design inter.	.14271	.03200	4.46	.000
x_4 – # Technologies	−.04273	.01225	−3.49	.001
x_5 – # Reg. and Std.	−.001185	.00067	−1.77	.076
x_6 – # Subsystems	.26399	.03257	8.11	.000
x_7 – Ref. job?	−.58987	.03055	−19.31	.000
S = .472984	R^2 = 39.8%	R^2_{adj} = 39.4%		

Table 4. ANOVA table.

Analysis of variance					
Source	DF	SS	MS	F	p
Regression	7	187.366	26.767	119.65	.000
Residual error	1268	283.669	.224		
Total	1275	471.034			

below the normal .050 alpha threshold and although x_2 and x_5 are marginally above the threshold, investigations with the ETO firms have concluded that these factors are valid. Hence, the model can be used for predictive purposes.

Step 4 – Given that the CF equation with coefficients is available, the RSA can be run. In order to generate the processing times, a second historical data-set consisting of (74) jobs was gathered from Firm A and represents the input data that would be used by the RSA. Recall that the regression formulation is based on p time periods before the current time period. Therefore, the data from the 1276 distinct jobs used in Step 1 iare from a time period preceding this set of 74 jobs. This condition simulates actual operating conditions which would be experienced by practitioners. Equation (1) is embedded in an MS Excel® spreadsheet and is evaluated for each of the 74 jobs to compute the predicted processing times. Examples of the input data and predicted process time values are given in Table 5.

Table 5. Example of input data for RegSPT and output.

Order No.	x_1	x_2	x_3	x_4	x_5	x_6	x_7	LOG predicted process time	Predicted process time
3001774379	128	71	0	68	0	1	1	.59	3.90
3001912763	142	113	1	112	0	1	0	1.24	17.31
3002194681	30	6	0	6	0	1	0	.94	8.72
3002246903	2	0	1	0	0	0	0	.78	6.04

Table 6. Example of the RegSPT schedule.

Order No.	Arrival time hour	Predicted process time (days)
3001774379	0	3.90
3002267611	0	8.27
3002194681	0	8.72
3002341332	0	12.36
3001912763	0	17.31
3002380038	7	2.35
3002400367	7	4.32
3002390220	7	8.03
3002454575	14	8.16
3002452423	14	8.50
3002447003	14	8.58
3002450199	14	11.18
3002491099	21	4.35
3002472412	21	6.44
3002488807	21	7.98

Step 5 – Referring to Table 6, the next step is to group the jobs based on their arrival time and then sequence the jobs in each group in ascending order of their processing time (SPT). The run time to sort the 74 jobs is extremely fast and was completed in less than 5 s.

6. Methods

Given the previous section's description of RegSPT as an alternative to current method of sequencing for the ETO engineering environment, it remains to be proven whether the new methodology is better than current approaches typically used by ETO firms and popular literature approaches. Therefore, an analysis using simulation employing data from one of the five aforementioned partner firms, Firm A, was performed to study the performance and behavior of sequencing methods in a real-world ETO engineering department. A discrete simulation model of the engineering process was built, verified, and validated using the approach of Harrell, Ghosh, and Bowden (2004). The output of the model yields queue lengths, utilization, job cycle times, and many other data useful for analysis of the long-term behavior of the engineering environment. In particular, the cycle times were used to determine job completion times. In order to provide a 95% percent confidence interval with a three-day half-length using the standard deviation of lateness (SDL) value based on an initial run of 30 replications, it was determined that 53 replications of the experiment would be required. Due dates for input into the model were calculated using the approach of Grabenstetter and Usher (2014). The process steps in the model follow those shown in Figure 2 and the processing times for those steps utilize industry values. The arrival patterns form the basis for testing the performance of RegSPT against popular sequencing methods.

Three popular sequencing approaches from the literature were chosen for the analysis: SPT, MDD, and EDD. Baker and Trietsch (2009) find that SPT performs quite robustly for a variety of objective functions (see Table 1) and, in particular, in the presence of imperfect information which is an advantage in the ETO environment. In addition, it is easy for practitioners to implement and understand. However, as can be observed, this rule requires foreknowledge of a job processing time. Hence, in order to simulate the operation of SPT in practice at an ETO firm, it is necessary to develop a method that ETO firms would most likely use to predict processing times using the limited data (material costs) that is available at the beginning of step A3. To accomplish this task, it is important to know that the firms will likely only implement a method which is simple and robust (O'Neil, 1991; Stoop & Wiers, 1996). Therefore, it was decided to use a simple material-based approach similar to what is currently utilized in industry. As such, jobs are grouped into three classes based upon perceived difficulty as predicted by the job material content (Grabenstetter & Usher, 2013). To accomplish this task, the 1276 jobs in the data-set were grouped by the three classes and an average process time by class determined. These average process times are then used to help drive SPT. In the literature, the release time r_j, is another variable that normally appears as a part of sequencing formulations. However, in this case study, it is assumed universally that the release time equals zero. Therefore, this term drops out of the formulations.

The MDD method chooses the assigned job due date (d_j) or the soonest time that the job could be completed ($t + p_j$), whichever time is later.

$$d'_j = \max\{d_j, t + p_j\} \tag{2}$$

The same approach for processing time prediction used previously for SPT is employed here to get the values of p_j.

EDD is the method of choice among ETO firms and has been reported to perform robustly. It is simple, easy to understand and implement, and does not require foreknowledge of job processing times. In order to sequence the jobs via the EDD method, within each arrival period, the jobs are simply arranged in ascending order according to their due dates. If ties are encountered, the jobs are ordered arbitrarily.

In order to judge the performance of the sequencing methods, it is necessary to define the performance metrics used. This study employs the criteria used by other published ETO studies such as Hicks and Pongcharoen (2006, 2009), Hicks and Braiden (2000), Hicks et al. (2007), and Song et al. (2002, 2006). These criteria are:

- Mean lateness (ML): lateness is calculated as the job completion time minus the due date. As such it measures the deviation, early or tardy from the due date.
- Mean absolute lateness (MAL): defined as the absolute value of the job lateness.
- SDL: used to add a measure of variability.

7. Effectiveness of the RegSPT method

The test results show the RegSPT method to be provide robust performance for the studied metrics. Referring to the summary statistics in Table 7, RegSPT produces the best performance in all three categories. In particular, the value for MAL and the standard deviation of ML are both statistically lower when compared to the closest competitor: EDD. Since the data for neither of these metrics is normally distributed, Levene's test is used for variance testing and the Mann–Whitney Test is used for median testing. Referring to Table 8, given that the p-value is less than the .05 alpha value, the ratio of the variances are considered to be unequal. Hence, the variance for RegSPT is significantly lower compared to EDD and Table 9 shows MAL for RegSPT to be likewise significantly different (lower) from that of EDD. Since lower values of absolute lateness are important for lean manufacturing (goal of zero), this is an important finding.

8. Implementation issues

RegSPT has been shown to provide more balanced performance compared to the existing sequencing methods and is easier to implement than the MDD and SPT approaches as there is no need to estimate a processing time for each job. In terms of tools required to implement the approach, all a company needs is a computer and standard software applications such as MS Excel® and possibly Minitab® (although Excel® statistical addons will work fine). RegSPT can be thought of as having two components: the CF regression equation and the sequencing algorithm. Because business conditions change over time, there is a need to regularly refresh the coefficients of the regression equation.

Table 7. Summary of lateness results for all methods.

Metric	EDD	MDD	SPT	REG SPT
MAL	10.55	10.68	29.83	7.84
SDL	11.33	12.12	24.49	7.11
AL	−7.06	−7.08	27.01	−7.09

Table 8. Statistical variance testing for ML.

Test and CI for Two Variances: EDD AVE lateness, Ave Lateness by

```
Method

Null hypothesis         Sigma(EDD AVE lateness by job) / Sigma(Ave
Lateness by job-REG SPT) = 1
Alternative hypothesis  Sigma(EDD AVE lateness by job) / Sigma(Ave
Lateness by job-REG SPT) not = 1
Significance level      Alpha = 0.05

Statistics

Variable                     N   StDev  Variance
EDD AVE lateness by job      74  11.331   128.400
Ave Lateness by job-REG SPT  74   7.111    50.573

Ratio of standard deviations = 1.593
Ratio of variances = 2.539

95% Confidence Intervals

                                    CI for
Distribution   CI for StDev       Variance
of Data            Ratio            Ratio
Normal         (1.264, 2.008)  (1.599, 4.031)
Continuous     (1.430, 2.123)  (2.045, 4.508)

Tests

Method                           DF1  DF2  Statistic  P-Value
F Test (normal)                   73   73       2.54    0.000
Levene's Test (any continuous)     1  146      32.22    0.000
```

Table 9. Statistical median testing for MAL.

Mann-Whitney Test and CI: EDD MAL by job, MAL by Job-REG SPT

```
                    N  Median
EDD MAL by job      74  10.397
MAL by Job-REG SPT  74   6.191

Point estimate for ETA1-ETA2 is 1.696
95.0 Percent CI for ETA1-ETA2 is (0.110,3.269)
W = 6058.0
Test of ETA1 = ETA2 vs ETA1 not = ETA2 is significant at 0.0368
```

This process is conducted offline to the everyday operation of RegSPT and given that most ETO firms use ERP systems, the historical data needed to compute the coefficients should be simple to access and export to a spreadsheet. The amount of time to accomplish this task depends on the firm's systems/data cleanliness and availability, but it should be on the order of several hours.

Next, given that the regression equation is up-to-date, the jobs which are to be run through the engineering step must be sequenced on a regular predetermined basis (typically weekly). In order to generate the input data required for the sequencing algorithm, the predicted process time for each job needs to be generated. In order to accomplish this task, personnel simply need to enter the aforementioned seven pieces of complexity information (x_i's) into the regression equation which can be encoded into a spreadsheet formula. At this point, it needs to be stated that the process of searching for these data is not an extra step that results from the use of RegSPT, but is actually already currently being performed as a normal part of the engineering review process. Because the job package is reviewed as a normal part of the engineering review process, the seven input data values can be quickly gathered and the RSA executed to generate the predicted processing times. The jobs are then sequenced in ascending order of predicted processing times. Once the input data are extracted, based upon discussions with the five firms studied, it is estimated that the entire sequencing process should typically take 15–20 min vs. the roughly one hour that it takes today. Overall, this algorithmic approach is quick, user-friendly, and runs using spreadsheet programs and tools that are well known by ETO practitioners.

9. Summary

The ETO environment drives complexity and variability which, in turn, provides a justification for administrative convenience. Administrative convenience leads personnel to choose methods which are familiar and easy. Unfortunately, per O'Neil (1991), administrative convenience also often pushes in a direction counter to best performance and, in this case, the use of ad hoc approaches that yield poor results. To combat administrative convenience, O'Neil states that simple, easy to understand robust heuristic approaches are needed and more likely to be implemented. In fact, Stoop and Wiers (1996) state that such simple and robust approaches are the most broadly used in practice. Therefore, research is needed which helps practitioners in terms of choosing the correct simple and robust scheduling rules for the ETO engineering environment. Yet, despite this need, there has been little sequencing research focused on the ETO engineering environment. Therefore, this research has focused on a more robust approach for enhancing scheduling in the ETO environment concentrating on what has been called the main bottleneck in the ETO environment, namely the engineering process.

The engineering process is distinct from traditional shop floor processes as its inputs and outputs are purely intellectual and not physical in nature. These and other issues have led to an overriding question: how can scheduling capabilities which co-exist with existing ERP systems be developed for the highly complex ETO engineering environment while, at the same time, remaining realistic, robust, acceptable, and easily implemented? Restated, the literature survey and the results of the study of the partner firms have shown that what is needed is a scheduling policy which focuses on the engineering process, is robust and relatively simple to implement and use, and is complementary to an ERP system.

As has been demonstrated, the RegSPT sequencing approach, as applied to the ETO engineering environment, meets all of the above goals. First, RegSPT is robust and simple. It requires only seven pieces of information in order to accurately predict the job processing time. These data are available in a typical job quotation package which is received from sales. Because the job package is reviewed as a normal part of the engineering review process, the seven data points can be quickly gathered and the RSA executed. In addition, the algorithm runs using a spreadsheet program which are extremely user-friendly and well known by ETO practitioners. Once the input data are extracted, it is estimated that the entire sequencing process should typically take 15–20 min vs. roughly one hour today.

Second, RegSPT is accurate. The industry and literature approaches were rigorously tested against one another and RegSPT exhibited the best performance in all categories compared to its next closest competitor, EDD. In particular, it achieved statistically better results for MAL (7.84 days vs. 10.55 days). Since lower values of absolute lateness are important for lean manufacturing, this is an important finding. In addition, RegSPT achieved a statistically significantly lower SDL compared to the same competitor (7.11 days vs. 11.33 days). Given the high amount of variability found in the ETO engineering environment, these results are fairly impressive. Lastly, it is known that ERP systems have the capability to export and import data to and from the spreadsheet software. Hence, since RegSPT can easily be implemented using spreadsheet programs, it is compatible with typical ERP systems. Therefore, there is no need to jettison an existing costly ERP system.

In terms of next steps, while RegSPT has been shown to be an improved method of sequencing in the ETO engineering environment, admittedly there are other possible approaches which can take advantage of the REG approach and perhaps the use of a complexity-driven regression might yield in an incremental improvement in performance. Also, an important part of the complexity-driven regressive approach is the determination of p, the number of time periods needed before data refresh. In this study, this is left up to the practitioner to determine, but further studies on an analytic method to determine the parameter p would be beneficial to the industrial community.

References

Baker, K. R. (1984). Sequencing rules and due-date assignments in a job shop. *Management Science, 30*, 1093–1104.

Baker, K. R., & Bertrand, J. V. M. (1981). A comparison of due date selection rules. *AIIE Transactions, 13*, 123–131.

Baker, K. R., & Trietsch, D. (2009). *Principles of sequencing and scheduling*. New York, NY: Wiley.

Brucker, P., Drexl, A., Moehring, R., Neumann, K., & Pesch, E. (1999). Resource-constrained project scheduling: Notation, classification, models, and methods. *European Journal of Operational Research, 112*, 3–41.

Carroll, D. C. (1965). *Heuristic sequencing of single and multiple component jobs* (Unpublished Ph.D. dissertation). Sloan School of Management, MIT, Cambridge, MA.

Coates, G., Ritchey, I., Duffy, H. B., Hills, W., & Whitfield, R. I. (2000). Integrated engineering environments for large complex products. *Concurrent Engineering: Research and Applications, 8*, 171–182.

Conway, R. W. (1965a). Priority dispatching and work-in-process inventory in a job shop. *Journal of Industrial Engineering, 16*, 123–130.

Conway, R. W. (1965b). Priority dispatching and job lateness in a job shop. *Journal of Industrial Engineering, 16*, 228–237.

Conway, R. W., & Maxwell, W. L. (1962). Network dispatching by shortest operation discipline. *Operations Research, 10*, 51–73.

Conway, R. W., Maxwell, W. L., & Miller, L. W. (1967). *Theory of Scheduling*. Reading, MA: Addison-Wesley.

Cutler, T. R. (2005). Are you ETO? The increasing importance of custom (engineer-to-order) manufacturers. *In MFG*, pp. 1–3.

Cutler, T. R. (2009). Special orders, engineer to order challenges for the industrial engineer. *Industrial Engineer, 41*, 36–38.

Elvers, D. A. (1973). Job shop dispatching using various due date setting criteria. *Production and Inventory Management, 14*, 62–69.

Elvers, D. A., & Taube, L. (1983). Time completion for various dispatching rules in a job shop. *Omega, 11*, 81–89.

Gere, W. S. (1966). Heuristics in job shop scheduling. *Management Science, 13*, 167–190.

Gosling, J., & Naim, M. M. (2009). Engineer to order supply chain management: A literature review and research agenda. *International Journal of Production Economics, 122*, 741–754.

Grabenstetter, D. H., & Usher, J. M. (2013). Determining job complexity in an engineer to order environment for due date estimation using a proposed framework. *International Journal of Production Research, 51*, 5728–5740.

Grabenstetter, D. H., & Usher, J. M. (2014). Developing due dates in an engineer-to-order engineering environment. *International Journal of Production Research, 52*, 6349–6361.

Harrell, C., Ghosh, B. K., & Bowden, R. O. (2004). *Simulation using ProModel* (2nd ed.). New York, NY: McGraw-Hill.

Hicks, C., & Braiden, P. M. (2000). Computer-aided production management issues in the engineer-to-order production of complex capital goods explored using a simulation approach. *International Journal of Production Research, 38*, 4783–4810.

Hicks, C., & Pongcharoen, P. (2006). Dispatching rules for production scheduling in the capital goods industry. *International Journal of Production Economics, 104*, 154–163.

Hicks, C., & Pongcharoen, P. (2009). Applying different control approaches for resources with high and low utilisation: A case study of the production of complex products with stochastic processing times. *International Journal of Technology Management, 48*, 202–218.

Hicks, C., McGovern, T., & Earl, C. F. (2000). Supply chain management: A strategic issue in engineer-to-order manufacturing. *International Journal of Production Economics, 65*, 179–190.

Hicks, C., McGovern, T., & Earl, C. F. (2001). A typology of UK engineer-to-order companies. *International Journal of Logistics: Research and Applications, 4*, 43–56.

Hicks, C., Song, D. P., & Earl, C. F. (2007). Dynamic scheduling for complex engineer to order products. *International Journal of Production Research, 45*, 3477–3503.

Kanet, J. J., & Hayya, J. C. (1982). Priority dispatching with operation due dates in a job shop. *Journal of Operations Management, 2*, 167–175.

Kutner, M. H., Nachtsheim, C. J., & Neter, J. (2004). *Applied linear regression models* (4th ed.). New York, NY: McGraw-Hill Irwin.

Little, D., Rollins, R., Peck, M., & Porter, J. K. (2000). Integrated planning and scheduling in the engineer to order sector. *International Journal of Computer Integrated Manufacturing, 13*, 545–554.

Miyazaki, S. (1981). Combined scheduling system for reducing job tardiness in a job shop. *International Journal of Production Research, 19*, 201–211.

Muhlemann, A. P., Lockett, A. G., & Farn, C. K. (1982). Job shop scheduling heuristics and frequency of scheduling. *International Journal of Production Research, 20*, 227–241.

O'Neil, P. (1991). *Performance of lot dispatching and scheduling algorithms through discrete event simulation*. IEEE/SEMI International Semiconductor Manufacturing Science Symposium (pp. 21–24). San Francisco, CA.

Pandit, A., & Zhu, Y. (2009). An ontology-based approach to support decision-making for the design of ETO (engineer to order) products. *Automation in Construction, 16*, 759–770.

Pongcharoen, P., Hicks, C., Braiden, P. M., & Stewardson, D. J. (2002). Determining optimum genetic algorithm parameters for scheduling the manufacturing and assembly of complex products. *International Journal of Production Economics, 78*, 311–322.

Pongcharoen, P., Hicks, C., & Braiden, P. M. (2004). The development of genetic algorithms for the finite capacity scheduling of complex products, with multiple levels of product structure. *European Journal of Operational Research, 152*, 215–225.

Qi, X., & Tu, F. S. (1998). Scheduling a single machine to minimize earliness penalties subject to the SLK due-date determination method. *European Journal of Operational Research, 105*, 502–508.

Salman, N., & Dogru, A. (2004). Design effort estimation using complexity metrics. *Journal of Integrated Design and Process Science, 8*, 83–88.

Song, D. P., Hicks, C., & Earl, C. F. (2001). Stage due date planning for multistage assembly systems. *International Journal of Production Research, 39*, 1943–1954.

Song, D. P., Hicks, C., & Earl, C. F. (2002). Product due date assignments for complex assemblies. *International Journal of Production Research, 76*, 4877–4895.

Song, D. P., Hicks, C., & Earl, C. F. (2006). An ordinal optimization based evolution strategy to schedule complex make to order products. *International Journal of Production Research, 44*, 4877–4895.

Stevenson, M., Hendry, L. C., & Kingsman, B. G. (2005). A review of production planning and control: The applicability of key concepts to the make-to-order industry. *International Journal of Production Research, 43*, 869–898.

Stoop, P. P. M., & Wiers, V. C. S. (1996). The complexity of scheduling in practice. *International Journal of Operations and Production Management, 16*, 37–53.

Weeks, J. K. (1979). A simulation study of predictable due dates. *Management Science, 25*, 363–373.

Worthen, B. (2002). Nestlé's ERP Odyssey. *CIO, 15*, 1–9.

10

Criteria definition and approaches in green supplier selection – a case study for raw material and packaging of food industry

Narges Banaeian[a]*, Hossein Mobli[a], Izabela Ewa Nielsen[b] and Mahmoud Omid[a]

[a]Department of Agricultural Machinery Engineering, Faculty of Engineering and Technology, University of Tehran, Karaj, Iran; [b]Department of Mechanical and Manufacturing Engineering, Aalborg University, Aalborg, Denmark

This paper formulates an integrated framework for deciding about the green supplier selection criteria in food supply chain and also proposes different methods that account for single and multiple sourcing of supplier selection. Green supplier selection relies on green criteria, so determination of suitable set of criteria will affect decision-maker results directly. In this research, an operational model including combination of general and environmental criteria is introduced for green supplier selection criteria in raw material and packaging of food industry. This model reviewed a literature on general and environmental criteria that help us to know and make a set of common green criteria. Afterward, weighting criteria and collecting of sub-criteria are done by an expert team using the Analytic Hierarchy Process and Delphi method. The expert team tried to propose important and practical sub-criteria which are well fitted to food industry. Finally, in the section of supplier selection, two kinds of qualitative and quantitative data are discussed when single and multiple sourcing are required, respectively. Fuzzy Grey Relational Analysis is proposed to ranking suppliers in presence of qualitative and imprecise data. Also linear programming is used to present a model which can select the best suppliers and allocate the orders among them optimally.

Keywords: green supplier selection; food industry; raw material; packaging supplier

1. Introduction

Enterprises realize that a large amount of direct and indirect profits can be obtained from effective and efficient supply chain management practices. Supplier selection has great impact on integration of the supply chain relationship. Effective and accurate supplier selection decisions are significant components for productions and logistics management in many firms to enhance their organizational performance.

The issue of supply chain (management of networks, decision-makings, scheduling, etc.) is the area of science that has been strongly developing since the 80s of the last century (Bocewicz, Wójcik, & Banaszak, 2012; Sitek & Wikarek, 2014). A definition of a supply chain can be proposed as: the global network of players involved in executing the production and distribution of a predefined good from raw materials through to delivery to satisfy a consumer demand (Chen, Simon, Reich-Weiser, & Woo, 2013). The aim of supply chain management is to increase sale, reduce costs, and take full advantage of business (Sitek & Wikarek, 2014). Green supply chain is similar while

*Corresponding author. Email: nrgsbanaei@ut.ac.ir

managers simultaneously consider environmental impacts and work to reduce those impacts over time. Namely according to Sirvastava (2007), Green Supply Chain Management (GSCM) can be defined as "integrating environmental thinking into supply chain management, including product design, material sourcing and selection, manufacturing process, delivery of the final product to the consumers as well as end-of-life management of the product after its useful life." GSCM is to eliminate or minimize waste in the form of energy, emission, hazardous, chemical, and solid waste (Olugu, Wong, & Shaharoun, 2011).

Deciding whether to make or buy a new product or service is the first decision that must be made in managing supply chain. It determines where and how the supply chain will be managed (Schoenfeldt, 2008). If you decide to buy, so next step of SCM implementation includes the purchase, movement, and storage of raw materials (Harland, 1996).

Having seen the significance of collaborative supply chain and centralized purchasing in environmental management performance, it is now obvious that deciding which suppliers to collaborate with and how to select suppliers is a very crucial decision for the organization performance.

Fundamentally, supplier selection is divided into two kinds as single sourcing and multiple sourcing. In single sourcing, one supplier can satisfy all the buyer's requirements, but in multiple sourcing, company needs to select a collection of selected suppliers and their allocations (Wikipedia contributors, 2014).

Decision-making on supplier selection is based on criteria, so definition and selection of criteria have an important role in your selection. Supplier selection is a critical managerial decision-making problem, which requires considering various qualitative and quantitative criteria (Toloo, 2014). In conventional supply chain, general criteria are used; general selection criteria ordinarily intend to cover issues such as quality, capacity in terms of finance, services, and equipment, quantity, responsiveness, and others. In GSCM, environmental criteria are joined to general; environmental supplier selection criteria arise from an organization inclination to respond to any existing trends in environmental issues related to business management and processes. A good combination of conventional and environmental criteria will show the exact image of green supplier selection.

Building accepted definitions and characteristics of these criteria before their implementation in decision models is necessary for the acceptance of the methodologies, and, most importantly, for the decisions that derive from this methodologies. Despite the large amount of literature to supplier selection, there is narrow literature on green supplier selection and especially criteria definition, most of which are qualitative. Also, the methods for identifying, defining, grouping, and filtering criteria would greatly enhance the research, development, and application of models and their criteria. These areas are important and necessary directions (Govindan, Rajendran, Sarkis, & Murugesan, in press). The selected criteria are systematically integrated from supplier selection papers (general criteria) and green supplier selection (environmental criteria).

Given this complexity, this study presents a compound approach, first capable to integrate Delphi and AHP for criteria selection and then, according to supplier's data and company's requirements, defines two different kinds of GSS: (1) single sourcing with qualitative supplier's data using FUZZY GRA to rank suppliers and select just one green supplier and (2) multiple sourcing with quantitative supplier's data using linear programming to select and allocate green suppliers of raw ingredients and packaging material food industries.

Generally, this study is organized as follows. Sections 2 and 3 explain greening food industry and reviews pertinent literature on green supplier selection of food industry. Next, Section 4 introduces the applied model and related methods. Section 5 describes the applicability of the proposed approach by food industry experts. Conclusions are finally drawn in Section 6, along with recommendations for future research.

2. Greening food industry

When compared to other, industries of food and agricultural products have some unique characteristics related to public health. Food supply chains can be distinguished into "fresh agricultural products" (e.g. vegetables or fruits) and "processed food products" (e.g. convenience food or soft drinks) (Apaiaha, Linnemanna, & van der Kooi, 2006) which here we focused on it. Resulting product-related issues include but are not limited to "shelf life constraints, variable quantity and quality due to biological variations, seasonality, random factors connected with weather and pests and other biological hazards" (van der Vorst, van Dijk, & Beulens, 2001).

Efforts have consequently been intensified to enable tracking and tracing of materials and ingredients in food supply chains, for example, food safety and traceability standards by the European Union (EU) requiring every ingredient to be traceable (EU, 2002). In supply chain management, to build strategic and strong relationships, food industries should select best suppliers by applying appropriate method and selection criteria.

Food companies are closely monitored by their customers who want to be confident that the food being purchased is safe. In this regard, they are usually asked to pass an independent audit to check whether good manufacturing practice, good hygiene practice, and HACCP (Hazard Analysis and Critical Control Points) principles are fully implemented and strictly followed in each step of the process (Losito, Visciano, Genualdo, & Cardone, 2011). Quality and safety of food ingredients and packaging are paramount to company success, so choosing a good supplier is a critical business decision. Generally, companies in food industry try to develop their supplier assessment form according to HACCP, ISO 9000, and other related regulation and special requirements.

Something that has become apparent is that in the near future, the design and operation of agricultural supply chains will be subject to more stringent regulations and closer monitoring, in particular, those for products destined for human consumption (Bosona & Gebresenbet, 2013). This implies that the traditional supply chain practices may be subject to revision and change. One of the aspects that may be the subject of considerable scrutiny is the supplier selection along the supply chains of food products (Gendron & Audet, 2012).

Beside the conventional factors like quality, safety, and hygienic factors, it should be noted that the food industry is one of the world's largest industrial sectors, and hence has large environmental impacts (Grunert, Hieke, & Wills, 2014). Greenhouse gas (GHG) emission, which has increased remarkably due to tremendous input and energy use, has resulted in global warming, perhaps the most serious problem that humankind faces today (Ilbery & Maye, 2005). With growing worldwide awareness of environmental protection, green food production has become an important issue for almost every producer and will determine the sustainability of producers in the long term. It is the responsibility of the food industry to develop and implement green supplier selection in the first step of green supply chain establishment.

3. Literature review

A food producer needs to buy raw ingredients and packaging material, so suppliers in food industries are divided in two groups. A number of literature reviews have been completed to summarize supplier selection criteria adopted and decision models published (Aissaoui, Haouari, & Hassini, 2007; Chai, Liu, & Ngai, 2013; Govindan et al., in press). Unfortunately, there are few studies which are focused on food supplier selection and none of them suggested environmental criteria. Some information about their criteria and techniques are summarized in Table 1.

In the field of package printing industry, Vachon and Klassen (2006) studied a comparative green supplier selection problem and determined green supplies in Canadian and United States Package Printing Industry. They purposed Green projects partnership as environmental criteria to select green supplier. Also Kumar, Hong, and Haggerty (2011) and Magdalena (2012) studied a global supplier selection process for packaging. Kumar et al. (2011) noted risk of disruption, volume and product flexibility, and innovation as specific criteria for packaging suppliers. This paper attempts to construct a practical model for purchase manager and decision-maker in green supplier selection of food industry. In the latest research of Banaeian, Nielsen, Mobli, and Omid (2014) on GSS in food industry, they used method of Data Envelopment Analysis (DEA) to rank and compare green suppliers, different criteria of price, and carbon footprint as inputs, and delivery and quality as outputs of models.

4. Material and method

4.1. General model (guideline)

Figure 1 shows the structured model. In this paper, we discussed widely about green supplier selection criteria and formed a selection model step by step:

Phase I

- Reviewing general & green supplier selection criteria
- Selection of relevant criteria for green supplier selection

Phase II

- Determining the weights of relevant criteria using AHP
- Determining branches for each relevant criterion and selecting the most effective branch based on Delphi group opinion in raw material and package suppliers

Phase III

- Identifying company requirements and supplier characteristics
- Decision-making for selection of best green supplier in case of single sourcing
- Modeling for selection and allocation of best green suppliers in case of multiple sourcing

Table 1. Supplier selection criteria in food industry.

Authors/Year	Technique	Criteria/Sub-criteria	
Cebi and Bayraktar (2003)	AHP+LGP	Logistic	Lead time, Supply lots, Flexibility, Delivery condition
		Technologic	Capacity, Involvement. Improvement efforts, problem solving
		Business	Reputation, Financial strength, Management,
		Relationship	Communication, Past experience, Sales representation
Shen, Liu, and Tzeng (2012)	DEMATEL+ANP	Cost	Unit cost, operating cost, warranty cost
		Delivery	Delivery performance, delivery speed, fill rate
		Quality	Product quality, steadiness of quality
		Service	Follow-up, process flexibility, professionalism, supply chain responsiveness
		Technology	Ability to solve technical problems, machinery, outlook for the future
		Company image	Attitude, relationship closeness, reputation and financial position
Prusak, Stefanow, Niewczas, and Sikora (2013)	AHP	Finance	Price, Price difference, Time limit for payment
		Assortment	Possible changes, Diversity
		Logistic	Buffer stock, Flexibility, Promptness of supply
		Service	Innovativeness, contact, Acceptable procedure, Customer service, On-line platform
		Quality	The use of SPC (Statistical process control), ISO standards, Technical support, Reaction on problem
Liao, Fu, Chen, and Chih (2012)	Fuzzy MSGP	Financial Stability Product quality Reputation in industry R&D capability Service quality	
Shen, Olfat, Govindan, Khodaverdi, and Diabat (2013)	DEA	Inputs	
		Synthetic cost, Geographical position	
		Outputs	
		Quantitative	Qualitative
		Product qualification ratio, Market occupancy ratio, Order changing received ratio, Deliver punctual Ratio, R&D success rate, R&D return rate, Total assets turnover ratio, Profit increasing ratio	Quality system, enterprise reputation, compatibility, upstream firm's management, traceability system, information level

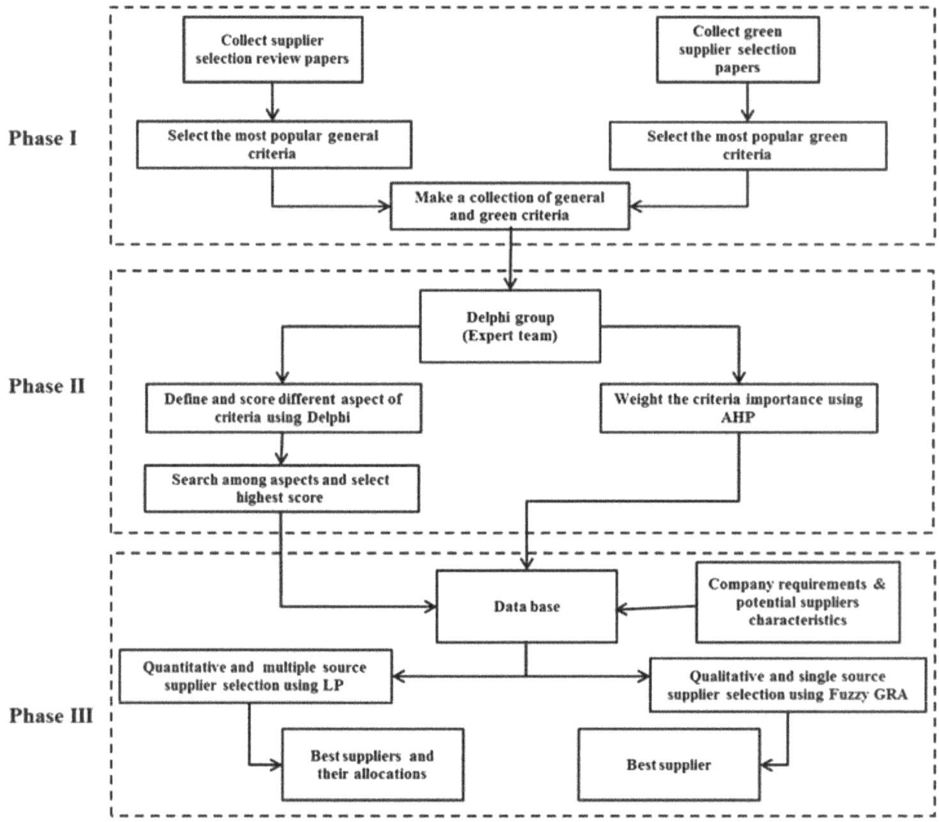

Figure 1. Structure of green criteria definition and supplier selection model.

4.2. Analytical hierarchy process

AHP is a structured technique for organizing and analyzing complex decisions based on paired comparisons of both projects and criteria (Saaty, 1990). This method compares weights in pairs and is more straightforward and easier to use for the decision-makers. The strength of the AHP method lies in its ability to construct complex, multi-person, multi-attribute, and multi-period problem hierarchically. Thus, it has been discussed in a wide variety of decision situations in fields such as business planning, resource allocation, priority setting, and selection among alternatives.

First, under the designed hierarchy model, pairwise comparisons are classically carried out by asking the decision-maker how valuable a criterion (C1) is when compared to another criterion (C2) with respect to overall goal. The verbal judgments of the decision-maker are then transformed into numerical values using the scale presented in Table 2, then a pairwise comparison matrix is constructed.

The pairwise comparison of criteria can be summarized in an $(n*n)$ evaluation matrix $A=(a_{ij})$ for n criteria, where every element $(i, j = 1,2,\ldots,n)$ is the quotient of weights of the criteria. This pairwise comparison can be shown by a square and reciprocal matrix, (see Equation (1)).

Table 2. Verbal judgments and numerical rate.

Verbal judgment of preference	Numerical rate
Equal importance	1
Weal importance of one over another	3
Essential or strong importance	5
Demonstrated importance	7
Absolute importance	9
Intermediate values between the two adjacent judgments	2,4,6,8

$$A = (a_{ij}) = \begin{pmatrix} 1 & a_{12} & \cdots & a_{1n} \\ a_{21} & 1 & \cdots & a_{2n} \\ \vdots & \vdots & \ddots & \vdots \\ a_{n1} & a_{n2} & \cdots & 1 \end{pmatrix}; \tag{1}$$

$$a_{ij} = \begin{Bmatrix} 1 & i = j \\ a_{ij},\ a_{ij}^{-1} & i \neq j \end{Bmatrix}; \tag{2}$$

In order to measure the subjective evaluation of the decision-maker, (Saaty, 1990) proposes a consistency index (CI). CIs are calculated for each pairwise comparison matrix and to check consistency using a random index (RI). If the calculated ratio is significantly small (10% or less), the priorities will accepted otherwise, the DM is asked to revise the pairwise comparisons.

4.3. *Delphi method*

Delphi is a method of popular survey technique that brings consensus of opinions among a set of experts or panelists (informed individuals) by maintaining the unanimity among them (Prusty et al., 2010). In this paper, Delphi method is used for describing sub-criteria based on food industry necessity and selects the most important sub-criteria. Each Delphi panelist was asked to answer two main questions:

First round – According to criteria collection, describe sub-criteria which are important for your supplier selection.

Second round – Score each sub-criteria between 1 (not important) and 9 (very important).

Finally, Delphi score is proposed by Listone and Turoff (2002) as illustrated in Equation (2):

$$\text{Delphi Score} = \frac{\text{Lowest Score} + \text{Highest Score} + (4 \times \text{Average Score})}{6} \tag{3}$$

4.4. *Qualitative and single source green supplier selection*

In decision-making process of supplier selection, the decision-makers try to gather as much information as possible through surveys, investigations, sampling, etc. To reach the aspired decision, sometimes obtaining all the quantitative information is impossible; therefore, decisions are often made in uncertain process, i.e. without complete or quantitative information. In these situations, decision-makers try to apply qualitative models.

Fuzzy theory and sets, first introduced by Zadeh (1965), have been recognized as a powerful tool for modeling systems with imprecise input data, such as the decision-maker opinions. Linguistics terms are used to express the variables by means of fuzzy sets and membership function. Triangular fuzzy number (TFN) has been the most popular form to present fuzzy numbers which represented with three points; $\tilde{A} = (l, m, u)$. A model based on subjective preferences of a decision-maker is not always accurate as it demands a great deal of knowledge, expertise, and experience (Li, Yamaguchi, & Nagai, 2008). This is where "fuzzy" approaches are used in MCDM problems. Incorporating fuzzy sets in decision-making models can produce more realistic outcomes as they aim to transform uncertain human knowledge into a mathematical formula (Zadeh, 1965).

Numerous MCDM techniques have been proposed (Chai et al., 2013) to tackle the problem of supplier selection in order to rank them in the presence of qualitative data. Grey Relational Analysis (GRA), originally proposed by Dong (1982), is applied to imprecise information in the form of interval values and also aims to show the degree of similarity or difference of development trends between an alternative and the reference (ideal) alternative (Kuo & Liang, 2011). Grey system theory has shown to be able to handle both incomplete information and ambiguities (Hu, Zhang, Hu, & Jiang, 2010).

Therefore, combination of Fuzzy logic and GRA is a good choice for single sourcing when quantitative and precise data are not available; so Fuzzy GRA would allow for effective qualitative assessment.

4.5. Quantitative and multiple source Green Supplier Selection

The incorporation of environmental criteria into supplier selection practices is gaining increasing importance. By increasing GSS studies and development of its concept, the applications of GSS in companies are growing quickly. Therefore in future, measuring of many quantitative criteria is possible. Future GSS researches can consider other important environmental criteria (like Carbon Footprint, GHG emissions, Life Cycle Assessment of products, etc.) which currently have been focused on. These kinds of environmental subjects will provide more practical results in GSS, provided that suppliers trace its environmental footprint accurately.

The purchaser only needs to make one decision, which supplier is the best. However, in multiple sourcing, purchasers need to decide which are the best suppliers and how much order should be placed on each selected supplier. It is based on firm's condition and depends on manager's decision directly. Linear programming is the easiest way which helps managers to select multiple source suppliers and their allocations, otherwise by changing constraints, managers are able to select single source. Linear programming (LP; also called linear optimization) is a method to achieve the best outcome (such as maximum profit or lowest cost) in a mathematical model whose requirements are represented by linear relationships. This method was proposed by Ghodsypour and O'Brien (1998) for the first time to indicate best suppliers and their allocations in multiple sourcing models.

5. Results

5.1. Criteria collection (Phase I)

One of the most significant business decisions faced by purchasing managers in a supply chain is the selection of appropriate suppliers while trying to satisfy multiple criteria based on price, quality, customer service, and delivery (Arikan, in press). Hence, supplier selection by its nature involves the need to trade-off multiple criteria, as well as the presence of both quantitative and qualitative data. A popular area of research in the field of supply chain was establishing the criteria for supplier selection and evaluation since the 1960s. If we look into the history of popular criterion identified in the literature, we can conclude three primary categories of emphasis: in the late 1970s and early 1980s, cost was the main focus; then in the early 1990s, cycle time and customer responsiveness were considered, and, finally, in the late 1990s, the focus shifted to flexibility. Now, environmental factors are a key issue which gives rise to the new paradigm of focusing on green supply chains (Haung & Keskar, 2007).

5.1.1. General criteria

Traditionally, companies only focus on their internal environmental measures of waste reduction, pollution control, and hazardous material control to comply with existing regulations and address their social responsibilities. These days, companies have started to realize that they have to do something more than these internal controls, so they are responsible for environmental concerns about supplier selection, product distribution, etc. (Chen et al., 2013).

Many previous studies on supplier selection and evaluation defined numerous evaluation criteria and selection models for supplier selection. In consideration of the criteria for supplier selection, the pathfinder work by Dickson (1966) has been one of the most cited studies.

Weber, Current, and Benton (1991) re-examined Dickson's work by reviewing published articles between 1966 and 1990. Twenty-three distinct criteria are identified in various supplier selection problems in Dickson (1966) and Weber et al. (1991) Among the 23 criteria identified for vendor selection, the product quality was ranked as most important, it was followed by on-time delivery, performance history of supplier and warranties and claimed policies, and so on.

In a later work, Cheraghi, Dadashzadeh, and Subramanian (2004) continued to extend these key players initial work to obtain the current perspective of supplier selection by analyzing articles published between 1990 and 2001. Cheraghi et al. (2004) provided an update of Dickson's seminal work with 13 more criteria. Also, Thiruchelvam and Tookey (2011) reviewed supplier selection studies and comprise the selection criteria between two periods of 1966–2001 and 2001–2010.

Previous studies indicate that quality, price, and delivery are general (traditional) key criteria that consistently appear for vendor selection (Wu, Zhang, Wu, & Olson, 2010).

Chang, Chang, and Wu (2011) used DEMATEL to find influential factors in selecting SCM suppliers and found that "technology ability", "stable delivery of goods", "lead-time", and "production capability" criteria are more influential than other evaluation criteria. They also suggested extending the scope of this study and exploring the addition of a green supply chain in future studies.

Table 3 shows summary results of review paper on most important general criteria on supplier selection. Often each of these important criteria contains several sub-criteria which are described completely in past literature.

5.1.2. Environmental criteria

The past few years have led researchers to investigate the environmental concepts in management and supply chains. We comprehensively collected the literature associated with the descriptors "Green/Environmental supplier selection," "vendor selection," and

Table 3. Most important general criteria on conventional supplier selection.

Authors	The most important criteria as a result of review papers
Dickson (1966)	Quality, delivery, performance history
Weber et al. (1991)	Price, delivery, facilities and capacity, geographic location, technology capability
Cheraghi et al. (2004)	Quality, delivery, price, repair service, technical capability
Ho et al. (2010)	Quality, delivery, price/cost, manufacturing capability, service, management, technology
Thiruchelvam and Tookey (2011)	Quality, delivery, price

from academic databases including Science Direct, Emerald, Springer-Link Journals, IEEE Xplore, Academic Search Premier, and World Scientific Net. After a methodological decision analysis of all collected articles, we reviewed 59 international journal articles published to 2014. This review of literature provides the basis for identification and definition of green supplier selection criteria.

As you can see in Figure 2, most of empirical studies on green supplier selection are concentrated on electronic (42%) and automotive industries (19%).

So for exact studies on other industries, specialized consultant is necessary to choose suitable criteria. Unfortunately, agriculture and food industry is neglected from the vision of researchers. Regarding to high environmental impacts on agriculture and food industry (Pluimers, Kroeze, Bakker, Calla, & Hordijik, 2000), it is necessary to pay attention more on this field.

Based on reviewing papers, widespread set of criteria were purposed. Unfortunately, lack of consistent studies about basic green criteria caused widespread number of green criteria in each study. Jabbour and Jabbour (2009) concluded there is no standard for criteria or for use on the part of analyzed companies. Nonetheless, evolving criteria have

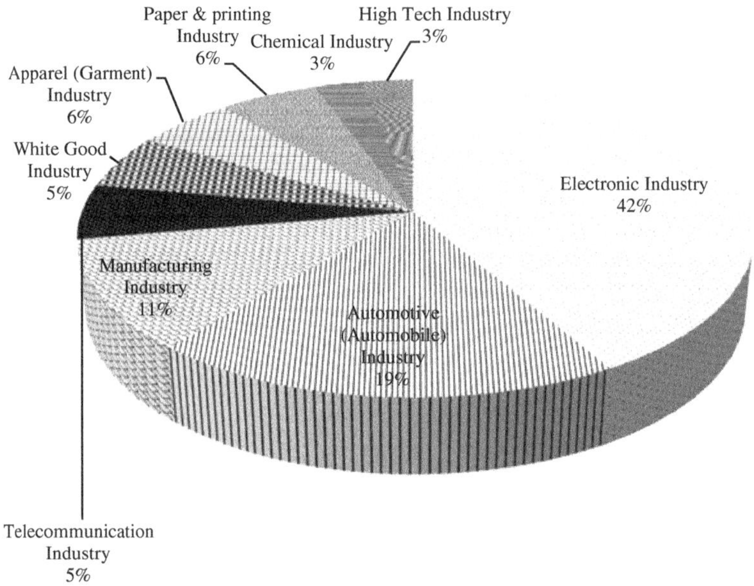

Figure 2. GSS position among different industries.

emerged due to competitive market as number of suppliers has increased over the years. Therefore, inception of new criteria is essential to narrow and qualify not only deserving but exceptional suppliers. Also under the current rapid GSCM transformation edge, reliability and flexibility of each supplier are considered as key contributing factors (Chang et al., 2011).

Govindan et al. (in press) reviewed 33 papers on green supplier and introduced relevant selection methods and criteria. The most popular approach in green supplier selection was AHP followed by ANP, DEA, also EMS was the most popular environmental criteria (Table 4).

In two comprehensive review paper (Govindan et al., in press; Nielsen, Banaeian, Golinska, Mobli, & Omid, 2014), researcher collected environmental criteria on supplier selection literature (Table 4). The most important green criteria derived from 57 review articles are: environmental management system (EMS) (20 frequencies), green image (8 frequencies), environmental competences (7 frequencies), design for environment (5 frequencies), environmental improvement costs (5 frequencies), and environmental performance (5 frequencies).

5.2. Criteria selection (Phase II)

According to Figure 1, Phase II investigate calculate the weights of criteria collected in previous section using expert team, also sub-criteria which are more important and influential in edible oil production are introduced.

5.2.1. Expert team

Due to many factors to be considered, purchasing managers might choose an easier path of evaluating based on certain important criteria and ignoring the rest. A larger number of attributes also contribute to the complications of being assigned with consistent and meaningful weights. The selection decision also becomes more difficult as the number of business rules that must be considered increases.

The Expert team is applied first to weighting criteria, second to define related sub-criteria and third for determination of most effective sub-criteria. Handfield, Walton, Sroufe, and Melnyk (2002) developed an expert team of environmental experts and noted that for better results in future studies, group should include purchasing managers who may not be directly familiar with environmental metrics. A good combination of experts will help to ensure that the tool is practical from an actual user's standpoint (Handfield et al., 2002). In order to obtain a rational framework, the model was developed through an expert team (10 people) described in Table 5. However, experts had a variety of titles, but were responsible for materials management and waste reduction activities within food processing organizations.

5.2.2. Weighting criteria

Weight of each criterion can be concluded from the paired comparison between the criteria using the AHP method, which represents the importance of criteria, is determined by means of a nine-point Likert scale.

The AHP method is often combined with Fuzzy Linear Programming in the decision-making process (Cebi & Bayraktar, 2003; Tektas & Aytekin, in press, etc.); the difference is that in relevant studies, AHP is used for estimating criteria situation in each supplier, but here, we used it for weighting criteria and tried to use quantified data for each supplier.

A questionnaire, which is based on the format proposed by Humphreys, Wong, and Chan (2003), ought to be carried out to determine the number of criteria. According to verbal judgement (Table 1), pairwise comparison of criteria are done by expert team. Finally using AHP method, criteria weights in raw material and package are calculated and shown in Table 6.

5.2.3. Sub-criteria description

The most relevant criteria are selected from reviewed papers (Cheraghi et al., 2004; Dickson, 1966; Ho et al., 2010; Thiruchelvam & Tookey, 2011; Weber et al., 1991) on supplier selection criteria (quality, price, and finally service and delivery). Main green criteria is selected based on reviewing green supplier selection papers (EMS), as (Govindan et al., in press) determined EMS as the most popular criteria on green supplier selection.

Since the lack of sufficient information about green supplier selection in food industries, expert team (Delphi group) determines sub-criteria based on the nature of industry. They also emphasized that in a practical model, measures should be relatively easy to assess, as commodity strategy development teams often do not have access to all the requisite information. Finally, each criterion is described by several sub-criteria as mentioned follow (Table 7). For example in criteria A, Capital and financial power of supplier company (A1) proposed raw material price (A2) and Transportation cost to the geographical location (A3) were the most important sub-criteria from the financial point of view.

Quality for food package material depends on durability, shelf life, interaction with content, density, etc., but based on expert opinion, the highest difference between package suppliers is providing precision and intact items to reduce leakage and waste rate. Also preserving nutritive and qualitative characters (Shelf life) is determined as quality factor for raw ingredients.

5.2.4. Conclusive criteria selection

Determining sub-criteria will help the Delphi group to decide more precision. The group was asked to score each sub-criteria for raw material and packaging supplier separately (It can be seen in Table 8). Delphi results of financial criteria shows that, capital and financial power of supplier in selection of raw ingredients and proposed price in

Table 4. Frequency of most important environmental criteria on conventional supplier selection.

	Govindan et al. (in press)	Nielsen et al. (2014)
Duration	1996–2011	1996–2014
Number of reviewed paper	33	57
Criterion		
Environmental Management System	11	20
Green Image	4	8
Environmental Competences	3	6
Design for environment	3	5
Environmental Improvement Costs	2	5
Environmental Performance	3	5

selection of packaging material were the most important sub-criteria in financial aspect. On-time delivery (lead time) and quality are most important in service and qualitative criteria in both raw ingredients and packaging materials. Mean Delphi score shows that there are some differences between raw ingredients and packaging materials from the aspect of EMS criteria. Environmental prerequisite and environmental friendly material are the most important for the raw ingredients and packaging materials, respectively. The greenest concerning subject in choosing packaging material is easily recycling capability; but in the food ingredient, training special courses for staff training are the most important criteria. For suppliers claiming to be green, the most tangible activity is awarding at least one environmental certificate or labels (like ISO14000, eco-friendly label and carbon footprint label, etc.), so we consider this term as first provision for green supplier selection.

5.3. Supplier selection (Phase III)

After identification of criteria and their weight (the main goal of this paper), decision-makers have to decide about supplier selection methodology. According to Sections 4.3 and 4.4, methods are different and should select based on company requirements and data characteristics.

5.3.1. Ranking and supplier selection by Fuzzy GRA

Fuzzy GRA works as follows. Assume that m potential suppliers are determined, n selected criteria (here based on the previous section, four selected sub-criteria in Table 8), and k decision-makers (here based on Table 5, 10 decision-makers). According to Table 9, the appropriate linguistic ratings can determine for suppliers with respect to criteria ($\tilde{x}_{ij}$) as TFN. It means that each decision-maker has to rank importance degree of n criteria in m suppliers (Overall $n \times m \times k$ linguistic variables) for raw material and packaging supplier separately.

Table 5. Summary of Delphi group.

Expert occupation	Experience	Total expert
Active member of Food Industry Association (president)	Professional and academic	4
Purchase manager in food industry	Professional and academic	4
Environmental technician	Professional and academic	2

Table 6. Criteria weight in raw material and package.

	Raw material	Package
Financial (w_1)	0.04	0.09
Service (w_2)	0.21	0.12
Quantitative (w_3)	0.65	0.51
EMS (w_4)	0.10	0.28

Table 7. Sub-criteria description based on expert team opinion.

Criteria	Sub-criteria
A- Financial	A1- Capital and financial power of supplier company A2- Proposed raw material price A3- Transportation cost to the geographical location (Availability)
B- Delivery & service	B1- Communication System (willing to trade, attitudes, acceptance of procedures and flexibility) B2- On time delivery (lead time) B3- After sales service (policy, quality assurance, and damage ratings) B4- Production capacity
C- Qualitative	C1- Quality (the supplier ability to access the quality characteristics) C2- Operational control (reporting, quality control, inventory control, research and development) C3- Expert labor, technical capabilities and facilities C4- Business experience and Position among competitors
D- Environmental management system	D1- Environmental prerequisite (environmental staff training) D2- Environmental planning (program to reduce environmental impacts, green research and development) D3- Environmental friendly material (low waste: easily recycling and reuse capability) D4- Environmental friendly technology (emission of pollutant: CO_2 equivalent, VOC, BOD and COD content and etc.)

Table 8. Mean Delphi score of experts.

		Delphi score	
Criteria	Sub-criteria	Raw ingredients	Packaging material
A	A1	24.5	21.5
	A2	22.5	23.5
	A3	21	20.5
B	B1	19.8	18.8
	B2	27.2	26.6
	B3	19.5	21.5
	B4	20.5	22.7
C	C1	30.9	18.9
	C2	11.3	11.2
	C3	14.6	16.4
	C4	15.4	17.3
D	D1	6.11	9
	D2	5.5	7.3
	D3	4.4	12.3
	D4	5.8	8.3

In the initial step of the selected multi-criteria supplier selection methods, aggregated ratings of suppliers (decision matrix) are constructed. The aggregated fuzzy rating $\tilde{r}_{ij}$ of alternative A_i under criterion C_j is:

$$\tilde{x}_{ij} = \frac{l}{k}\left[\tilde{x}_{ij}^{1} + \tilde{x}_{ij}^{2} + \cdots + \tilde{x}_{ij}^{k}\right] \quad (4)$$

The aggregated decision matrix can then be constructed as follows:

$$\tilde{X} = \begin{matrix} A_1 \\ A_2 \\ \vdots \\ A_m \end{matrix} \begin{bmatrix} \tilde{x}_{11} & \tilde{x}_{12} & \cdots & \tilde{x}_{1n} \\ \tilde{x}_{21} & \tilde{x}_{22} & \cdots & \tilde{x}_{2n} \\ \vdots & \vdots & \vdots & \vdots \\ \tilde{x}_{m1} & \tilde{x}_{m2} & \cdots & \tilde{x}_{mn} \end{bmatrix}; \tag{5}$$

Decision matrix ($\tilde{X}$) is converted into the normalized decision matrix $\tilde{R}$. Given $\tilde{r}_{ij} = (l_{ij}, m_{ij}, u_{ij})$, the normalized performance rating can be calculated as (Gumus, Yayla, Celik, & Yildiz, 2013; Zhang & Liu, 2011):

$$\tilde{r}_{ij} = \left(l_{ij}/u_j^+, m_{ij}/u_j^+, u_{ij}/u_j^+\right), i = 1, \ldots, m, \quad j = 1, \ldots, n \tag{6}$$

$$\text{where } u_j^+ = \max_i\{u_{ij}\} \quad \forall ii = 1, \ldots, m \tag{7}$$

The reference series can be determined from (Gumus et al., 2013):

$$\tilde{R}_0 = [\tilde{r}_{01}, \tilde{r}_{02}, \tilde{r}_{03}, \tilde{r}_{04}] \text{ where } \tilde{r}_{0j} = \max(\tilde{r}_{ij})\, j = 1, \ldots, 4 \tag{8}$$

The distance matrix is then established. The distance δ_{ij} between the reference value and each comparison value is computed from Equation (7) of TFNs. The distance between $\tilde{A}$ and $\tilde{B}$ is defined as follows (Krohling & Campanharo, 2011; Zhang & Liu, 2011):

$$d(\tilde{A}, \tilde{B}) = \sqrt{\frac{1}{3}\left[(l_1 - l_2)^2 + (m_1 + m_2)^2 + (u_1 - u_2)^2\right]} \tag{9}$$

The grey relational coefficient (ξ_{ij}) is defined as (Gumus et al., 2013):

$$\xi_{ij} = \frac{\delta_{\min} + \rho\delta_{\max}}{\delta_{ij} + \rho\delta_{\max}}, \ \delta_{\max} = \max(\delta_{ij}), \ \delta_{\min} = \min(\delta_{ij}) \text{ and } \rho \text{ resolving coefficient } \rho \in [0, 1] \tag{10}$$

Defuzzification formula (Definition 4) is applied to $\tilde{w}_j$; then the grey relational grade γ_i is estimated by the relation (Gumus et al., 2013):

$$\gamma_i = \sum_{j=1}^{n} w_j \xi_{ij}, \quad i = 1, \ldots, m \tag{11}$$

Table 9. Linguistic variables for supplier ratings.

Number	Linguistic variables (degree of importance)	TFNs
1	Very Poor (VP)	(0, 1, 2)
2	Poor (P)	(1, 2, 3)
3	Medium Poor (MP)	(2, 3.5, 5)
4	Fair (F)	(4, 5, 6)
5	Medium Good (MG)	(5, 6.5, 8)
6	Good (G)	(7, 8, 9)
7	Very Good (VG)	(8, 9, 10)

where w_j is the weight of the jth criterion, $\sum_{j=1}^{n} w_j = 1$ which are presented in Table 5. The alternatives can now be ranked in accordance with the value of grey relational grade. The bigger the value, the better would be the alternatives.

5.3.2. Supplier selection and allocation by programming

An environmental objective function was added to the proposed model by Ghodsypour and O'brien (2001). Thus, our model is a multi-objective model considering four objectives of financial, quantity, service, and EMS issues for green supplier selection in food industry. The presented model can select the best suppliers and allocate the orders among them optimally too. Consequently, one objective function and five constrains are considered to calculate Total Purchase Value (TPV). The elements of model and constraints are presented below:

All suppliers have limited capacities and minimum order quantity (13, 9).
Total order quantity must be equal the amount of demand (14).
Summation of supplier's capacity must be above the amount of demand (15).
Since managerial matter, also there is a limitation on number of selected suppliers (16).
The parameters of the proposed model are defined as follows:

$$\text{Max TPV} = \sum_{i=1}^{n} \left[Y_i Z_i \left(w_1 F_i^* + w_2 S_i + w_3 Q_i^* + w_4 E_i^* \right) \right] * X_i \tag{12}$$

$$ST : X_i \leq C_i Y_i i = 1, 2, \ldots, n \tag{13}$$

$$\sum_{i=1}^{n} X_i = D \tag{14}$$

$$\sum_{i=1}^{n} C_i \geq D \tag{15}$$

$$n_{\min} \leq \sum_{i=1}^{n} Y_i \leq n_{\max} \tag{16}$$

$$X_i \geq Q_{i\min} Y_i i = 1, 2, \ldots, n \tag{17}$$

Base on criteria selection process, the model includes four objective functions to optimize the four criteria for green supplier selection: financial, service, qualitative, and EMS of suppliers are expressed as follows:

$$F_i^* : \begin{cases} \text{for raw ingredients} = \frac{F_i}{\text{Max}F}; F_i : \text{Capital of supplier} \\ \text{for packaging materials} = \frac{\text{Min}F}{F_i}; F_i : \text{Proposed price by supplier } i \end{cases} \tag{18}$$

$$E_i^* : \begin{cases} \text{for raw ingredients} = \dfrac{\text{Number of trained staff}}{\text{Number of total staff}} \\ \text{for packaging material} = \dfrac{\text{Amount of collected package from consumer}}{\text{Total amount of production}} \end{cases} \tag{19}$$

$$Q_i^* : \begin{cases} \text{for raw ingredients} = \frac{Q_i}{\text{Max } Q}; Q_i = \text{Shelf life of items provided by supplier } i \\ \text{for packaging material} = \text{percent of intact items delivered by supplier } i \end{cases}$$

$S_i = \frac{S_i}{\text{Max}} S$; S_i: Lead time (day) items provided by the supplier i

n: Number of suppliers

D: The amount of demand for the period

X_i: Order quantity for the supplier i

$Q_{i\ \min}$: Minimum order quantity of the ith supplier

C_i: The capacity of the supplier i

Y_i: 1 if the supplier is selected; 0 otherwise

Z_i: 1 if the supplier award environmental certificate; 0 otherwise

$W_{1,2,3,4}$: Represent the weight of each criterion using the AHP method as described completely in previous section which is different for raw ingredients and packaging materials.

6. Conclusion

GSS literature reviewed in this paper showed that most of papers are focused on automotive, electronics, and manufacturing industries. The principal benefit of this paper is to provide an extended GSS guideline (from the definition of criteria to GSS) to help decision-makers in various industries often neglected in previous GSS studies.

Food and agricultural products have some unique characteristics related to public health, which has not been properly investigated in GSS literature. Green purchase is an inclusive concept and people may attach different meanings to it, especially among multifarious number of environmental management actions in food industry. So in this study, an expert team tried to define easy understanding criteria and sub-criteria and also select easy access ones in a food company.

As the main goal, this study tried to show the single platform in order to choice criteria and sub-criteria in GSS and calculate the weights of criteria, applied in an edible oil company. In addition, an example of different ways to GSS (single and multiple sourcing) is also briefly illustrated.

In this paper, general and environmental criteria are systematically selected: general and green supplier selection criteria are reviewed comprehensively, and relevant criteria for green supplier selection are selected. In order to obtain a rational framework, the model was developed through an expert team (including Active members of Food Industry Association, purchase managers and environmental technicians) to find most suitable criteria and their weights. The weights of relevant criteria using AHP are determined, also branches for each relevant criterion and selecting the most effective branch based on Delphi group opinion in raw material and package suppliers are calculated.

In the section of supplier selection, two kinds of qualitative and quantitative data are discussed when single and multiple sourcing are required, respectively. Fuzzy GRA is proposed to ranking suppliers in presence of qualitative and imprecise data. Also programming is used to present a model which can select the best suppliers and allocate the orders among them optimally.

There are certain limitations for this study that can provide opportunities for further research in this area. While it is believed that the presented model provides operational value in criteria selection, there are also further points in supplier selection that can be included. Application of single and multiple sourcing of GSS (phase III) will be study in future papers. Additional research is required to identify and define more clearly specific criteria in different products and industries.

Acknowledgments

The financial support provided by University of Tehran is duly gratefully acknowledged. Thanks and appreciations of authors also go to colleagues in developing the research and people who have willingly helped us with their abilities.

References

Aissaoui, N., Haouari, M., & Hassini, E. (2007). Supplier selection and order lot sizing modeling: A review. *Computers & Operations Research, 34*, 3516–3540.

Apaiaha, R. K., Linnemanna, A. R., & van der Kooi, H. J. (2006). Exergy analysis: A tool to study the sustainability of food supply chains. *Food Research International, 39*(1), 1–11.

Arikan, F. (in press). An interactive solution approach for multiple objective supplier selection problem with fuzzy parameters. *Journal of Intelligent Manufacturing*. doi:10.1007/s10845-013-0782-6

Banaeian, N., Nielsen, I. E., Mobli, H., & Omid, M. (2014). Green supplier selection in edible oil production by a hybrid model using Delphi method and Green Data Envelopment Analysis (GDEA). *Management and Production Engineering Review, 5*, 3–8.

Bocewicz, G., Wójcik, R., & Banaszak, Z. (2012). Cyclic scheduling for supply chain network. *Trends in practical applications of agents and multiagent systems* (pp. 39–47). Springer Berlin Heidelberg.

Bosona, T., & Gebresenbet, G. (2013). Food traceability as an integral part of logistics management in food and agricultural supply chain. *Food control, 33*, 32–48.

Cebi, F., & Bayraktar, D. (2003). An integrated approach for supplier selection. *Logistics Information Management, 16*, 395–400.

Chai, J., Liu, J. N. K., & Ngai, E. W. T. (2013). Application of decision-making techniques in supplier selection: A systematic review of literature. *Expert Systems with Applications, 40*, 3872–3885.

Chang, B., Chang, C.-W., & Wu, C.-H. (2011). Fuzzy DEMATEL method for developing supplier selection criteria. *Expert Systems with Applications, 38*, 1850–1858.

Chen, Y., Simon, R., Reich-Weiser, C., & Woo, J. (2013). Green supply chain. In D. A. Donrfeld (Ed.), *Green manufacturing: Fundamental and applications* (pp. 83–105). Springer.

Cheraghi, S. H., Dadashzadeh, M., & Subramanian, M. (2004). Critical success factors for supplier selection: An update. *Journal of Applied Business Research, 20*, 91–108.

Dickson, G. W. (1966). An analysis of vendor selection system and decisions. *Journal of Purchasing, 2*, 5–17.

Dong, J. L. (1982). Control problems of grey systems. *Systems and Controls Letters, 5* 288–294.

EU. (2002). Regulation (EC) no 178/2002 of the European parliment and of the council of 28 January 2002. *Official Journal of the European Commiunities, L31*, 1–24.

Gendron, C., & Audet, R. (2012). Key drivers of the food chain. In J. L. Boye & Y. Arcand (Eds.), *Green technologies in food production and processing* (pp. 23–39). Food Engineering Series. New York, NY: Springer.

Ghodsypour, S. H., & O'brien, C. (2001). The total cost of logistics in supplier selection, under conditions of multiple sourcing, multiple criteria and capacity constraints. *International Journal of Production Economics, 73*, 15–27.

Ghodsypour, S. H., & O'Brien, C. (1998). A decision support system for supplier selection using an integrated analytic hierarchy process and linear programming. *International Journal of Production Economics, 56–57*, 199–212.

Govindan, K., Rajendran, S., Sarkis, J., & Murugesan, P. (in press). Multi criteria decision making approaches for green supplier evaluation and selection: A literature review. *Journal of Cleaner Production*. doi:10.1016/j.jclepro.2013.06.046

Grunert, K. G., Hieke, S., & Wills, J. (2014). Sustainability labels on food products: Consumer motivation, understanding and use. *Food Policy, 44*, 177–189.

Gumus, A. T., Yayla, A. Y., Celik, E., & Yildiz, A. (2013). A combined fuzzy-AHP and Fuzzy-GRA methodology hydrogen energy storage method selection in Turkey. *Energies, 6*, 3017–3032.
Handfield, R., Walton, S. V., Sroufe, R., & Melnyk, S. A. (2002). Applying environmental criteria to supplier assessment: A study in the application of the analytical hierarchy process. *European Journal of Operational Research, 141*, 70–87.
Harland, C. M. (1996). Supply chain management, purchasing and supply management, logistics, vertical integration, materials management and supply chain dynamics. *Blackwell encyclopedic dictionary of operations management*. Oxford, UK: Blackwell.
Haung, S. H., & Keskar, H. (2007). Comprehensive and configurable metrics for supplier selection. *International Journal of Production Economics, 105*, 510–523.
Ho, W., Xu, X., & Dey, P. K. (2010). Multi-criteria decision making approaches for supplier evaluation and selection: A literature review. *European Journal of Operational Research, 202*, 16–24.
Hu, L., Zhang, C., Hu, C., & Jiang, G. (2010). Use of grey system for assessment of drinking water quality: A case study of Jiaozuo City, China. In S. Liu & J. Y. L. Forrest (Eds.), *Advance in Grey Systems* (pp. 469–480). Berlin Heidelberg, Germany: Springer.
Humphreys, P. K., Wong, Y. K., & Chan, F. T. S. (2003). Integrating environmental criteria into the supplier selection process. *Journal of Materials Processing Technology, 138*, 349–356.
Ilbery, B., & Maye, D. (2005). Food supply chains and sustainability: Evidence from specialist food producers in the Scottish/English borders. *Land Use Policy, 22*, 331–344.
Jabbour, A. B. L., & Jabbour, C. J. (2009). Are supplie selection criteria going green? Case studies of companies in Brazil. *Industrial Managemant & Data Systems, 109*, 477–495.
Krohling, R. A., & Campanharo, V. C. (2011). Fuzzy TOPSIS for group decision making: A case study for accidents with oil spill in the sea. *Expert Systems with Applications, 38*, 4190–4197.
Kumar, S., Hong, Q. S., & Haggerty, L. N. (2011). A global supplier selection process for food packaging. *Journal of Manufacturing Technology Management, 22*, 241–260.
Kuo, M. S., & Liang, G. S. (2011). Combining VIKOR with GRA techniques to evaluate service quality of airports under fuzzy environment. *Expert Systems with Applications, 38*, 1304–1312.
Li, G. D., Yamaguchi, D., & Nagai, M. A. (2008). A grey-based rough decision-making approach to supplier selection. *The International Journal of Advanced Manufacturing Technology, 36*, 1032–1040.
Liao, C. N., Fu, Y. K., Chen, Y. C., & Chih, L. (2012). Applying Fuzzy-MSGP approach for supplier evaluation and selection in food industry. *African Journal of Agricultural Research, 5*, 726–740.
Listone, H. A., & Turoff, M. (2002). *The Delphi Method: Techniques and Applications. College of Computing Sciences*. Newark, NJ: New Jersey Institute of Technology.
Losito, P., Visciano, P., Genualdo, M., & Cardone, G. (2011). Food supplier qualification by an italian large-scale-Distributor: Auditing system and non-conformances. *Food Control, 22*, 2047–2051.
Magdalena, R. (2012). Supplier selection for food industry: A combination of Taguchi loss function and fuzzy analytical hierarchy process. *The Asian Journal of Technology Management V, 5*, 13–22.
Nielsen, I. E., Banaeian, N., Golinska, P., Mobli, H., & Omid, M. (2014). Green supplier selection criteria: From a literature review to a flexible framework for determination of suitable criteria. In P. Golinska (Ed.), *Logistic operations, supply chain management and sustainability* (pp. 79–100). Springer.
Olugu, E. U., Wong, K. Y., & Shaharoun, A. M. (2011). Development of key performance measures for the automobile green supply chain. *Resource, Conservation and Recycling, 55*, 567–579.
Pluimers, J. C., Kroeze, C., Bakker, E. J., Calla, H., & Hordijik, L. (2000). Quantifying the environmental impact of production in agriculture and horticulture in The Netherlands: Which emissions do we need to consider? *Agricultural Systems, 66*, 167–189.
Prusak, A., Stefanów, P., Niewczas, M., & Sikora, T. (2013). Application of the AHP in evaluation and selection of suppliers, 57th EOQ congress quality renaissance – co-creating a viable future, June 17–20, 2013, Tallinn, Estonia.
Prusty, S. K., Pratap, K. J., Mohapatra, C. K., & Mukherjee (2010). GOS tree (Goal–Objective–Strategy tree) approach to strategic planning using a fuzzy-Delphi process: An application to the Indian Shrimp Industry. *Technological Forecasting and Social Change, 77*, 442–456.

Saaty, T. L. (1990). How to make a decision: The analytic hierarchy process. *European Journal of Operational Research, 48*, 9–26.

Schoenfeldt, T. I. (2008). Make or buy: The first step in supply chain management. In P. O'Mara (Ed.), *A practical application of supply chain management principles* (pp. 53–59). Milwaukee, WI: American Society for Quality Press.

Shen, J. L., Liu, Y. M., & Tzeng, Y. L. (2012). The cluster-weighted DEMATEL with ANP method for supplier selection in food industry. *Journal of Advanced Computational Intelligence and Intelligent Informatics, 16*, 567–575.

Shen, L., Olfat, L., Govindan, K., Khodaverdi, R., & Diabat, A. (2013). A fuzzy multi criteria approach for evaluating green supplier's performance in green supply chain with linguistic preferences. *Resources, Conservation and Recycling, 74*, 170–179.

Sirvastava, S. K. (2007). Green supply-chain management: A state-of-the-art literature review. *International Journal of Management Reviews, 9*, 53–80.

Sitek, P., & Wikarek, J. (2014). Hybrid solution framework for supply chain problems. Distributed computing and artificial intelligence (DCAI 2014). *Book Series: Advances in Intelligent Systems and Computing, 290*, 11–18.

Tektas, A., & Aytekin, A. (in press). Supplier selection in the international environment: A comparative case of a Turkish and an Australian company. *IBIMA Publishing*. doi:10.5171/2011.598845

Thiruchelvam, S., & Tookey, J. E. (2011). Evolving trends of supplier selection criteria and methods. *International Journal of Automotive and Mechanical Engineering, 4*, 437–454.

Toloo, M. (2014). Selecting and full ranking suppliers with imprecise data: A new DEA method. *The International Journal of Advanced Manufacturing Technology, 74*, 1141–1148.

Vachon, S., & Klassen, R. D. (2006). Green project partnership in the supply chain: The case of the package printing industry. *Journal of Cleaner Production, 14*, 661–671.

van der Vorst, J. G. A. J., van Dijk, S. J., & Beulens, A. J. M. (2001). Supply chain design in the food industry. *The International Journal of Logistics Management, 12*, 73–86.

Weber, C., Current, J., & Benton, W. (1991). Vendor selection criteria and methods. *European Journal of Operation Research, 50*, 2–18.

Wikipedia contributors, S. C. M. (2014). Wikipedia, The Free Encyclopedia. Retrieved 30 June, 2014, from http://en.wikipedia.org/wiki/Supply_chain_management#cite_note-11

Wu, D. D., Zhang, Y., Wu, D., & Olson, D. L. (2010). Fuzzy multi-objective programming for supplier selection and risk modeling: A possibility approach. *European Journal of Operational Research, 200*, 774–787.

Zadeh, L. A. (1965). Fuzzysets. *Inform Control, 8*, 338–353.

Zhang, S. F., & Liu, S. Y. (2011). A GRA-based intuitionistic fuzzy multi-criteria group decision making method for personnel selection. *Expert Systems with Applications, 38*, 11401–11405.

11

A review of turning of hard steels used in bearing and automotive applications

Suha Karim Shihab*, Zahid A. Khan, Aas Mohammad and Arshad Noor Siddiquee

Department of Mechanical Engineering, Jamia Millia Islamia (A Central University), MMA Jauhar Marg, New Delhi 110025, India

Hard turning is a recent technology that involves machining of hard steels using modern machine tools. Hard machining presents challenges in terms of selection of tool insert with improved tool life and high-precision machining. Turning of hardened steels using single-point cutting tool has got considerable interest among manufacturers of ball bearings, automotive, gear, and die industry. Hard turning in the automotive industry and bearing applications typically has a number of potential benefits over traditional form grinding including lower equipment costs, shorter setup time, and fewer process steps which in turn provides high flexibility and ability to cut complex geometries. Moreover, the hard turning process is usually carried out without coolant/lubricant and thus, the problem of storage, handling, and disposal of cutting fluid is eliminated, and at the same time, it probably favors the health of operators. This paper presents an overview of the past research hard turning using hard turning tools such as PCBN, cubic boron nitride, Ceramics, Carbide, etc. Major hard turning cutting materials and effect of hard turning process parameters on cutting forces, heat generation during cutting, surface finish and surface integrity, and tool wear have been discussed in light of the findings of the past research.

Keywords: hard turning; hard tool materials; cutting parameters

1. Introduction

High-hardness materials include various hardened alloy steels, tool steels, case-hardened steels, super alloys, nitrided steels, hard-chrome coated steels, and heat-treated powder metallurgical parts. Finishing of hardened steel, (e.g. through hardened AISI 52100 steel for bearing applications, and case hardened steel 16MnCr5 for automotive gears and shafts) using hard turning using super hard cutting tools (PCBN, cubic boron nitride (CBN), Ceramics, and Carbide) was early recognized by the automotive industry as a means of manufacturing of precisely finished transmission components (Davim, 2011).

If hard turning is applied to the manufacture of complex parts, manufacturing costs can be reduced up to 30%, and US industries exploited the advantages of hard turning for an annual gain of up to $6 billion (Huang, Chou, & Liang, 2007). A qualitative comparison of the capabilities of hard turning and grinding processes in terms of workpiece quality, process flexibility, dimension and shape accuracy, etc. has been made by M'Saoubi, Outeiro, Chandrasekaran, Dillon, and Jawahir (2008).

*Corresponding author. Email: suhakarim_10@yahoo.com

In the past, many studies have been carried out to explore different facets of the hard turning of alloy steel. Thiele and Melkote (1999) investigated the effects of tool edge geometry of CBN tools and workpiece hardness (45, 52, and 60 HRC) on the surface roughness and cutting forces in the finish hard turning of AISI 52100 steel. Their study showed that large edge hones result in higher forces in the axial, radial, and tangential directions as compared to those tools with small edge hone. Further, the effect of workpiece hardness on the axial and radial components of force was found to be significant, particularly for large edge hones. Ramesh, Melkote, Allard, Riester, and Watkins (2005) examined the differences in structure and properties of white layers formed during machining of hardened AISI 52100 steel (62 HRC) at different cutting speeds. Their results indicated that the grain sizes of white layers formed were considerably smaller than the grain sizes of the bulk. They also observed that white layers generated at higher machining speeds are coarser than those generated at lower speeds. Diniz, Gomes, and Braghini (2005) studied the hardened steel turning (SAE 01 steel with 58 ± 2 HRC hardness) with continuous, interrupted, and semi-interrupted cutting using two different kinds of CBN cutting tools (high CBN content material and CBN content material). Their main conclusion pointed out that the low CBN content tool is more suitable for continuous and semi-interrupted surfaces, whereas the high CBN content tool presents a slightly longer life during interrupted cutting. Umbrello, Rizzuti, Outeiro, Shivpuri, and M'Saoubi (2008) developed hardness-based flow stress and fracture models for machining AISI H13 (50-60 HRC) tool steel, which can be applied for a wide range of work material hardness using FEM. They implemented these models in a viscoplastic non-isothermal finite element model of the orthogonal cutting process and validated the results of these models using the data available in the literature. Umbrello, Ambrogio, Filice, and Shivpuri (2008) determined residual stresses distribution and optimal cutting conditions during hard turning of AISI 52100 bearing steel using the hybrid model based on the artificial neural networks (ANNs) and finite element method (FEM). Adesta, Riza, Hazza, Agusman, and Rosehan (2009) investigated tool wear of cermet tools and also surface roughness under different rake angles (0, −3, −6, −9, and −12) and different cutting speeds (1000 and 300 m/min). Their experimental results showed that increase in negative rake angles causes higher wear, shorter duration of tool life, and poor surface finish. Further, high cutting speed turning gave shorter tool life, high wear rate but finer surface finish than conventional one. Gaitonde, Karnik, Figueira, and Davim (2009a) analyzed the effects of depth of cut and machining time on surface roughness and tool wear during turning of high chromium AISI D2 cold work tool steel (59/61 HRC) using ceramic inserts with TiN coating (CC650, CC650WG, and GC6050WH). From their analysis, it was observed that the CC650WG wiper insert performs better with reference to surface roughness and tool wear. They also observed that the surface roughness is minimal at lower values of depth of cut and machining time in case of CC650 and CC650WG inserts machining, while the minimum surface roughness occurs at 0.4 mm depth of cut for hard turning with GC6050WH insert. Finally, they found that the CC650 conventional insert is useful in reducing the cutting force. Fnides, Yallese, Mabrouki, and Rigal (2009) investigated the effect of three cutting parameters on surface roughness in turning of X38CrMoV5-1 hardened steel treated at 50 HRC using mixed ceramic tool. Their results revealed that the effect of feed rate on surface roughness is more significant than cutting speed, whereas the depth of cut is not significant. Moreover, the highest level of cutting speed, 180 m/min, the lowest level of feed rate, 0.08 mm/rev, and the highest level of depth of cut, 0.45 mm, resulted in larger amount of material removal and also a good surface finish. Thamizhmanii and Hasan

(2010) evaluated the performance of CBN and PCBN tools based on flank wear and cutting forces during machining of AISI 440 C hard martensitic stainless steel. They observed high cutting forces during machining and they suggested that this may be due to heat and flank wear combinations. In addition, they also found that lower cutting forces lead to low flank wear and provide good dimensional accuracy of the work material including low surface roughness. Meyer, Köhler, and Denkena (2012) studied the effect of geometrical contact conditions between the tool (WBN560 with a CBN content of 56%) and the workpiece (AISI 5115 steel) on the cutting forces and tool wear during hard turning. Their results showed that the geometrical contact conditions have a significant influence on the forces, tool, and workpiece load as well as on the tool wear behavior. Moreover, they suggested that by applying a load-specific tool and process design, the tool operation behavior and the productivity of the process can be enhanced in a beneficial way.

2. Major hard turning cutting tool materials

Material developments for the cutting tool, one of the most critical elements in metal cutting, have always been characterized by an increase in wear resistance to machine harder, tougher, or chemically reactive materials. For example, superhard materials such as ceramic and CBN were one of the main keys to enable the hard turning technology to be an alternative to grinding processes. Correlation between chemical, physical, and mechanical characteristics of cutting tool materials and their performances in cutting operations is therefore, a key issue for both tool manufacturers and users (Suresh, Basavarajappa, and Samuel, 2012).

Cutting tools must simultaneously withstand big mechanical loads and high temperatures. Temperature in the chip/tool interface reaches more than 700 °C in some cases such as turning of high hardness alloy steel (Davim, 2011). Additionally, the friction between tool and removed chip, on one hand, and tool against the new machined surface, on the other, is very severe (Davim, 2011). Bearing this in mind, the following main factors should be considered for a good tool design and post-manufacturing (Davim, 2011):

- Chemical and physical stability of cutting-tool substrate material must be maintained at high temperatures.
- Material hardness must be maintained to high temperatures suffered at the chip/tool interface.
- Tool material has to present high resistance for the abrasion and adhesion wear.
- Tool material must present enough toughness to avoid fracture, especially when operation to perform implies interrupted cutting.

The main tool materials that are used for hard turning include sintered carbides, ceramics (e.g. Al_2O_3 or Si_3N_4 etc.), and extra-hard materials (e.g. PCD, PCBN etc.). The selection of appropriate tool material is vital for process efficiency and depends on the accuracy and surface finish required. Within each type, hundreds of different grades are available from various tool material, cutting insert, and tool manufacturers. Therefore, the selection of appropriate tool-material grade is one of the most important tasks in hard turning in terms of obtaining efficient and stable machining process. The insert shape, tool holders, and optimal machining regime just add more complications to this multivariable optimization problem. Knowledge, as understanding the essence of hard machining, and experience are prerequisites for success (Davim, 2011). The studies carried out by the researchers using cutting tool of the above-mentioned materials are describe in the following sections:

2.1. Sintered carbide (hardmetal)

Sintered carbide tools, also known as hardmetal tools or cemented carbide tools, are made by a mixture of tungsten carbide micrograins with cobalt at high temperature and pressure. Tantalum, titanium, or vanadium carbides can also be mixed in small proportions. This type of material in the straight grade or in the coated grades is mostly used today for hard machining and high-speed machining.

A sintered tungsten carbide also includes TiC (carbide with hardness 3200 HV) and in some cases TiCN, but they typically have a nickel–chrome binder. New grades with TaNbC and MoC increase the tool-edge strength against the cyclic impacts which finds typical application in milling. Tungsten carbide is very stable with respect to chemical and thermal aspects of machining, and is very hard as well. In most cases, cemented carbide degradation starts from the cobalt binder and the tungsten carbide–cobalt cohesion (Davim, 2011).

Aslan (2005) investigated the wear behavior of different cutting tools such as TiCN coated tungsten carbide, TiCN + TiAlN coated tungsten carbide, TiAlN coated cermet, mixed ceramic with Al_2O_3 + TiCN, and CBN, in end milling of X210 Cr12 cold-work tool steel hardened to 62 HRC. The results of his investigations indicated that the TiAlN coated carbide and cermet tools perform slightly better than TiCN coated carbide tool which can be attributed to better high-temperature properties of TiAlN compared to TiCN. Further, he also found that CBN tool exhibits the best cutting performance in terms of both flank wear and surface finish and the highest volume of metal removal is obtained with CBN tool. Arsecularatne, Zhang, and Montross (2006) studied wear and tool life for tungsten carbide WC cutting tool during machining of AISI 4142 steel and AISI 1045 steel, and also for PCBN cutting tool during machining of AISI 52100 with two different hardness values of 60HRC and 62HRC. It was concluded from their study that the most dominant tool wear mechanism for WC is diffusion and for PCBN is chemical wear. Lima, Avila, and Abrao (2007) performed a study which is concerned with continuous turning of AISI 4340 steel hardened from 250 up to 525 HV using coated carbide tools. They assessed machining forces, tool life, and wear mechanisms. The results of their study indicated that the machining force components increase with the work material hardness, however, the cutting force components decrease slightly as the work hardness increase from 250 up to 345 HV. Further, tool wear was lower when machining the workpiece with a hardness value of 345 HV as compared to machining of the workpiece possessing hardness of 250 HV steel. Finally, it was observed that the abrasion is the principal wear mechanism and catastrophic failure takes place when attempting to machine the 525 HV steel. Chowdhury and Dhar (2011) investigated the tool wear and surface roughness for varying cutting parameters under dry and Minimum Quantity Lubrication (MQL) environment while turning hardened medium carbon steel using coated carbide insert. The results of their study indicated that the application of MQL technique significantly helps to obtain better performance of coated carbide insert in comparison to dry condition. Hwang and Lee (2010) studied the effect of cutting parameters (nozzle diameter, cutting speed, feed rate, and depth of cut) on cutting forces and surface roughness of carbon steel AISI 1045 using coated carbide inserts under MQL and wet turning conditions. They found that cutting speed and depth of cut show opposite effects on the cutting force and therefore, they suggested that cutting conditions should be set under a clear standard because the optimal combination of cutting parameters could be different depending on machinability. Moreover, cutting speed and depth of cut showed opposite effects on surface roughness in MQL turning and it produced better surface roughness compared to wet turning. They also observed that the MQL turning provides more advantages than wet turning.

2.2. *Ceramics*

Ceramics are very hard and refractory materials, withstanding more than 1500 °C without chemical decomposition. These features recommend them to be used for the machining of metals at high cutting speeds and in dry machining conditions. Ceramic tools are based primarily on alumina (Al_2O_3), silicon nitride (Si3N4), and sialon (a combination of Si, Al, O, and N). Alumina tools can contain titanium, magnesium, chromium, or zirconium oxides distributed homogeneously into the alumina matrix to improve toughness (Davim, 2011)

Several researches have been done to study the effect of hard turning on different types of ceramics tools. Luo, Liao, and Tsai (1999) in their research revealed the wear behavior of ceramic and CBN tools in the turning of AISI 4340 hardened alloy steels. It was found from their study that the main wear mechanism for the CBN tools is the abrasion of the binder material by the hard carbide particles of the workpiece and for the ceramic tools it is adhesive wear and abrasive wear. They also found that there is a protective layer formed on the chip–tool interface which plays an important role in wear behavior of CBN and ceramics tools. Kumar, Durai, and Sornakumar (2003) studied the machinability of EN 24 steel (HRC 40 and HRC 45) using two types of ceramic cutting tool materials namely, Ti[C, N] mixed alumina ceramic cutting tool and zirconia toughened alumina ceramic cutting tool. It was found from their study that the performance of ceramic cutting tools is good in the machining of hardened steel and Ti[C, N] mixed alumina ceramic cutting tool produces the best surface finish. Gaitonde, Karnik, Figueira, and Davim (2009b) analyzed the effects of depth of cut and machining time on machinability aspects such as machining force, power, specific cutting force, surface roughness, and tool wear, with the use of second-order mathematical models during turning of high chromium AISI D2 cold work tool steel with CC650, CC650WG, and GC6050WH ceramic inserts. From their analysis, it was revealed that the CC650WG wiper insert performs better with reference to surface roughness and tool wear, while the CC650 conventional insert is useful in reducing the machining force, power, and specific cutting force. Elmunafi (2012) evaluated the performance of wiper coated ceramic tool when turning ASSAB DF-3 grade hardened steel with hardness 55 HRC. Their study showed that the effects of cutting speed and feed rate on the tool life are statistically significant. They also found that the effects of cutting speed and feed rate on the surface roughness are statistically significant, specifically, at high cutting speed and low feed rate. Accordingly, wiper inserts were able to produce better surface finish. Maňková, Kovac, Kundrak, and Beňo (2011) studied the influence of cutting parameters on accompanying phenomena during hard turning of hardened steel with hardness of HRC 55 using mixed oxide ceramic inserts (70% Al_2O_3 + 30% TiC). They concluded that hard turning has many advantages in comparison to other processes in machining of hardened steels. Fnides, Yallese, Mabrouki, and Rigal (2011) stated the effect of cutting parameters on cutting force components in hard turning of AISI H11 hot work tool steel (50 HRC) using a mixed ceramic tool. They found that the depth of cut is the dominant factor affecting cutting force components. Additionally, the feed rate influences tangential cutting force more than radial and axial forces. Finally, the cutting speed affects radial force more than tangential and axial forces. Zhao, Yuan, and Zhou (2010) investigated the cutting performance, failure mode, and mechanism of the Al_2O_3-based composite ceramic tool material reinforced with WC microparticles and TiC nanoparticles in both continuous and intermittent turning of hardened AISI1045 steel. Their study revealed that the tool life of the Al_2O_3/WC/TiC ceramic tool increases when the cutting speed increases to 170 m/min. They also observed that the longer tool life of the Al_2O_3/WC/TiC composite ceramic tool is attributed to its synergistic strengthening/toughening mechanisms induced by the WC microparticles and TiC nanoparticles.

2.3. Extra-hard materials

PCD and PCBN are extra-hard materials. There are several grades in the PCD and PCBN groups. PCD is suitable for tools focused on machining abrasive non-ferrous metals, plastics, and composites. PCBN finds applications in the machining of hardened tool steels and hard cast irons. PCD plates are obtained by a high temperature and pressure process where synthetic diamond grains are sintered with cobalt. CBN is a polymorph boron-nitride-based material. It possesses high mechanical properties due to its crystalline structure and its covalent link. It has been industrially produced since 1957, starting from hexagonal boron nitride put under high pressures (8 GPa) and temperatures (1500 °C). With a lower hardness (<4500 HV) than diamond (>9000 HV), CBN is the second-hardest synthetic material. The CBN grains are sintered together with a binder to form a composite, PCBN. The size, shape, and ratio of CBN/binder define the different PCBN grade (Davim, 2011).

The literature reveals that many researchers have suggested effect of hard turning on CBN and PCBN cutting tools (Chowdhury and Dhar 2011; Diniz et al., 2005; Elmunafi 2012; Lima et al., 2007). In addition, many studies have been conducted to investigate the performance of CBN and PCBN tool in machining of various hard materials. Oliveira, Diniz, and Ursolino (2009) investigated the performance of PCBN and alumina-based ceramic reinforced with silicon carbide tools during turning of AISI 4340 steel with 56 HRC under continuous and interrupted cutting. The results of their study indicated that the longest tool life is achieved using PCBN in continuous turning, but similar tool longevity is attained in interrupted turning using both PCBN and ceramic. Also, the roughness values were lower for both continuous and interrupted surfaces when PCBN tools were used. Sahin (2009) compared the tool life of mixed alumina ceramic with that of Al_2O_3 (70%):TiC (30%) matrix and CBN cutting tools during machining of AISI 52100 steel. They showed that the CBN cutting tool has the best performance than that of ceramic-based cutting tool. Poulachon, Moisan, and Jawahir (2001) studied various modes of wear and damage of PCBN cutting tool under different loading conditions during turning of AISI 52100 steel in order to establish a reliable wear model. Their results led to the conclusion that the main wear mechanism of the PCBN tools is abrasion and it depends not only on the chemical composition of the PCBN, and the nature of the binder phase, but also on the hardness value and above all on the microstructure of the machining work material. They also found that use of TiN coated PCBN improves the tool-wear and hence the tool-life by reducing the diffusion wear between workpiece and tool rake face. Diniz, Ferreira, and Filho (2003) investigated the influence of cutting speed under three cutting conditions: dry cutting, wet cutting, and minimum volume of oil i.e. MVO (oil flow of 10 ml/h) on CBN tool wear in turning of AISI 52100 hardened steel. Their results revealed that dry and MVO cuttings produce most of the time, similar values of flank wear which is always smaller than the values obtained during wet cutting. They also observed that highest cutting speed for three cutting conditions gives the smallest value of surface roughness. Poulachon, Bandyopadhyay, Jawahir, Pheulpin, and Seguin (2003) examined the influence of the microstructure of different hardened tool steels (AISI D2 cold work steel, AISI H11 hot work steel, 35NiCrMo16 hot work steel, and AISI 52100 bearing steel) on the wear of PCBN cutting tools under dry turning. They observed a large variation in tool wear in the machining of these steels. Further, they found that the flank grooves are correlated to the microstructure of these steels, namely the presence of various carbides. Benga and Abrao (2003) studied the machinability of hardened 100Cr6 bearing steel (62-64 HRC) under dry turning using mixed alumina, whisker reinforced alumina, and PCBN inserts. As far as tool life is concerned, they obtained best results with the PCBN compact, followed by the mixed alumina tool at low feed rates and by the whisker reinforced alumina when feed rate was increased. Poulachon, Bandyopadhyay, Jawahir, Pheulpin, and Seguin (2004) investigated the tool wear mechanisms of CBN cutting tools in finish turning of the hardened steels (AISI D2 cold work steel, AISI H11 hot work steel, 35NiCrMo16 hot work steel, and AISI 52100bearing steel), treated at 54 HRC. They observed a large variation in tool-wear rate in machining of these steels. They also found that the generated tool flank grooves are correlated with the hard carbide content of the workpieces. In addition, they also performed a crater wear study in which they found the appearance of an adhesive third body which could induce a chemical wear in the tool. Kurt and Seker (2005) investigated the effects of chamfer angle of PCBN cutting tools on the cutting forces and the cutting tool stresses, principal, and Von Mises stress, in finishing hard turning of AISI 52100 bearing steel. Their results revealed a great influence of the chamfer angle on the cutting forces and tool stress. Based on their work, they concluded that critical chamfer angle is 20° in finishing hard turning of AISI 52100 bearing steel. Galoppi, Filho, and Batalha (2006) tested different types of CBN inserts with wiper geometry, coated with TiAlN and TiN as well as with no coated ones, during hard turning of tempered DIN 100Cr6 steel. Their results revealed significant cracking on the coating surface layer of both TiAlN and TiN coated tools which resulted in the removal of the surface hard coating by cracking. Further, they also found that the cratering that arises on insert surfaces is similar to that found on the no coated inserts. Finally, they observed that the inserts with a wiper geometry presents longer tool life compared with no coated tools. Remadna and Rigal (2006) explored the parameters of hard steel turning (alloyed steel 52HRC) with a CBN tool. They analyzed the correlation between tool wear and the direction of cutting force. Their results showed that the cutting forces increase gradually with increasing of cutting distance and tool flank wear. They also revealed that the wear of CBN does not directly or appreciably affect the manufactured surface. More, Jiang, Brown, and Malshe (2006) measured tool wear and evaluated machining performance of CBN–TiN coated carbide inserts and PCBN compact inserts in turning of AISI 4340 hardened steel. They observed that the flank wear is mainly due to abrasive actions of the martensite which is present in the hardened AISI 4340 alloy. Further, they found that the crater wear of the CBN–TiN coated inserts is less than that of the PCBN inserts because of the lubricity of TiN capping layer on the CBN–TiN coating. Arsecularatne, Zhang, Montros, and Mathew (2006) investigated the machinability of AISI D2 steel of hardness 62 HRC using PCBN tools. They found that the most feasible feeds and speeds fall in the ranges 0.08–0.20 mm/rev and 70–120 m/min, respectively, and that most of the tested PCBN tools reach the end of life mainly due to flank wear. Moreover, the highest acceptable values of tool life and volume of material removal were obtained at the lowest speed (70 m/min), indicating that this speed is more suitable for machining the selected tool/ work material combination. They also found that the highest feed results in the highest volume of material removal and lower feeds results in higher tool life values. They suggested that the most appropriate feeds for AISI D2 steel are 0.14 mm/rev for finishing operations and 0.20 mm/rev for roughing operations. Lin, Liao, and Wei (2008) studied tool wear mechanisms in turning of AISI 4340 alloy steels by CBN tools which contained 50% and 45% of CBN, respectively, and also TiC based-binder, under varying cutting speeds. It was observed from their results that at low cutting speed, the binder of the hard particles of the cutting tool is removed from the substrate due to high cutting force and the tool wear is mainly due to abrasion. Their study revealed formation of a protective layer, resulting from the diffusion of the bond material of the cutting tool, on the chip–tool interface when the cutting speed was increased. The layer so formed works as a diffusion barrier and reduces tool wear rate which results in the prolonged life of the CBN tool. They also found that at further higher cutting speed, (i) cutting temperature becomes the dominant factor instead of the cutting force which causes inhomogeneous shear strain, and a transition from continuous chip to saw-tooth chip, (ii) the friction force increases because of the very irregular chip–tool contact and it results in the removal of the protective layer, and (iii) the bond between tool particles gets weakened due to serious diffusion between the work material and the cutting tool. Subsequently, hard particles of the tool get detached from it and the tool life is reduced. Özel (2009) investigated the influence of different edge micro-geometry of PCBN tools on forces, stresses, friction, and tool wear during turning of alloy steel AISI 4340. On the basis of the results of their study, they made the following conclusions: (i) variable micro-geometry insert edge design reduces the heat generation and stress concentration along the tool cutting edge significantly which improves the tool life, (ii) variable micro-geometry insert cutting edge induces less plastic strain on the machined workpiece which improves surface integrity, and (iii) variable micro-geometry insert edge design reduces tool wear depth and wear rate. Yallese, Chaoui, Zeghib, Boulanouar, and Rigal (2009) investigated the behavior of a CBN tool during hard turning of 100Cr6-tempered steel. They observed that the cutting speeds ranging from 90 to 220 m/min is the most interesting cutting conditions for the system CBN7020-100Cr6. However, they found that beyond 280 m/min the machining system becomes unstable and produces significant sparks and vibrations after only a few minutes of work. Further, they found optimal productivity at the speed of 120 m/min with an acceptable tool flank wear below 0.4 mm. Godoy and Diniz (2011) made a comparison between the performance of CBN and ceramic tools in continuous and interrupted cutting of AISI 4340 steel with 56 HRC of hardness. They concluded that in both continuous and interrupted cutting, the CBN tools exhibit a much better performance with respect to both tool life and workpiece surface roughness than the ceramic tools. Zawada-Tomkiewicz (2011) evaluated the surface finish in a continuous dry turning of EN 41Cr4 low chromium alloy steel with 58 HRC using PCBN coated and uncoated tools. He observed that the hard turning with PCBN tools can produce a very smooth and uniform surface. Also, in the case of the PCBN coated tool material, the wedge's wear was almost unnoticeable. Further, the random part of the generated surface was significantly greater for the PCBN coated cutting tool material. Deepakkumar and Sadaiah (2011) investigated the effects of cutting speed, feed rate, and different cutting environments (dry, wet, and MQL) on the surface roughness and tool wear during turning of AISI 4340 steel using CBN insert. They analyzed the effect of cutting parameters and determined its optimum condition by using ANOVA technique. They found that increase in the cutting speed and feed rate results in decrease in R_a value. In addition, he observed the R_a value for dry and wet turning as 1.2 and 1.1 μm, respectively, while in the case of turning in MQL condition, the R_a value was 0.9 μm. Finally, the results of their study indicated that as cutting speed and feed rate increase, the tool wear value increases.

3. Effect of process parameters in hard turning:

Machining parameters such as cutting speed, feed rate, and depth of cut do affect the production costs and product quality. Thus, it is important to use optimization technique to determine optimal levels of these parameters so as to reduce the production costs and to achieve the desired product quality simultaneously. One of the main objectives in the optimization of a turning process is minimization of the cost of production and maximization of the production rate while keeping the quality of machined parts as per design specifications. The cost of machining is strongly related to the material removal rate. The material removal rate for a turning operation is given by the product of cutting parameters (cutting speed (*Vc*), feed rate (*f*), and depth of cut (*d*)). Therefore, if an increase in productivity is desired then an increase in these three cutting parameters is required. But, there are limits to these cutting parameters since they also have an effect on the tool life, tool wear, surface quality, cutting forces, cutting temperature, etc. Keeping this in view, many researchers have investigated effect of these parameters pertaining to the hard turning. The following sections present the findings of some of the research studies involving these parameters with reference to hard turning.

3.1. Studies on cutting forces

The forces acting on the tool are important aspect of machining. Knowledge of the cutting forces is needed for the estimation of power requirements, the adequately rigid design of machine tool elements, tool-holders, and fixtures, for vibration free operations.

Many force measurement devices like dynamometers have been developed which are capable of measuring tool forces with increasing accuracy. Power consumed in metal cutting is largely converted into heat near the cutting edge of tool, and many of the economic and technical problems of machining are caused directly or indirectly by this heating action (Sharma, Dhiman, Sehgal, & Sharma, 2008).

By measuring the cutting forces, one is able to understand the cutting mechanism such as the effects of cutting variables on the cutting force, the machinability of the workpiece, the process of chip formation, chatter, and tool wear. It has been observed that engineering calculations used for obtaining the force values give some errors when compared to experimental measurements of the forces. The cutting force in even unsteady state conditions is affected by many parameters and the variation of cutting force with time has a typical characteristic. Cutting forces can be resolved into three components i.e. the radial thrust force (Fx), feed force (Fy), and tangential cutting force (Fz).Usually, the tangential cutting force is the largest of the three components, though in finish turning the radial thrust force is often larger, while the feed force is minimal. The findings of some of the research studies pertaining to the effect of cutting parameters on cutting forces are presented below.

The cutting forces increase drastically when machining materials with hardness higher than about 45 HRC (Davim, 2011). Kurt and Seker (2005) investigated the effects of chamfer angle on the cutting forces and the stresses on the PCBN cutting tools in finishing hard turning of AISI 52100 bearing steel. They found that the chamfer angles (20° and 30°) have a great influence on the passive cutting forces and Von Mises tool stresses distribution. Lee (2011) developed a theoretical model to predict cutting forces for machining AISI 4140-hardened materials (45 HRC) that contain more than 0.58% carbon. His predicted values of cutting forces from the model were found to be in good agreement with those measured from an experiment of hard machining of AISI 4140 steel heat-treated. Panzera, Souza, Rubio, Abrao, and Mansur (2012) investigated the effect of the cutting parameters (cutting speed, feed rate, and depth of cut) on the cutting force components during dry turning AISI 4340 steel using coated carbide inserts. The results of their study indicated that the three components of the turning force decrease slightly as cutting speed increases and also they increase linearly with feed rate and depth of cut. Further, the results of analysis of variance (ANOVA) revealed that the three components of the force are not significantly affected by cutting speed but they are significantly affected by feed rate and depth of cut. Aouici, Yallese, Fnides, and Mabrouki (2010) studied the influence of the cutting parameters (cutting speed, feed rate, and depth of cut) on cutting force components, surface roughness, temperature in the cutting zone, and tool life during turning of AISI H11 steel treated at 50 HRC using CBN tool (57% CBN and 35% Ti(C, N)). Their results indicated that (i) tangential cutting force is very sensitive to the variation in cutting depth, (ii) thrust force is dominating compared to both others cutting forces, (iii) surface roughness is very sensitive to the variation of the feed rate, and (iv) the temperature is greatly influenced by the cutting speed.

3.2. Studies on heat generation and cutting temperature

Most of the energy in the cutting process is converted into heat. This heat is generated by plastic deformation and friction at the tool–chip and the tool–workpiece interfaces. The generation of the heat during machining increases the temperature in the cutting zone which affects the strength, hardness, wear resistance, and life of the cutting tool and causes difficulty in controlling the dimensional accuracy and surface integrity. The temperature also causes thermal damage to the workpiece and affects its properties and service life.

The temperature in the cutting zone is affected by the cutting parameters. In addition, it also depends on the properties of the workpiece material, as well as on the physical properties of the tool. Therefore, considerable attention has been paid to the measurement and prediction of the temperatures at the tool, chip, and workpiece in the metal cutting (Aouici et al., 2010; Özel, 2009). The cutting tools that are used for machining should possess adequate hot hardness to withstand elevated temperatures generated at high-speed condition. Under these conditions, most tool materials generally lose their hardness resulting in the weakening of the interparticle bond strength and consequently, tool wear gets accelerated as reported by Ezugwu, Bonney, and Yamane (2003). Amritkar, Prakash, and Kulkarni (2012) performed machining of SAE 8620 material at various cutting speed and feed rate using uncoated tungsten carbide tool. They designed and developed a simple and economical technique of temperature measurement i.e. tool-work thermocouple setup for the measurement of the cutting temperature for which they calibrated the setup in order to establish a relationship between obtained e.m.f. during machining and the cutting temperature. They used regression analysis for establishing the relationship between temperature and the generated voltage.

They evaluated the performance of the setup for the different material like EN19, EN31, mild steel, SS 304, and SAE 8620 using uncoated tungsten carbide tool. Their obtained results confirmed that the setup is having better accuracy and good repeatability. Further, they observed that the tool-work thermocouple technique is the best method for measuring the average chip–tool interface temperature during metal cutting. Sutter, Faure, Molinari, Ranc, and Pina (2003) studied the effects of the cutting speed and depth of cut on the temperature profile at the chip during an orthogonal machining of 42 CrMo 4 steel using standard carbide tools TiCN coated. They performed the machining with a gas gun. It was found from their results that the temperature at the chip increases with the increase in both cutting speed as well as the depth of cut. Ren, Yang, James, and Wang (2004) determined the cutting temperatures during hard turning of high chromium hardfacing materials using PCBN tools. They found that the average cutting temperatures ranged from 600 to 700 °C which increased with higher cutting speed and feed rate. Abukhshim, Mativenga, and Sheikh (2005) used FEA to estimate the amount of heat flowing into the cutting tool in high-speed turning of AISI 4140 high strength alloy steel using uncoated cemented carbide. Their results showed that the maximum temperature at the tool–chip contact increases with cutting speed but not linearly and this could be attributed to the trend of the heat fraction flowing into the tool. Sutter and Ranc (2007) measured the temperature during machining of two steels i.e. C15 and 42CrMo4 for a range of cutting speed around 15–65 m/s. Their results showed that the increase in cutting speed from 10 to 65 m/s maximizes temperature at the chip continuously. List, Sutter, and Bouthiche (2012) predicted the interface cutting temperature and its relation with the crater wear mechanism. Their work was focused on the domain of the high-speed machining above 20 m/s. They analyzed the mechanical and thermal parameters that influenced the temperature distribution at the tool rake face.

3.3. *Studies on surface finish and surface integrity*

The surface integrity of a machined surface is defined in terms of residual stresses, surface roughness, microhardness, etc. Surface roughness and dimensional accuracy play an important role in the performance of a machined component. High cutting forces and high localized temperatures may dramatically affect the surface integrity, often resulting in the development of high tensile residual stresses in the machined surfaces. Residual stress on the machined surface and the subsurface is known to influence the service quality of a component, such as fatigue life, tribological properties, and distortion. Therefore, it is essential to predict and control it for enhanced performance as suggested by several researchers (Adesta et al., 2009; Aouici et al., 2010; Benga & Abrao, 2003; Chowdhury & Dhar, 2011; Diniz et al., 2003; Elmunafi, 2012; Fnides et al., 2009; Gaitonde et al., 2009; Godoy & Diniz, 2011; Kumar et al., 2003; More et al., 2006; Poulachon et al., 2001; Ramesh et al., 2005; Sharma et al., 2008; Thamizhmanii & Hasan, 2010; Thiele & Melkote, 1999; Yallese et al., 2009; Zawada-Tomkiewicz, 2011). In addition to the research studies cited above, Hua et al. (2005) analyzed the effects of cutting edge geometry, workpiece hardness, and cutting parameters, such as cutting speed and feed rate on subsurface residual stress in hard turning of AISI 52100 bearing steel. It was revealed from their analysis that hone edge plus chamfer cutting edge and aggressive feed rate help to increase both compressive residual stress and penetration depth. Moreover, using medium hone radius (0.02–0.05 mm) plus chamfer was good for keeping tool temperature and cutting force low, while obtaining desired residual stress profile. Rech et al. (2008) provided a comprehensive characterization of residual stresses that were developed in dry turning of a hardened AISI 52100 bearing steel using PCBN tools. For better understanding of the experimental results, they investigated the generated residual stresses in hard turning or in 'hard turning + belt finishing' by two complementary ways: an experimental X-ray diffraction characterization after each step of the process, and a finite element model of the belt finishing operation. Additionally, they explored the sensitivity of some parameters such as the lubrication and the indentation force during belt finishing. They observed that the belt finishing process improves the surface integrity by the induction of strong compressive residual stresses in the external layer and by a great improvement of the surface roughness. In addition, they also found that among the process parameters of the belt finishing technique, the lubrication is a key parameter to get compressive stresses. Caruso, Umbrello, Outeiro, Filice, and Micari (2011) examined the effects of the tool cutting-edge geometry, workpiece hardness, cutting speed, and microstructural changes (white and dark layers) on the residual stresses in dry orthogonal hard machining of AISI 52100 steel using PCBN tool inserts. Their results showed that tool geometry, workpiece hardness, and cutting parameters significantly affect the surface residual stress, maximum compressive residual stress below the machined surface, and its location. Moreover, microstructural analysis showed that thermally-induced phase transformations have a significant impact on the magnitude and location of this maximum compressive residual stress peak.

On the other hand, a number of investigations have been carried out in the past to assess the surface roughness that could be achieved with hard turning in comparison to grinding. Asilturk and Akkus (2011) investigated the effects of cutting speed, feed rate, and depth of cut on surface roughness during dry turning of AISI 4140 (51 HRC) with coated carbide cutting tools. Their results indicated that the feed rate has the most significant effect on the surface roughness. In addition, the effects of two factor interactions of the feed rate-cutting speed and depth of cut-cutting speed appear to be important. Aouici, Yallese, Chaoui, and Mabrouki (2012) investigated the effects of cutting speed, feed rate, workpiece hardness, and depth of cut on surface roughness in the hard turning of AISI H11 steel (hardened to (40; 45; and 50) HRC) using CBN which is essentially made of 57% CBN and 35% TiCN. Their results showed that both the feed rate and workpiece hardness have statistically significant effect on the surface roughness. Further, the best surface roughness was achieved at the lower feed rate and the highest cutting speed. Umbrello et al. (2011) investigated the effects of cryogenic coolant on surface integrity in orthogonal machining of hardened AISI 52100 bearing steel using chamfered CBN tool inserts. The results of their study showed that the use of cryogenic coolant significantly affects the surface integrity and improves product's functional performance. Grzesik, Żak, Prażmowski, Storch, and Pałka (2012) explored the effect of cryogenic cooling on the surface integrity produced in hard turning of low alloy 41Cr4 steel with hardness of 57 ± 2 HRC using low content CBN tools containing about 60% CBN. They confirmed that hard machining produces surfaces with acceptable surface roughness and, in some cases, with attractive service properties. They also observed that cryogenic hard cutting operations can partly eliminate grinding operations in cases when white layer is not produced. Abhang and Hameedullah (2012) determined the optimum cutting parameters (cutting speed, feed rate, tool nose radius, and concentration of solid–liquid lubricants (MQL)) for the multiperformance characteristics (surface roughness and chip thickness) in the turning of EN-31 steel by using the grey relational analysis coupled with factorial design. They concluded that gray relational analysis coupled with factorial design can effectively be used to obtain the optimal combination of cutting parameters. Additionally, they observed that the surface roughness and the chip thickness in the turning process can be improved effectively through this approach. Finally, they found that the minimum surface roughness and smallest chip thickness are 9.83 μm and 0.32 mm, respectively, which are obtained at optimal conditions at: 1200 rpm cutting speed; 0.06 mm/rev feed rate; 0.8 mm nose radius; and concentration of solid–liquid lubricant (10% boric acid + SAE-40 base oil. Islam (2013) investigated the effect of cooling method, blank size, and work material on the dimensional accuracy characteristics (diameter error and circularity) and surface roughness of turned parts under three turning conditions (dry, flood, and MQL) for different materials (aluminum (AISI 6061), mild steel (AISI 1030), and alloy steel (AISI 4340)). He analyzed the results by applying three methods: traditional analysis, Pareto ANOVA, and Taguchi method. He noticed that cooling method, blank size, and work material have demonstrated considerable effects on the dimensional accuracy characteristics, whereas the effect of these parameters on surface roughness is relatively low. He also revealed that while work material has the greatest effect on diameter error and surface roughness, the major contributor to circularity is blank size. Finally, dimensional accuracy and surface roughness of different work materials were influenced differently by the cooling methods, and in most cases the best result was achieved by MQL.

3.4. Studies on the tool wear

During machining, the cutting tools are subjected to severe forces and temperature which may cause tool wear and therefore, it is necessary to study and predict the tool wear during machining for the effective design of cutting tools and determination of cutting conditions that will lead to the formulation of the tool change strategies. Intensive research studies pertaining to the tool wear have been carried out in the past century which has contributed greatly to the understanding of the factors responsible for the tool wear and also the tool wear mechanisms. Many researchers focused their studies on the prediction of the tool wear during hard turning (Adesta et al., 2009; Arsecularatne et al., 2006, Arsecularatne et al., 2006; Aslan, 2005; Chowdhury & Dhar, 2011; Diniz et al., 2003; Elmunafi, 2012; Fnides et al., 2009; Gaitonde et al., 2009; Galoppi et al., 2006; Godoy and Diniz, 2011; Huang et al., 2007; Kumar et al., 2003; Lima et al., 2007; Lin et al., 2008; Luo et al., 1999; Meyer et al., 2012; More et al., 2006; Oliveira et al., 2009; Özel, 2009; Poulachon et al., 2001, 2003, 2004; Remadna & Rigal, 2006; Sahin, 2009; Suresh et al., 2012; Thamizhmanii & Hasan, 2010; Yallese et al., 2009; Zawada-Tomkiewicz, 2011; Zhao et al., 2010). In addition to the references cited before for the tool wear study, Grzesik and Zalisz (2008) investigated the wear phenomenon of the mixed ceramic tips during dry hard turning of AISI 5140 steel (60 HRC). They performed finishing cuts under varying feed rate, constant cutting speed of 100 m/min and small depth of cut of 0.2 mm. It was observed from their results that depending on the mechanical and thermal conditions generated on the wear zones, the wear mechanisms involve abrasion, fracture, plastic flow, material transfer, and tribochemical. El Hakim, Abad, Abdelhameed, Shalaby, and Veldhuis (2011) studied the performance of different tool materials, PCBN (CBN + TiN), TiN coated PCBN (CBN + TiN), mixed alumina ceramic (Al_2O_3 + TiC), coated tungsten carbide (TiN coated over a multilayer coating (TiC/TiCN/Al_2O_3)) in the machining of medium hardened steel AISI T15 HSS. Their results indicated that the mixed alumina ceramic and coated carbide tool materials have longer tool life than PCBN tools when they machined the selected workpiece material.

Chinchanikar and Choudhury (2013) investigated the effect of workpiece hardness, cutting parameters, and type of coating for coated cemented carbide inserts on flank wear during turning of hardened AISI 4340 steel at different levels of hardness. The results of their study revealed that flank wear is dominant wear form for CVD coated tool and crater wear is dominant wear form for PVD coated tool. Further, they found that abrasion and adhesion are the main causes for wear of CVD coated tool and abrasion, adhesion and diffusion leads to the wear of PVD coated tool. Gaitonde et al. (2009) studied the relationships between the cutting parameters (cutting speed, feed rate, and machining time) on tool wear. They used RSM to analyze the effects of process parameters on machinability during turning of high chromium AISI D2 cold work tool steel using CC650WG wiper ceramic inserts. They found that the maximum tool wear occurs at a cutting speed of 150 m/min for all values of feed rate and the tool wear increases with the increase in machining time. Dogra, Sharma, Sachdeva, Suri, and Dureja (2011) made comparison between the performance of CBN inserts with coated carbide and cryogenically treated coated/uncoated carbide inserts in terms of flank wear during finish turning of hardened AISI H11 steel (48–49 HRC). They indicated that the flank wear of CBN is lower than that of other inserts.

4. Summary

Hard machining including hard turning is widely used by several manufacturing industries such as ball bearings, automotive, gear, and die-making industries, since it offers numerous advantages when compared with traditional methodology based on finish-grinding operation after heat treatment of workpieces. This technology possesses immense potential to machine very hard materials to produce near net shape components and also to contribute to a great extent to the sustainable manufacturing. In the recent past, this technology has created interest among researchers and attracted their attention, and consequently, many research studies have been conducted and the results have been reported which are available in the literature. This paper has presented an overview focusing mainly on turning of hardened steels that are used by ball bearings, automotive, gear, and die-making industries. On the basis of the research findings reported in the available literature reviewed and presented in this paper, following conclusions can be drawn:

- Hard turning offers a number of potential benefits over traditional form grinding, including lower equipment costs, shorter setup time, fewer process steps, greater part geometry flexibility, and elimination of the use of cutting fluid.
- During hard turning, the cutting tool is subjected to heavy mechanical loads and also it is exposed to very high temperature due to excessive heat generation, and therefore, cutting tools made of superhard materials such as coated cemented carbide, ceramic, PCD, CBN, etc. must be used as they perform well under severe machining conditions.
- The cutting tool geometry, workpiece hardness, and cutting parameters significantly affect the cutting forces, surface residual stress, surface roughness, surface integrity, tool wear, and tool life.
- The information gained through study and prediction of the tool wear during machining can be used as a basis for the effective design of cutting tools and determination of cutting conditions that will lead to the formulation of the tool change strategies.

- The complex phenomena involved in hard turning can be studied through simulation and modeling using techniques such as FEM, ANN, etc. and the results of the models can be validated with experimental results.

Thus, it is concluded that the information gathered through extensive literature review has been presented in a modular way in this paper. The review has been organized in terms of role of machining parameters on machining of hard steel, cutting force, heat generation, and temperature evolution during machining, surface integrity, and tool wear during hard machining, etc. The information presented are immensely useful to the researchers in identifying solutions to the several machining problems to mention some: (i) to identify strategies with regard to tool edge geometry, cutting parameters, etc. for specific work material hardness so as to obtain better surface integrity and surface finish; and (ii) to identify prevalent wear mechanisms and appropriate tool material for specified machining situations.

References

Abhang, L. B., & Hameedullah, M. (2012). Determination of optimum parameters for multi-performance characteristics in turning by using grey relational analysis. *The International Journal of Advanced Manufacturing Technology, 63*, 13–24.

Abukhshim, N. A., Mativenga, P. T., & Sheikh, M. A. (2005). Investigation of heat partition in high speed turning of high strength alloy steel. *International Journal of Machine Tools and Manufacture, 45*, 1687–1695.

Adesta, E. Y. T., Riza, M., Hazza, M., Agusman, D., & Rosehan. (2009). Tool wear and surface finish investigation in high speed turning using cermet insert by applying negative rake angles. *European Journal of Scientific Research, 38*, 180–188.

Amritkar, A., Prakash, C., & Kulkarni, A. P. (2012). Development of temperature measurement setup for machining. *World Journal of Science and Technology, 2*, 15–19.

Aouici, H., Yallese, M. A., Chaoui, K., & Mabrouki, T. (2012). Analysis of surface roughness and cutting force components in hard turning with CBN tool: Prediction model and cutting conditions optimization. *Measurement, 45*, 344–353.

Aouici, H., Yallese, M. A., Fnides, B., & Mabrouki, T. (2010). Machinability investigation in hard turning of AISI H11 hot work steel with CBN tool. *MECHANIKA, 86*, 71–77.

Arsecularatne, J. A., Zhanga, L. C., & Montross, C. (2006). Wear and tool life of tungsten carbide, PCBN and PCD cutting tools. *International Journal of Machine Tools & Manufacture, 46*, 482–491.

Arsecularatne, J. A., Zhang, L. C., Montross, C., & Mathew, P. (2006). On machining of hardened AISI D2 steel with PCBN tools. *Journal of Materials Processing Technology, 171*, 244–252.

Asilturk, I., & Akkus, H. (2011). Determining the effect of cutting parameters on surface roughness in hard turning using the Taguchi method. *Measurement, 44*, 1697–1704.

Aslan, E. (2005). Experimental investigation of cutting tool performance in high speed cutting of hardened X210 Cr12 cold-work tool steel (62 HRC). *Materials and Design, 26*, 21–27.

Benga, G. C., & Abrao, A. M. (2003). Turning of hardened 100Cr6 bearing steel with ceramic and PCBN cutting tools. *Journal of Materials Processing Technology,* 143–144, 237–241.

Caruso, S., Umbrello, D., Outeiro, J. C., Filice, L., & Micari, F. (2011). An experimental investigation of residual stresses in hard machining of AISI 52100 steel. *Procedia Engineering, 19*, 67–72.

Chinchanikar, S., & Choudhury, S. K. (2013). Investigations on machinability aspects of hardened AISI 4340 steel at different levels of hardness using coated carbide tools. *International Journal of Refractory Metals and Hard Materials, 38*, 124–133.

Chowdhury, N. T., & Dhar, N. R. (2011). Experimental analysis and modeling of tool wear and surface roughness in hard turning under minimum quantity lubricant environment. Proceedings of the International Conference on Industrial Engineering and Operations Management, Malaysia, 326–330.

Davim, J. P. (2011). *Machining of hard materials*. London: Springer-Verlag. ISBN 1849964505, 9781849964500. doi:10.1007/978-1-84996-450-0

Deepakkumar, H. P., & Sadaiah, M. (2011). Investigations on finish turning of AISI 4340 steel in different cutting environments by CBN insert. *International Journal of Engineering Science and Technology, 3*, 7690–7706.

Diniz, A. E., Ferreira, J. R., & Filho, F. T. (2003). Influence of refrigeration/lubrication condition on SAE 52100hardened steel turning at several cutting speeds. *International Journal of Machine Tools & Manufacture, 43*, 317–326.

Diniz, A. E., Gomes, D. M., & Braghini, A., Jr. (2005). Turning of hardened steel with interrupted and semi-interrupted cutting. *Journal of Materials Processing Technology, 159*, 240–248.

Dogra, M., Sharma, V. S., Sachdeva, A., Suri, N. M., & Dureja, J. S. (2011). Performance evaluation of CBN, coated carbide, cryogenically treated uncoated/coated carbide inserts in finish-turning of hardened steel. *The International Journal of Advanced Manufacturing Technology, 57*, 541–553.

El Hakim, M. A., Abad, M. D., Abdelhameed, M. M., Shalaby, M. A., & Veldhuis, S. C. (2011). Wear behavior of some cutting tool materials in hard turning of HSS. *Tribology International, 44*, 1174–1181.

Elmunafi, M. H. S. (2012). *Performance evaluation of wiper coated ceramic tool when hard turning tool steel* (Dissertation thesis). University Technology of Malaysia.

Ezugwu, E. O., Bonney, J., & Yamane, Y. (2003). An overview of the machinability of aeroengine alloys. *Journal of Materials Processing Technology, 134*, 233–253.

Fnides, B., Yallese, M. A., Mabrouki, T., & Rigal, J.-F. (2009). Surface roughness model in turning hardened hot work steel using mixed ceramic tool. *MECHANIKA, 77*, 68–73.

Fnides, B., Yallese, M. A., Mabrouki, T., & Rigal, J.-F. (2011). Application of response surface methodologyfor determining cutting force model in turning hardened AISI H11 hot work tool steel. *Sadhana, 36*, 109–123.

Gaitonde, V. N., Karnik, S. R., Figueira, L., & Davim, J. P. (2009a). Machinability investigations in hard turning of AISI D2 cold work tool steel with conventional and wiper ceramic inserts. *International Journal of Refractory Metals & Hard Materials, 27*, 754–763.

Gaitonde, V. N., Karnik, S. R., Figueira, L., & Davim, J. P. (2009b). Analysis of machinability during hard turning of cold work tool steel (Type: AISI D2). *Materials and Manufacturing Processes, 24*, 1373–1382.

Galoppi, G. de S., Filho, M. S., & Batalha, G. F. (2006). Hard turning of tempered DIN 100Cr6 steel with coated and no coated CBN inserts. *Journal of Materials Processing Technology, 179*, 146–153.

Godoy, V. A. A. de, & Diniz, A. E. (2011). Turning of interrupted and continuous hardened steel surfaces using ceramic and CBN cutting tools. *Journal of Materials Processing Technology, 211*, 1014–1025.

Grzesik, W., Żak, K., Prażmowski, M., Storch, B., & Pałka, T. (2012). Effects of cryogenic cooling on surface layer characteristics produced by hard turning. *Archives of Materials Science and Engineering, 54*, 5–12.

Grzesik, W., & Zalisz, Z. (2008). Wear phenomenon in the hard steel machining using ceramic tools. *Tribology International, 41*, 802–812.

Hua, J., Shivpuri, R., Cheng, X., Bedekar, V., Matsumotob, Y., Hashimoto, F., & Watkins, T. R. (2005). Effect of feed rate, workpiece hardness and cutting edge on subsurface residual stress in the hard turning of bearing steel using chamfer + hone cutting edge geometry. *Materials Science and Engineering: A, 394*, 238–248.

Huang, Y., Chou, Y. K., & Liang, S. Y. (2007). CBN tool wear in hard turning: A survey on research progresses. *The International Journal of Advanced Manufacturing Technology, 35*, 443–453.

Hwang, Y. K., & Lee, Ch. M. (2010). Surface roughness and cutting force prediction in MQL and wet turning process of AISI 1045 using design of experiments. *Journal of Mechanical Science and Technology, 24*, 1669–1677.

Islam, M. N. (2013). Effect of additional factors on dimensional accuracy and surface finish of turned parts. *Machining Science and Technology, 17*, 145–162.

Kumar, A. S., Durai, A. R., & Sornakumar, T. (2003). Machinability of hardened steel using alumina based ceramic. *International Journal of Refractory Metals & Hard Materials, 21*, 109–117.
Kurt, A., & Seker, U. (2005). The effect of chamfer angle of polycrystalline cubic boron nitride cutting tool on the cutting forces and the tool stresses in finishing hard turning of AISI 52100 steel. *Materials and Design, 26*, 351–356.
Lee, T. H. (2011). Development of a theoretical model to predict cutting forces for hard machining. *International Journal of Precision Engineering and Manufacturing, 12*, 775–782.
Lima, J. G. de, de Avila, R. F., & Abrao, A. M. (2007). Turning of hardened AISI 4340 steel using coated carbide inserts. *Proceedings of the Institution of Mechanical Engineers, Part B: Journal of Engineering Manufacture, 221*, 1359–1366.
Lin, H. M., Liao, Y. S., & Wei, C. C. (2008). Wear behavior in turning high hardness alloy steel by CBN tool. *Wear, 264*, 679–684.
List, G., Sutter, G., & Bouthiche, A. (2012). Cutting temperature prediction in high speed machining by numerical modelling of chip formation and its dependence with crater wear. *International Journal of Machine Tools and Manufacture*, 54–55, 1–9.
Luo, S. Y., Liao, Y. S., & Tsai, Y. Y. (1999). Wear characteristics in turning high hardness alloy steel by ceramic and CBN tools. *Journal of Materials Processing Technology, 88*, 114–121.
M'Saoubi, R., Outeiro, J. C., Chandrasekaran, H., Dillon, O. W., Jr, & Jawahir, I. S. (2008). A review of surface integrity in machining and its impact on functional performance and life of machined products. *International Journal of Sustainable Manufacturing, 1*, 203–236.
Maňková, I., Kovac, P., Kundrak, J., & Beňo, J. (2011). Finite element analysis of hardened steel cutting. *Journal of production engineering, 14*, 7–10.
Meyer, R., Köhler, J., & Denkena, B. (2012). Influence of the tool corner radius on the tool wear and process forces during hard turning. *The International Journal of Advanced Manufacturing Technology, 58*, 933–940.
More, A. S., Jiang, W., Brown, W. D., & Malshe, A. P. (2006). Tool wear and machining performance of CBN–TiN coated carbide inserts and PCBN compact inserts in turning AISI 4340 hardened steel. *Journal of Materials Processing Technology, 180*, 253–262.
Oliveira, A. J. de, Diniz, A. E., & Ursolino, D. J. (2009). Hard turning in continuous and interrupted cut with PCBN and whisker-reinforced cutting tools. *Journal of Materials Processing Technology, 209*, 5262–5270.
Özel, T. (2009). Computational modelling of 3D turning: Influence of edge micro-geometry on forces, stresses, friction and tool wear in PCBN tooling. *Journal of Materials Processing Technology, 209*, 5167–5177.
Panzera, T. H., Souza, P. R., Rubio, J. C. C., Abrao, A. M., & Mansur, T. R. (2012). Development of a three-component dynamometer to measure turning force. *The International Journal of Advanced Manufacturing Technology, 62*, 913–922.
Poulachon, G., Bandyopadhyay, B. P., Jawahir, I. S., Pheulpin, S., & Seguin, E. (2003). The influence of the microstructure of hardened tool steel workpiece on the wear of PCBN cutting tools. *International Journal of Machine Tools and Manufacture, 43*, 139–144.
Poulachon, G., Bandyopadhyay, B. P., Jawahir, I. S., Pheulpin, S., & Seguin, E. (2004). Wear behavior of CBN tools while turning various hardened steels. *Wear, 256*, 302–310.
Poulachon, G., Moisan, A., & Jawahir, I. S. (2001). Tool-wear mechanisms in hard turning with polycrystalline cubic boron nitride tools. *Wear, 250*, 576–586.
Ramesh, A., Melkote, S. N., Allard, L. F., Riester, L., & Watkins, T. R. (2005). Analysis of white layers formed in hard turning of AISI 52100 steel. *Materials Science and Engineering: A, 390*, 88–97.
Rech, J., Kermouche, G., Grzesik, W., Garcia-Rosales, C., Khellouki, A., & Garcia-Navas, V. (2008). Characterization and modelling of the residual stresses induced by belt finishing on a AISI52100hardened steel. *Journal of Materials Processing Technology, 208*, 187–195.
Remadna, M., & Rigal, J. F. (2006). Evolution during time of tool wear and cutting forces in thecase of hard turning with CBN inserts. *Journal of Materials Processing Technology, 178*, 67–75.
Ren, X. J., Yang, Q. X., James, R. D., & Wang, L. (2004). Cutting temperatures in hard turning chromium hardfacings with PCBN tooling. *Journal of Materials Processing Technology, 147*, 38–44.

Sahin, Y. (2009). Comparison of tool life between ceramic and cubic boron nitride (CBN) cutting tools when machining hardened steels. *Journal of materials processing technology, 209*, 3478–3489.

Sharma, V. S., Dhiman, S., Sehgal, R., & Sharma, S. K. (2008). Estimation of cutting forces and surface roughness for hard turning using neural networks. *Journal of Intelligent Manufacturing, 19*, 473–483.

Suresh, R., Basavarajappa, S., & Samuel, G. L. (2012). Some studies on hard turning of AISI 4340 steel using multilayer coated carbide tool. *Measurement, 45*, 1872–1884.

Sutter, G., Faure, L., Molinari, A., Ranc, N., & Pina, V. (2003). An experimental technique for the measurement of temperature fields for the orthogonal cutting in high speed machining. *International Journal of Machine Tools and Manufacture, 43*, 671–678.

Sutter, G., & Ranc, N. (2007). Temperature fields in a chip during high-speed orthogonal cutting – An experimental investigation. *International Journal of Machine Tools and Manufacture, 47*, 1507–1517.

Thamizhmanii, S., & Hasan, S. (2010). Relationship between flank wear and cutting force on the machining of hard Martensitic stainless steel by super hard tools. Proceedings of the World Congress on Engineering, London, U.K, 3, 2185–2190.

Thiele, J. D., & Melkote, S. N. (1999). Effect of cutting edge geometry and workpiece hardness on surface generation in the finish hard turning of AISI 52100 steel. *Journal of Materials Processing Technology, 94*, 216–226.

Umbrello, D., Rizzuti, S., Outeiro, J. C., Shivpuri, R., & M'Saoubi, R. (2008). Hardness-based flow stress for numerical simulation of hard machining AISI H13 tool steel. *Journal of Materials Processing Technology, 199*, 64–73.

Umbrello, D., Ambrogio, G., Filice, L., & Shivpuri, R. (2008). A hybrid finite element method–artificial neural network approach for predicting residual stresses and the optimal cutting conditions during hard turning of AISI 52100 bearing steel. *Materials and Design, 29*, 873–883.

Umbrello, D., Pu, Z., Carusoa, S., Outeiro, J. C., Jayal, A. D., Dillon, O. W., Jr, & Jawahir, I. S. (2011). The effects of cryogenic cooling on surface integrity in hard machining. *Procedia Engineering, 19*, 371–376.

Yallese, M. A., Chaoui, K., Zeghib, N., Boulanouar, L., & Rigal, J.-F. (2009). Hard machining of hardened bearing steel using cubic boron nitride tool. *Journal of Materials Processing Technology, 209*, 1092–1104.

Zawada-Tomkiewicz, A. (2011). Analysis of surface roughness parameters achieved by hard turning with the use of PCBN tools. *Estonian Journal of Engineering, 17*, 88–99.

Zhao, J., Yuan, X., & Zhou, Y. (2010). Cutting performance and failure mechanisms of an Al_2O_3/WC/TiCmicro- nano-composite ceramic tool. *International Journal of Refractory Metals and Hard Materials, 28*, 330–337.

Appendix

Major findings of the researches presented in this review are summarized in the following table.

Author(s)	Tool material	Cutting parameters	Remarks
Thiele and Melkote (1999)	CBN tools	Tool edge geometries, workpiece hardness, and feed rate	Large edge hones of CBN tools result in higher forces in the axial, radial, and tangential directions as compared to the tools with small edge hone during turning AISI 52100 steel
Ramesh et al. (2005)	CBN	Cutting speed	The grains of white layers are smaller than the grains of the bulk material Machining at higher speeds results in the formation of coarser grains
Diniz et al. (2005)	CBN	Two different kinds of CBN cutting tools (high CBN content material and CBN content material), cutting speed, three cutting operations continuous, semi-interrupted surfaces interrupted cutting)	Low CBN content tool is more suitable for continuous and semi-interrupted surfaces, whereas the high CBN content tool presents a slightly longer life during interrupted cutting
Adesta et al. (2009)	Cermet tools	Rake angles, cutting speed	Increase in negative rake angles causes higher wear, shorter duration of tool life, and poor surface finish. Further, high cutting speed turning gave shorter tool life, high wear rate but finer surface finish than conventional one
Gaitonde et al. (2009)	Ceramic inserts with TiN coating (CC650, CC650WG, and GC6050WH)	Depth of cut and machining time	CC650WG wiper insert performs better with reference to surface roughness and tool wear. The surface roughness is minimal at lower values of depth of cut and machining time in case of CC650 and CC650WG inserts machining, while the minimum surface roughness occurs at 0.4 mm depth of cut for hard turning with GC6050WH insert. Finally, the CC650 conventional insert is useful in reducing the cutting force
Fnides et al. (2009)	Mixed ceramic tool(insert CC650)	Cutting speed, feed rate, depth of cut	Statistical models of surface roughness criteria were developed

Meyer et al. (2012)	CBN content of 56%)	Geometrical contact conditions	It was suggested that by applying a load-specific tool and process design, the tool operation behavior and the productivity of the process can be enhanced in a beneficial way
Aslan (2005)	TiCN coated tungsten carbide, TiCN + TiAlN coated tungsten carbide, TiAlN coated cermet, mixed ceramic with Al_2O_3 + TiCN and CBN	Different cutting tool materials	TiAlN coated carbide and cermet tools perform slightly better than TiCN coated carbide tool which can be attributed to better high-temperature properties of TiAlN compared to TiCN CBN tool exhibits the best cutting performance in terms of both flank wear and surface finish and the highest volume of metal removal is obtained with CBN
Lima et al. (2007)	Coated carbide	Workpiece hardness, cutting speed, feed rate, depth of cut, cutting time	The abrasion is the principal wear mechanism and catastrophic failure takes place when attempting to machine the AISI 4340 steel with 525 HV
Chowdhury and Dhar (2011)	Coated carbide insert	Tool geometry	The application of MQL technique significantly helps to obtain better performance of coated carbide insert in comparison to dry condition
Hwang and Lee (2010)	Coated carbide insert	Cutting speed, Feed rate, Depth of cut, Supplied air pressure, Nozzle diameter	Turning of AISI 1045 with coated carbide inserts under MQL results in better surface finish than wet turning
Luo et al. (1999)	CBN Ceramic Carbide P10	Cutting speed, feed rate, depth of cut	A protective layer formed on the chip–tool interface plays an important role in wear behavior of CBN and ceramics tools
Kumar et al. (2003)	Ti[C,N] mixed alumina ceramic and zirconia toughened alumina ceramic		Ti[C, N] mixed alumina ceramic cutting tool produces the best surface finish during turning of EN 24 steel (HRC 40 and HRC 45)

(Continued)

Appendix (*Continued*).

Author(s)	Tool material	Cutting parameters	Remarks
Gaitonde (2009)	CC650, CC650WG, and GC6050WH ceramic inserts	Cutting speed, feed rate, machining time	CC650WG wiper insert performs better with reference to surface roughness and tool wear, while the CC650 conventional insert is useful in reducing the machining force, power and specific cutting force during turning of high chromium AISI D2 cold work tool steel
Elmunafi (2012)	Wiper coated ceramic	Cutting speed, feed rate	Wiper inserts produce better surface finish when turning ASSAB DF-3 grade hardened steel with hardness 55 HRC
Maňková et al. (2011)	Mixed oxide ceramic inserts (70% Al_2O_3 + 30% TiC)	Cutting speed, feed rate	Ceramic cutting tools have an advantage in the machining of hard work piece materials at high speed
Fnides et al. (2011)	Mixed ceramic tool	Cutting speed, feed rate, depth of cut	Statistical models of cutting force components for turning of AISI H11 grade steel treated at 50 HRC, with a mixed ceramic tool (insert CC650) was developed
Zhao et al. (2010)	The Al_2O_3-based composite ceramic tool material reinforced with WC microparticles and TiC nanoparticles	Cutting speed, depth of cut	The tool life of the Al_2O_3/WC/TiC ceramic tool increases when the cutting speed increases to 170 m/min. The longer tool life of the Al_2O_3/WC/TiC composite ceramic tool is attributed to its synergistic strengthening/toughening mechanisms induced by the WC microparticles and TiC nanoparticles
Oliveira et al. (2009)	PCBN and alumina-based ceramic reinforced with silicon carbide tools	Continuous and interrupted surfaces	The longest tool life is achieved using PCBN in continuous turning, but similar tool longevity is attained in interrupted turning using both PCBN and ceramic. PCBN tools result in lower roughness for continuous and interrupted surfaces
Sahin (2009)	Mixed alumina ceramic with that of Al_2O_3 (70%):TiC (30%) matrix and CBN cutting tools	Cutting speed, feed rate, tool hardness	The CBN cutting tool has the best performance as compared to the ceramic based cutting tool

Poulachon et al. (2001)	PCBN cutting tool	Workpiece hardness	TiN coated PCBN improves the tool-wear and hence the tool-life by reducing the diffusion wear between workpiece and tool rake face.
Diniz et al. (2003)	CBN	Cutting speed, and under three cutting conditions: dry cutting, wet cutting, and minimum volume of oil i.e. MVO (oil flow of 10 ml/h)	Surface roughness in all three cutting conditions i.e. dry cutting, wet cutting, and minimum volume of oil cutting is the lowest at the highest cutting speed
Poulachon et al. (2003)	PCBN	Steel workpiece with different hardness	A large variation in tool wear occurs in the machining of these steels
Benga and Abrao (2003)	Mixed alumina, whisker reinforced alumina, and PCBN inserts	Cutting speed, feed rate	At low feed rates PCBN compact tool gives best results in terms of the tool life followed by the mixed alumina tool. However, as feed rate is increased tool life of the whisker reinforced alumina tool is better
Poulachon et al. (2004)	CBN	Different steels (AISI D2 cold work steel, AISI H11 hot work steel, 35NiCrMo16 hot work steel and AISI 52100bearing steel), cutting speed, and feed rate	The generated tool flank grooves are correlated with the hard carbide content of the workpieces
Kurt and Seker (2005)	PCBN	Chamfer angle	Chamfer angle significantly affects the cutting forces and the tool stress
Galoppi et al. (2006)	Different types of CBN inserts with wiper geometry, coated with TiAlN and TiN as well as with no coated ones	Cutting speed, and feed rate	CBN insert with a wiper geometry presents longer tool life compared to no coated tools
Remadna and Rigal (2006)	CBN	Cutting speed	The wear of CBN does not directly or appreciably affect the manufactured surface
More et al. (2006)	CBN–TiN coated carbide inserts and PCBN compact inserts	Cutting speed, and feed rate	The crater wear of the CBN–TiN coated inserts is less than that of the PCBN inserts because of the lubricity of TiN capping layer on the CBN–TiN coating
Arsecularatne et al. (2006)	PCBN tools	Cutting speed, and feed rate	The highest acceptable values of tool life and volume of material removal is obtained at low speed

(*Continued*)

Appendix (*Continued*).

Author(s)	Tool material	Cutting parameters	Remarks
Lin et al. (2008)	CBN tools which contained 50% and 45% of CBN	Cutting speed	At high cutting speed a protective layer is formed on the chip–tool interface which prolongs life of the CBN tool. With further increase in the cutting speed (i) cutting temperature becomes the dominant factor instead of the cutting force which causes inhomogeneous shear strain, and a transition from continuous chip to saw-tooth chip, (ii) the friction force increases because of the very irregular chip–tool contact and it results in the removal of the protective layer, and (iii) the bond between tool particles gets weakened due to serious diffusion between the work material and the cutting tool. Subsequently, hard particles of the tool get detached from it and the tool life is reduced
Özel (2009)	PCBN	Different edge micro-geometry, cutting speed, and feed rate	Micro-geometry insert edge design improves surface integrity, the tool life, and reduces tool wear depth and wear rate
Yallese et al. (2009)	CBN	Cutting speed, feed rate, and depth of cut	At very high cutting speed the machining system becomes unstable and produces significant sparks and vibrations after only a few minutes of work and causes tool wear
Godoy and Diniz (2011)	CBN and ceramic tools	Continuous and interrupted cutting, cutting speed	The CBN tools exhibit a much better performance with respect to both tool life and workpiece surface roughness than the ceramic tools
Zawada-Tomkiewicz (2011)	PCBN coated and uncoated tools		Turning with PCBN tools can produce a very smooth and uniform surface
			The wedge's wear is almost unnoticeable for PCBN coated tool material

Deepakkumar and Sadaiah (2011)	CBN	Cutting speed, feed rate, depth of cut, and different cutting environments (dry, wet, and MQL)	The tool wear increases with increase in the cutting speed and feed rate
Panzera et al. (2012)	Coated carbide inserts	Cutting speed, feed rate, and depth of cut	The three components of the turning force decrease slightly as cutting speed increases and also they increase linearly with feed rate and depth of cut
Aouici et al. (2010)	CBN tool (57% CBN and 35% Ti(C, N))	Cutting speed, feed rate, and depth of cut	Tangential cutting force is very sensitive to the variation in cutting depth Thrust force is dominating compared to both others cutting forces Surface roughness is very sensitive to the variation of the feed rate The temperature is greatly influenced by the cutting speed
Amritkar et al. (2012)	Uncoated tungsten carbide tool	Cutting speed and feed rate	The tool-work thermocouple technique is the best method for measuring the average chip–tool interface temperature during metal cutting
Sutter et al. (2003)	Carbide tools TiCN coated	Cutting speed and depth of cut	The temperature at the chip increases with the increase in both cutting speed as well as the depth of cut
Ren et al. (2004)	PCBN		The average cutting temperature increases with increase in both the cutting speed and feed rate
Abukhshim et al. (2005)	Uncoated cemented carbide	Cutting speed	The maximum temperature at the tool–chip contact increases with cutting speed but not linearly and this could be attributed to the trend of the heat fraction flowing into the tool

(Continued)

Appendix (*Continued*).

Author(s)	Tool material	Cutting parameters	Remarks
Rech et al. (2008)	PCBN	Cutting speed, feed, depth of cut, tool geometry (all fixed values), and subsequent belt finishing	Belt finishing process improves the surface integrity by the induction of strong compressive residual stresses in the external layer and due to reasonable improvement in the surface roughness Lubrication used in the belt finishing technique is a key parameter to get compressive stresses to improve surface roughness
Caruso et al. (2011)	PCBN	Tool geometery, workpiece hardness, cutting speed	Tool geometry, workpiece hardness, and cutting parameters significantly affect the surface residual stress
Asilturk and Akkus (2011)	Coated carbide	Cutting speed, feed rate, and depth of cut	The feed rate significantly affects the surface roughness
Aouici et al. (2012)	CBN which is essentially made of 57% CBN and 35% TiCN	Workpiece hardness, cutting speed, feed rate, and depth of cut	The feed rate and workpiece significantly affect the surface roughness
Umbrello et al. (2011)	CBN	Under dry and cryogenic conditions, cutting speed	Cryogenic coolant significantly affects the surface integrity and improves product's functional performance
Grzesik et al. (2012)	CBN tools containing about 60% CBN	Under dry and cryogenic conditions, feed rate	Hard machining produces surfaces with acceptable surface roughness and, in some cases, with attractive service properties. In addition, cryogenic hard cutting operations can partly eliminate grinding operations in cases when white layer is not produced
Abhang and Hameedullah (2012)	CNMA 120404, CNMA 120408, CNMA 120412, and diamond-shaped carbide	Cutting speed, feed rate, tool nose radius, and concentration of solid–liquid lubricants (MQL)	The surface roughness and the chip thickness in the turning process can be improved effectively through application of MQL

Islam (2013)		Cooling method, blank size, and work material	The effect of these parameters on surface roughness is relatively low
			Dimensional accuracy and surface roughness of different work materials get influenced differently by the cooling methods, and in most cases the best result is achieved by MQL
Grzesik and Zalisz (2008)	Mixed ceramic	Feed rate	The wear mechanisms involve abrasion, fracture, plastic flow, material transfer and tribochemical
El Hakim et al. (2011)	PCBN (CBN + TiN), TiN coated PCBN (CBN + TiN), mixed alumina ceramic (Al_2O_3 + TiC), coated tungsten carbide (TiN coated over a multilayer coating (TiC/TiCN/Al_2O_3))	Tool materials	The mixed alumina ceramic and coated carbide tool materials have longer tool life than PCBN tools
Chinchanikar and Choudhury (2013)	Coated cemented carbide inserts	Workpiece hardness, cutting speed, feed rate, depth of cut, and type of coating for coated cemented carbide inserts	Flank wear is dominant wear form for CVD coated tool and crater wear is dominant wear form for PVD coated tool
Gaitonde et al. (2009)	CC650WG wiper ceramic inserts	Cutting speed, feed rate, and machining time	The maximum tool wear occurs at a cutting speed of 150 m/min for all values of feed rate and the tool wear increases with the increase in machining time
Dogra et al. (2011)	CBN with coated/uncoated carbide	Cutting speed under cryogenic cooling	The flank wear of CBN is lower than that of other inserts

12

Stabilizing wave attenuation effects in turning process

Chigbogu Godwin Ozoegwu*

Department of Mechanical Engineering, Nnamdi Azikiwe University, PMB 5025, Awka, Nigeria

Efforts have been made and continue to be made to understand the self-excited vibrations of machine tools called regenerative chatter. Theory of stabilizing wave attenuation effects in machining process presented in another work is expanded in this work to postulate the behaviour of turning stability lobes to changes in material, process and structural parameters. The wave attenuation theory simply stated that rise in attenuating forces suppresses chatter instability. Analysis in this work showed that rise in tool natural frequency ω_n, damping ratio ξ or feed speed v causes a rise in attenuating forces thus suppressing chatter. As expected from the wave attenuation theory harder workpiece materials with higher cutting coefficients C will better resist wave attenuation, and thus exhibit more chatter instability. Numerical verification of postulations is given.

Keywords: turning process; wave attenuation; chatter; turning point critical force; stability

1. Introduction

Regenerative vibrations are the self-excited vibrations in machine tools that are referred to as secondary chatter in literature (Wiercigroch & Budak, 2001). Regenerative chatter is a much more detrimental problem to the machining industry than the primary chatter phenomena that stem from frictional, thermo-mechanical and mode-coupling effects. Some of the unwanted effects of regenerative chatter are noise pollution in the work environment, poor surface quality, aggravated tool wear, reduced productivity and waste of materials and energy. Two of the most important reviews of machine tool chatter research (Quintana & Ciurana, 2011; Siddhpura & Paurobally, 2012) noted that regenerative chatter is not yet fully understood even though it has received most attention of all chatter phenomena studied over a century. Regenerative chatter is caused by Regenerative effects which are effects of waviness on a machined surface that are in turn caused by response of the machine tool system to disturbances. The waves of consecutive turns are normally out of phase causing cutting force variation that excites the tool. If cutting process parameters are stable, then cutting force variation decays because phase difference of consecutive turns die out with time. The opposite is the case when cutting process parameters constitute an unstable combination. Before regenerative chatter was distinguished and understood as stemming from self-excitation effects it was originally thought by Arnold (1946) as stemming from negative damping. Some of the standard early works (Merritt, 1965; Tobias, 1961; Tobias & Fishwick, 1958) on regenerative chatter considered a single degree of freedom orthogonal turning process vibrating in

*Email: chigbogug@yahoo.com

the feed direction only. Analysis of non-linear regenerative chatter was introduced by Hanna and Tobias (1974). It is known that the so-called Hopf bifurcation (lose of stability) in turning operation occurs when a pair of complex conjugate characteristic roots crosses from the left-half plane to right-half plane of the complex plane. More recently non-linear analysis has been strengthened by the experiment of Shi and Tobias (1984) and analysis of Stepan and Kalmar-Nagy (1997) by revealing that Hopf bifurcation of turning operation is subcritical. In addition to chatter frequencies of Hopf bifurcation type, damped natural frequency can also be spotted in the frequency response spectrum of high-speed turning tool due to the effects of tool-workpiece contact loss. Damped natural frequency has been experimentally confirmed to exist in the frequency response spectrum of high-speed milling (Insperger, Stepan, Bayly, & Mann, 2003). Experimental data of chatter and other vibrations in the turning and milling processes have been characterized using multi-scale methods (Litak, Syta, & Rusinek, 2011; Sen, Litak, Syta, & Rusinek, 2013) and flicker-noise spectroscopy (Litak, Polyakov, Timashev, & Rusinek, 2013). The theory of stabilizing wave attenuation effects in machining process was presented by Ozoegwu and Omenyi (2013). The theory posits that some of the force of interaction between the tool and workpiece acts normal to surface waves, thus depressing or flattening them in a way as to attenuate the chatter–causing regenerative effects. The attenuating force is either compressive when there is a continuous cutting edge-workpiece engagement or impulsive when the active pass of the cutting edge is highly interrupted as in high-speed machining. The theory further posits that any variable of the attenuating force that varies in a way as to increase the amplitude of the attenuating force will suppress chatter instability. In Ozoegwu and Omenyi (2013), the wave attenuation theory was able to explain why chatter stability rises with rise in feed, fall in number of teeth and fall in hardness of workpiece material. The theory of stabilizing wave attenuation effects is being expanded in this work to cover the turning process.

2. The cutting force, dynamics and stability of regenerative turning

Figure 1 represents an orthogonal turning process in which the cutting edge of the tool is perpendicular to the feed motion. In the physical configuration the workpiece is clamped in the chuck of a rotating spindle while the tool is fed in the workpiece. Regenerative chatter is the self-excited vibration in machining due to the effects of waviness (regenerative effects) on a machined surface resulting from disturbed dynamic interaction between the tool and the workpiece. The present turn of the workpiece has waviness that is not in phase with those of the last turn resulting chip thickness variation, hence cutting force variation that powers vibration which subsequently builds up to chatter if cutting parameter combination is unfavourable. As shown in Figure 1, a tool with modal parameters; m the mass, c the equivalent viscous damping coefficient and k the stiffness is fed into the workpiece at a speed v in $x-$ direction. The response of the tool $x(t)$ to the regenerative cutting force variation is seen from Figure 1 to be governed by the differential equation

$$m\ddot{x}(t) + c[\dot{x}(t) - v] + k[x(t) - vt] + F_x(t) = 0 \tag{1}$$

The cutting force $F_x(t)$ has the empirical form (Insperger, 2002; Stepan, Szalai, & Inspeger, 2004; Tlusty, 2000);

$$F_x(t) = Cwf_a^{\gamma} \tag{2}$$

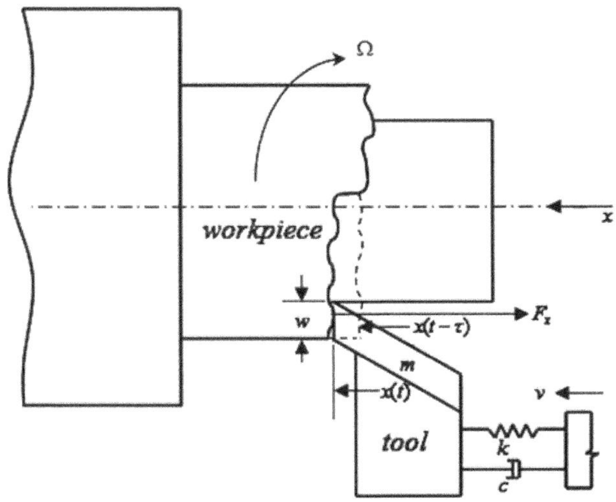

Figure 1. Mechanical model of orthogonal turning.

where C is the cutting coefficient, w is the depth of cut, f_a is the actual feed rate defined as difference of present and one period delayed position of tool; $f_a = x(t) - x(t-\tau)$ and γ is the feed exponent of the cutting force that has typical values 1, .8 and 3/4. With the use of Equation (2), Equation (1) is re-arranged to give

$$m\ddot{x} + c\dot{x} + kx = cv + kvt - Cw(x - x_\tau)^\gamma. \tag{3}$$

where $x(t) = x$ and $x(t-\tau) = x_\tau$. The response of the tool is a linear combination of prescribed feed motion vt, static deflection $x_t(t) = -(Cwf^\gamma)/k = -[Cw(v\tau)^\gamma]/k$ and perturbation $z(t)$ such that Equation (3) becomes.

$$m\ddot{z} + c\dot{z} + kz = Cw(v\tau)^\gamma - Cw[v\tau + (z - z_\tau)]^\gamma. \tag{4}$$

The linearized Taylor's series expansion Equation (4) about $v\tau$ gives

$$m\ddot{z} + c\dot{z} + kz = -hw(z - z_\tau). \tag{5}$$

where $h = C\gamma(v\tau)^{\gamma-1}$. Recognizing that the natural frequency and damping ratio of the tool are given in terms of modal parameters k, m and c, respectively as $\omega_n = \sqrt{k/m}$ and $\xi = c/2\sqrt{mk}$ gives the modal form

$$\ddot{z} + 2\xi\omega_n\dot{z} + \left(\omega_n^2 + \frac{hw}{m}\right)z = \frac{hw}{m}z_\tau. \tag{6}$$

Equation (6) is the linear dynamical model for the regenerative turning tool. With the substitutions $y_1 = z$ and $y_2 = \dot{z}$ made, Equation (6) could be put in state differential equation form

$$\underline{y} = \mathbf{A}\mathbf{y} + \mathbf{B}\mathbf{y}_\tau \tag{7}$$

where $\mathbf{A} = \begin{bmatrix} 0 & 1 \\ -\left(\omega_n^2 + \frac{hw}{m}\right) & -2\xi\omega_n \end{bmatrix}$, $\mathbf{B} = \begin{bmatrix} 0 & 0 \\ \frac{hw}{m} & 0 \end{bmatrix}$, $\mathbf{y} = \begin{Bmatrix} y_1 \\ y_2 \end{Bmatrix}$, $\mathbf{y}_\tau = \begin{Bmatrix} y_{1,\tau} \\ y_{2,\tau} \end{Bmatrix}$ and $y_{i,\tau} = y_i(t-\tau)$ for $i = 1$ and 2. A trial solution of form $y(t) = \boldsymbol{K}e^{\lambda t}$ put in Equation (7) to give the characteristic equation

$$|\lambda\mathbf{I} - \mathbf{A} - \mathbf{B}e^{-\lambda\tau}| = 0. \tag{8}$$

On the stability lobes of turning process the pair of maximum-magnitude eigenvalues of Equation (8) are on the imaginary axis; $\lambda = \pm i\omega$ given the pair of equations

$$-\omega^2 + \omega_n^2 + \frac{hw}{m}(1 - \cos\omega\tau) = 0 \tag{9}$$

$$2\xi\omega_n\omega + \frac{hw}{m}\sin\omega\tau = 0. \tag{10}$$

These are solved simultaneously to give the critical parameter combinations

$$\Omega_c = \frac{60}{\tau_c} = \frac{30\omega}{j\pi - tan^{-1}\left(\dfrac{\omega^2 - \omega_n^2}{2\xi\omega_n\omega}\right)}, \qquad j = 1, 2, 3, \tag{11}$$

$$w_c = \frac{m}{2h}\left\{\frac{(\omega^2 - \omega_n^2)^2 + 4\xi^2\omega_n^2\omega^2}{\omega^2 - \omega_n^2}\right\}. \tag{12}$$

The critical speed Ω_c and critical depth of cut w_c constitute the pair of stability equations for turning process. The details of deriving Equations (11) and (12) from Equations (9) and (10) are seen in other works (Ozoegwu, 2011; Ozoegwu & Omenyi, 2012).

3. The stabilizing wave attenuation effects in turning process

What follows is geared towards establishing the wave attenuation theory for orthogonal turning process. The cutting force model used in this work is seen from Figure 1 to be normal to the machined surface waves and thus, it is reasonably true that the magnitude of attenuating force will be proportional to cutting force $F_x(t)$ in the manner; $|F_{at}| = \alpha|F_x(t)|^\beta$ where α and β are both positive real numbers. Equations (11) and (12) that jointly describe the stability lobes of turning process also defines the critical cutting force F_c. Being that $\alpha, \beta \geq 0$ the values $\alpha = 1$ and $\beta = 1$ can be adopted without loss of generality such that F_c becomes used as a measure of attenuating force on a stability lobe. This is possible because sign of the partial derivative $\frac{\partial}{\partial\varepsilon}|F_{at}| = \alpha\beta|F_x(t)|^{\beta-1}\frac{\partial}{\partial\varepsilon}|F_x(t)|$ is solely determined by the sign of the partial derivative $\frac{\partial}{\partial\varepsilon}|F_x(t)|$ where ε is a parameter of F_{at}. Since a turning process on stability lobe is stable in the Liapunov sense, then the critical cutting force F_c is seen from Equation (2) to be given by

$$F_c = Cv^\gamma w_c \tau_c^\gamma \tag{13}$$

where in light the specific force variation $h = C\gamma(v\tau)^{\gamma-1}$ and Equations (11) and (12), τ_c and w_c are as given as

$$\begin{aligned} \tau_c &= \frac{60}{\Omega_c} = \frac{2}{\omega}\left\{j\pi - \tan^{-1}\left(\frac{\omega^2 - \omega_n^2}{2\xi\omega_n\omega}\right)\right\} \\ w_c &= \frac{mv^{1-\gamma}}{2C\gamma}\tau_c^{1-\gamma}\left\{\frac{(\omega^2 - \omega_n^2)^2 + 4\xi^2\omega_n^2\omega^2}{\omega^2 - \omega_n^2}\right\} \end{aligned} \tag{14}$$

It is of interest here to investigate the effect of variation of the passive parameters ω_n and ξ and the prescription parameter v on F_c and interpret the result in light of the wave attenuation theory. The most mathematically amenable is the critical cutting force at the turning points of the stability lobes. The stability lobes generated using Equation (14) are normally given as w_c as a function of Ω_c. The turning point (minimum point) for the *j*th lobe is derived in details in (Ozoegwu, Omenyi, Achebe & Uzoh, 2012) the non-dimensionalized cutting parameter space to be located at

$$\begin{aligned} \tilde{\Omega}_{ct}^{j} &= \frac{\pi \Omega_{ct}^{j}}{30\omega_n} = \frac{\pi\sqrt{2\xi+1}}{\{j\pi - \tan^{-1}(1/\sqrt{2\xi+1})\}} \\ \tilde{W}_{ct} &= \frac{w_{ct}h}{m\omega_n^2} = 2\xi(\xi+1) \end{aligned} \tag{15}$$

It is clear from Equations (14) and (15) in light of the specific force variation $h = C\gamma(v\tau)^{\gamma-1}$ that the critical cutting parameters at the turning point of *j*th lobe is at

$$\tau_{ct}^{j} = \frac{2}{\pi\sqrt{2\xi+1}}\left\{j\pi - \tan^{-1}\left(\frac{1}{\sqrt{2\xi+1}}\right)\right\} \tag{16}$$

$$w_{ct} = \frac{2^{2-\gamma}m\omega_n^{\gamma+1}}{\gamma C v^{\gamma-1}}(\xi^2+\xi)(2\xi+1)^{\frac{1}{2}(\gamma-1)}\left\{j\pi - \tan^{-1}\left(\frac{1}{\sqrt{2\xi+1}}\right)\right\}^{(1-\gamma)} \tag{17}$$

When Equations (16) and (17) are inserted in Equation (13) the turning point critical force reads

$$F_{ct} = \frac{4mv\omega_n}{\gamma}(\xi^2+\xi)(2\xi+1)^{-\frac{1}{2}}\left\{j\pi - \tan^{-1}\left(\frac{1}{\sqrt{2\xi+1}}\right)\right\} \tag{18}$$

It is seen from Equation (18) that the magnitude of the turning point critical force F_{ct} will rise with rise in either feed v or natural frequency ω_n since both partial derivatives; $\frac{\partial F_{ct}}{\partial v}$ and $\frac{\partial F_{ct}}{\partial \omega_n}$ are positive. This causes a rise in the attenuating forces which by the wave attenuation theory means more flattening of the regenerative effects and suppression of chatter instability. This point is verified by generating stability charts of 15 lobes ($j = 1$ to 15) for a system with $m = .0431\,\text{kg}$, $\gamma = .75$, $C = 3.5 \times 10^7\,\text{Nm}^{-7/4}$ and damping ratio $\xi = .02$. The natural frequencies $\omega_n = 1000, 3000,\ 5700$ and $7000\,\text{rad s}^{-1}$ are considered at fixed feed $v = .0025\,\text{ms}^{-1}$ in generating Figure 2. The feed speeds $= .0008333,\ .0025,\ .08$ and $.6\,\text{ms}^{-1}$ are considered for a turning tool with natural frequency $\omega_n = 5700\,\text{rad s}^{-1}$ in generating Figure 3.

The effect of variation of damping ratio ξ on the turning point critical force F_{ct} is not obvious from Equation (18) except on further investigation. This further investigation involves determining and checking the sign of the derivative $\frac{\partial F_{ct}}{\partial \xi}$. The derivative is

$$\frac{\partial F_{ct}}{\partial \xi} = \frac{4mv\omega_n}{\gamma(2\xi+1)}\left\{\frac{\xi^2+\xi}{2\xi+2} + \frac{3\xi^2+3\xi+1}{\sqrt{2\xi+1}}\left[j\pi - \tan^{-1}\left(\frac{1}{\sqrt{2\xi+1}}\right)\right]\right\} \tag{19}$$

Since machine tools usually have low damping ratio that lie in the range $\xi \approx .005 - .02$ (Insperger, 2002), the derivative $\frac{\partial F_{ct}}{\partial \xi}$ as given in Equation (19) is positive meaning that F_{ct} rises with rise in damping ratio ξ. Based on the wave attenuation theory a rise in ξ will suppress chatter being that rise in F_{ct} which is a measure of the attenuating force improves flattening of the waves. To verify this point damping ratios ξ= .005, .01, .015 and .02 are considered at fixed natural frequency $\omega_n = 5700\,\text{rad s}^{-1}$ and feed speed $v = .0025\,\text{ms}^{-1}$ in generating Figure 4. It is visible from Figure 4 that rise in ξ only improves the domain of delay independent stability of the turning stability chart. This

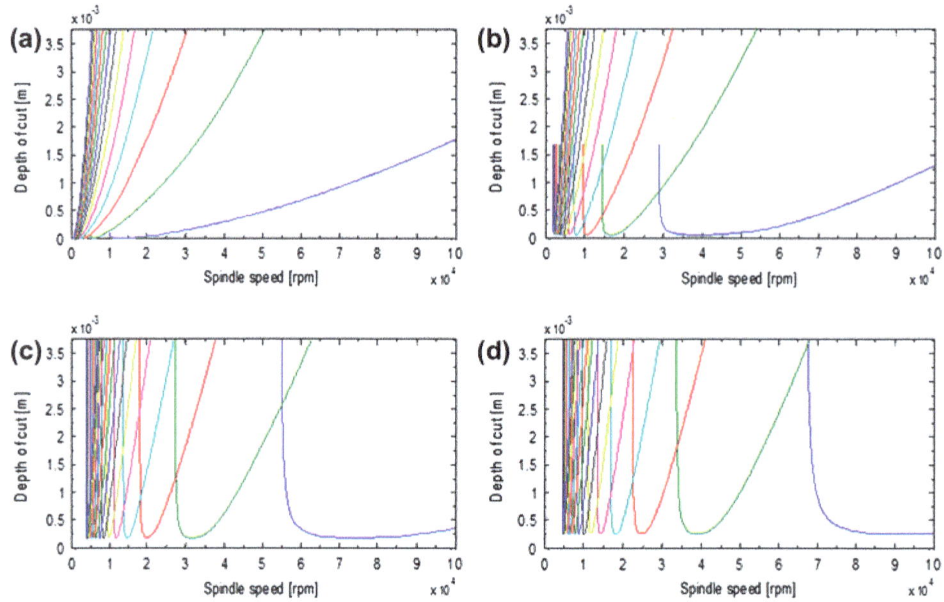

Figure 2. Stability charts of turning tools with same $m = .0431$, $\gamma = .75$, $C = 3.5 \times 10^7\,\mathrm{Nm}^{-7/4}$, $\xi = .02$ and $v = .0025\,\mathrm{ms}^{-1}$ when natural frequencies are (a) 1000 rpm, (b) 3000 rpm, (c) 5700 rpm and (d) 7000 rpm. It is seen that in conformity with the wave attenuation theory that the chatter instability (union of all domain above the stability lobes) is suppressed with rise in natural frequency.

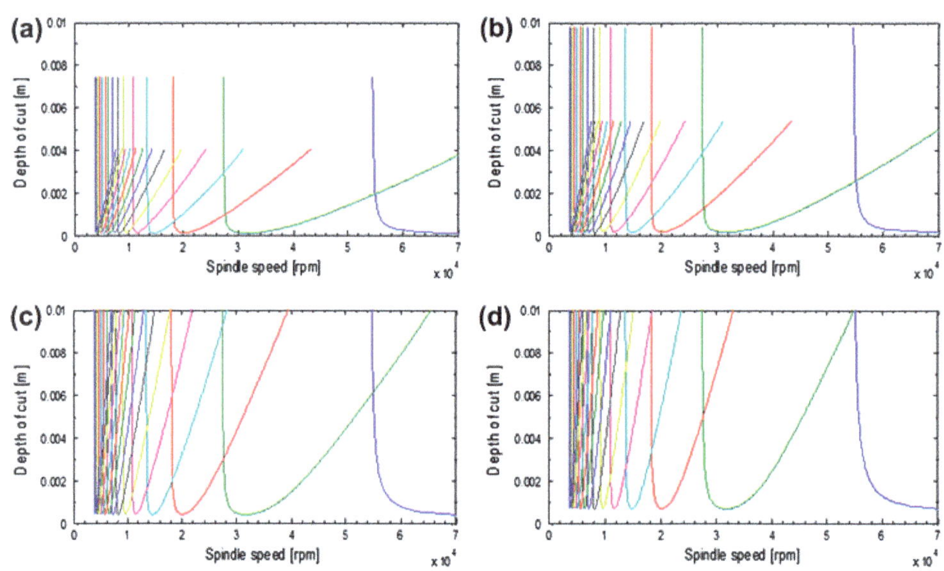

Figure 3. Stability charts of a turning tool with $m = .0431$, $\gamma = .75$, $C = 3.5 \times 10^7\,\mathrm{Nm}^{-7/4}$, $\xi = .02$ and $\omega_n = 5700$ rpm when feed speeds are (a) $.0008333\,\mathrm{ms}^{-1}$ (b) $.0025\,\mathrm{ms}^{-1}$ (c) $.08\,\mathrm{ms}^{-1}$ (d) $.6\,\mathrm{ms}^{-1}$. It is seen that in conformity with the wave attenuation theory that the chatter instability (union of all domain above the stability lobes) is suppressed with rise in feed speed.

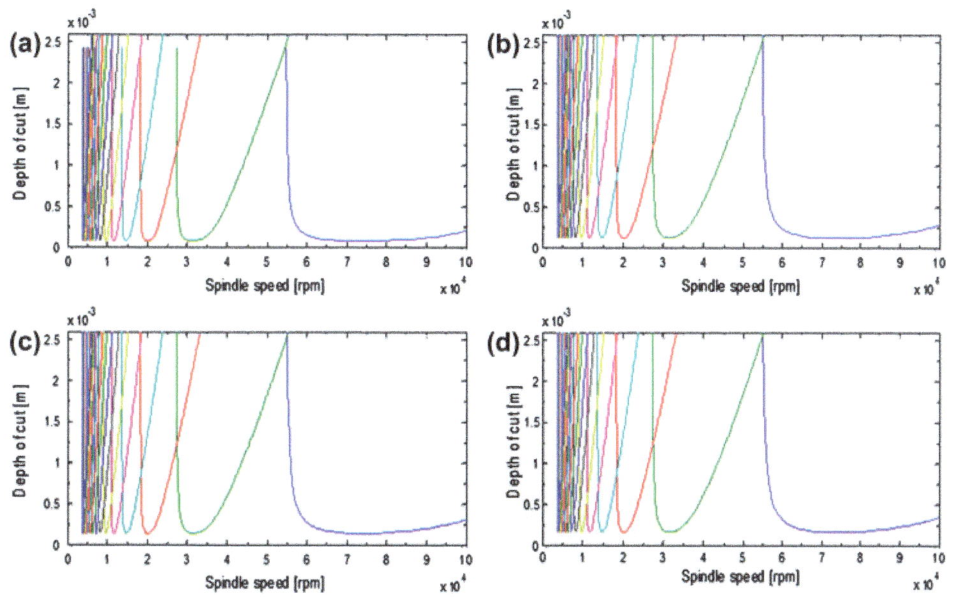

Figure 4. Stability charts of turning processes with $\omega_n = 5700$, $v = .0025\,\mathrm{ms}^{-1}$ and $C = 3.5 \times 10^7\,\mathrm{Nm}^{-7/4}$ and damping ratios (a) .005, (b) .01, (c) $\xi = .015$ and (d) $\xi = .02$. The domain of delay-independent stability (the domain in which a fixed depth of cut line does not cross any of the lobes) rises with rise in ξ.

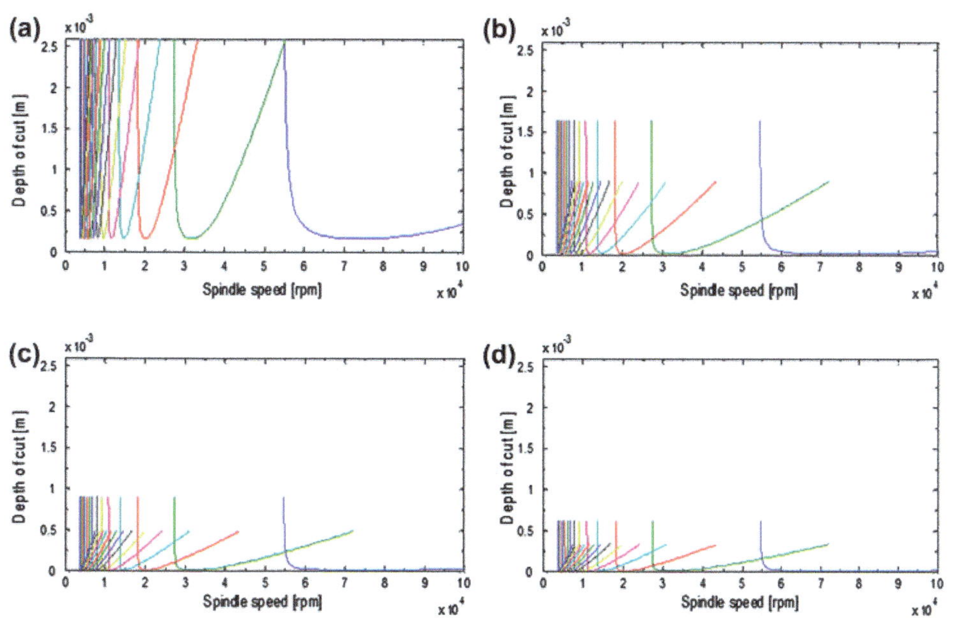

Figure 5. Stability charts of turning with $m = .0431$, $\gamma = .75$, $\xi = .02$ and $v = .0025\,\mathrm{ms}^{-1}$ and $\omega_n = 5700$ for the cutting coefficients (a) $C = 3.5 \times 10^7\,\mathrm{Nm}^{-7/4}$ (b) $C = 2.0667 \times 10^8\,\mathrm{Nm}^{-7/4}$ (c) $C = 3.7834 \times 10^8\,\mathrm{Nm}^{-7/4}$ (d) $C = 5.5 \times 10^8\,\mathrm{Nm}^{-7/4}$. It is seen that size of stable domain falls with rise in C suggesting that stabilizing effects of wave attenuation is resisted more at higher C.

point was noted in Ozoegwu et al. (2012). One major consequence of the wave attenuation theory pointed in Ozoegwu and Omenyi (2013) is that the waves of harder and more heat resistant materials will better resist the attenuating forces, and thus enhance chatter instability. Thus, wave attenuation effect is expected to favour chatter stability in cutting of some light materials like composites and magnesium but this favour will considerably diminish when machining materials of hard machinability like stainless steel and inconel. Numerical verification of this point is presented in Figure 5 for a turning tool with parameters; $m = .0431$ kg, $\gamma = .75$, $\xi = .02$ and $v = .0025$ ms^{-1} and $\omega_n = 5700$ rad s^{-1} machining four workpiece materials with different cutting coefficients C.

4. Conclusion

The theory of stabilizing wave attenuation effects is developed for the turning process. The theory generally states that some of the force of interaction between the tool and workpiece is flattening on regenerative waves thus attenuating and depopulating the chatter –causing regenerative effects. The flattening force is called the attenuating force. In this work a measure for the attenuating force in turning process is established as a function of tool modal parameters; natural frequency ω_n and damping ratio ξ and turning process parameter; feed speed v. It is shown that the attenuating force rises in response to rise of each of ω_n, ξ, v. This by the wave attenuation theory means more flattening of the regenerative effects and suppression of chatter instability. One major consequence of the wave attenuation theory is noted; the waves of harder and more heat-resistant materials will better resist the attenuating forces thus retaining chatter instability. All the above points deriving from the wave attenuation are verified numerically.

References

Arnold, R. N. (1946). The Mechanism of tool vibration in the cutting of steel. *Proceedings of institution of mechanical engineers, 54*, 261–284.

Hanna, N., & Tobias, S. (1974). A theory of nonlinear regenerative chatter. *ASME Journal of Engineering for Industry, 96*, 247–255.

Insperger, T. (2002). *Stability analysis of periodic delay-differential equations modelling machine tool chatter*. Budapest: Budapest University of Technology and Economics.

Insperger, T., Stepan, G., Bayly, P. V., & Mann, B. P. (2003). Multiple chatter frequencies in milling processes. *Journal of Sound and Vibration, 262*, 333–345.

Litak, G., Polyakov, Y. S., Timashev, S. F., & Rusinek, R. (2013). Dynamics of stainless steel turning: Analysis by flicker-noise spectroscopy. *Physica A: Statistical Mechanics and its Applications, 392*, 6052–6063.

Litak, G., Syta, A., & Rusinek, R. (2011). Dynamical changes during composite milling: Recurrence and multiscale entropy analysis. *The International Journal of Advanced Manufacturing Technology, 56*, 445–453.

Merritt, H. E. (1965). Theory of self-excited machine-tool chatter: Contribution to machine-tool chatter research – 1. *Journal of Engineering for Industry, 87*, 447–454.

Ozoegwu, C. G. (2011). *Chatter of plastic milling CNC machine*. Awka: Nnamdi Azikiwe University Awka.

Ozoegwu, C. G., & Omenyi, S. N. (2012). Stability characterization of a turning process. *Journal of Engineering and Applied Sciences, 8*, 68–75.

Ozoegwu, C. G., & Omenyi, S. N. (2013). Wave attenuation effects on the chatter instability of end-milling. *Noise Control Engineering Journal, 61*, 436–444.

Ozoegwu, C. G., Omenyi, S. N., Achebe, C. H., & Uzoh, C. F. (2012). Effect of modal parameters on both delay-independent and global stability of turning process. *Journal of Mechanical Engineering and Automation, 2*, 159–168.

Quintana, G., & Ciurana, J. (2011). Chatter in machining processes: A review. *International Journal of Machine tools and manufacture, 51*, 363–376.
Sen, A. K., Litak, G., Syta, A., & Rusinek, R. (2013). Intermittency and multiscale dynamics in milling of fiber reinforced composites. *Meccanica, 48*, 738–789.
Shi, H. M., & Tobias, S. A. (1984). Theory of finite amplitude machine tool instability. *International Journal of Machine Tool Design and Research, 24*, 45–69.
Siddhpura, M., & Paurobally, R. (2012). A review of chatter vibration research in turning. *International Journal of Machine tools and manufacture, 61*, 27–47.
Stepan, G., & Kalmar-Nagy, T. (1997). *Nonlinear regenerative machine tool vibrations.* Proceedings of DETC'97 1997 ASME Design Engineering Technical Conferences, Sacramento, CA, September 14–17.
Stepan, G., Szalai, R., & Inspeger, T. (2004). This is a chapter. In R. Radons & R. Neugebauer (Eds.), *Nonlinear dynamics of production systems: Nonlinear dynamics of high-speed milling subjected to regenerative effect* (pp. 111–127). Weinheim: Wiley-VCH.
Tlusty, J. (2000). *Manufacturing processes and equipment.* Abingdon: Prentice Hall, New Jersey.
Tobias, S. A. (1961). Machine tool vibration research. *International Journal of Machine Tool Design and Research, 1*, 1–14.
Tobias, S. A., & Fishwick, W. (1958). The chatter of lathe tools under orthogonal cutting conditions. *Transactions of ASME, 80*, 1079–1088.
Wiercigroch, M., & Budak, E. (2001). Sources of nonlinearities, chatter generation and suppression in metal cutting. *Philosophical Transactions of the Royal Society A: Mathematical, Physical and Engineering Sciences, 359*, 663–693.

13

End mill tools integration in CNC machining for rapid manufacturing processes: simulation studies

Muhammed Nafis Osman Zahid, Keith Case* and Darren Watts

Mechanical and Manufacturing Engineering, Loughborough University, Loughborough, Leicestershire LE11 3TU, UK

Computer numerical controlled (CNC) machining has been recognized as a manufacturing process that is capable of producing metal parts with high precision and reliable quality, whereas many additive manufacturing methods are less capable in these respects. The introduction of a new layer-removal methodology that utilizes an indexing device to clamp the workpiece can be used to extend CNC applications into the realm of rapid manufacturing (CNC-RM) processes. This study aims to improve the implementation of CNC machining for RM by formulating a distinct approach to integrate end mill tools during finishing processes. A main objective is to enhance process efficiency by minimizing the staircasing effect of layer removal so as to improve the quality of machined parts. In order to achieve this, different types of end mill tools are introduced to cater for specific part surfaces during finishing operations. Virtual machining simulations are executed to verify the method and the implications. The findings indicate the advantages of the approach in terms of cutting time and excess volume left on the parts. It is shown that using different tools for finishing operations will improve the capabilities of CNC machining for rapid manufacturing applications.

Keywords: end mill tools; CNC machining; rapid manufacturing

1. Introduction

In machining processes, it is crucial to ensure good surface finish quality as this will influence the performance of machined parts and production costs (Davim, 2001). Improving surface finish has become a major concern particularly in rapid manufacturing processes where the objective is to achieve final geometry directly. A major consideration in rapid manufacturing processes is that planning time needs to be substantially reduced so that overall time from design to finished part is also reduced. In additive processes (AM), relatively long fabrication times are compensated by the limited planning time required. In traditional computer numerical controlled (CNC) the opposite situation exists, and whilst machining can be very rapid, the planning process may be complex and time-consuming. In trying to use CNC for rapid applications, the main effort is in maintaining a simple approach to planning whilst possibly sacrificing some of the available machining efficiency.

An approach that has some of the advantages of additive manufacturing involves the use of an indexable fourth axis device to hold and clamp the workpiece onto the table

*Corresponding author. Email: K.Case@lboro.ac.uk

of a three-axis CNC milling machine (Figure 1). This approach allows a layer-removal process to take place at various rotations about one axis and complex shapes and features can be created by layer-based removal of material (Frank, Wysk, & Joshi, 2006). The indexable device permits rotation about one axis while also clamping the workpiece. Layers are removed from several orientations to reveal all surfaces of the part without refixturing (Frank et al., 2006). Determination of the necessary orientations is central to the process and this is achieved by visibility algorithms (a minimum set of orientations are sought where the surfaces are 'visible' from the direction of the tool).

Within each orientation, roughing and finishing operations are performed sequentially using different tool sizes and machining parameters. Machining starts with the first roughing operation that cuts until the furthest possible surface is reached or the workpiece is fully cut (Frank, 2007). Then, finishing operations are executed no further than the central axis of the cylindrical workpiece. In order to reduce effort in the planning tasks, feature recognition on machined parts is avoided. Therefore, the selection of a cutting tool becomes straightforward and is based on the smallest diameter and a flat end mill tool is most likely to be selected.

Despite the effectiveness of reducing complexities in the planning phase, the use of a single end mill tool in finishing processes leads to several drawbacks. First, due to the geometry of flat end mill cutters, a staircasing effect is visible on part surfaces particularly on contours and non-flat regions. This effect still can be seen even if very small cutting layer depths are adopted (Frank, Joshi, & Wysk, 2002). In addition, using small cutting depths will obviously increase cutting times considerably. Another limitation can be seen in the assessment of tool accessibility of part features. Analysis that focuses on machinability has resulted in some uncut material being left on the part due to inaccessibility of the cutting tool (Li & Frank, 2006) and inevitably this will affect dimensional accuracy. Figure 2 illustrates potential non-machined regions generated on a toy jack part. To date, no specific method has been developed to cater for the integration of different end mills in rapid machining processes. Therefore, this study attempts to show that using different end mill tools in finishing operations can improve surface finish by reducing the excess volume left on the parts. The next section of the paper will review the previous approach in cutting tools selection for CNC-RM processes. A methodology to integrate end mill tools is then described considering the implications for process planning and cutting operations. A framework for the simulation studies is established that is later used to generate the results to validate the proposed method. The findings

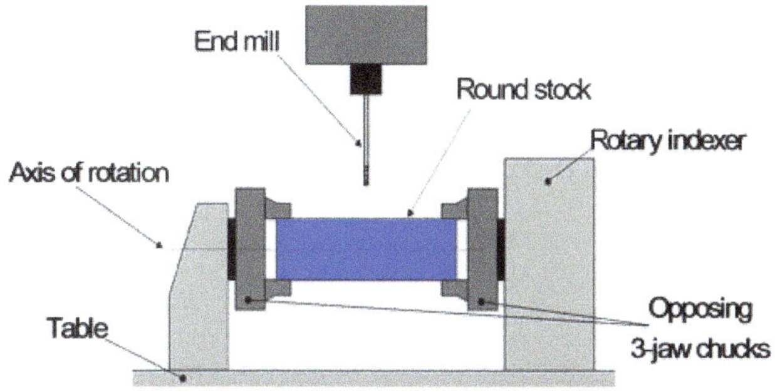

Figure 1. Setup for CNC-RP (Frank, Wysk, & Joshi, 2004).

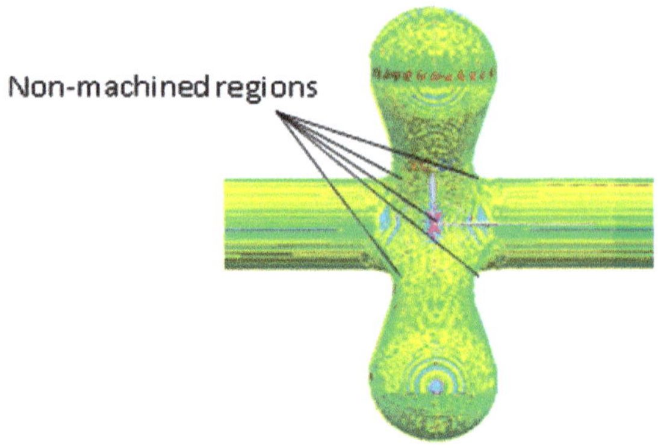

Figure 2. Possible non-machined regions using a flat end mill tool (Li & Frank, 2006).

are described further to highlight the achievements and limitations of the approach before drawing conclusions.

2. Cutting tools and machining operations

Initially, due to the objective of simplifying process planning, only a single cutting tool with the smallest possible diameter is used in CNC machining for rapid prototyping processes (Frank, 2003). Since the visibility analysis is based on 2D cross-sectional layers of part geometries, a flat end mill tool is usually selected to finish cut the part. Consequently, the machined parts exhibit the same staircase effect as presented in most of the AM processes. Figure 3 shows the staircase effect formed on contoured surfaces of a part. However, CNC machines are capable of reducing the layer thickness down to 0.0127 mm. Therefore, the step appearance is minimized within an acceptable range for making prototypes. From a planning perspective, the use of a single cutting tool has generalized the cutting areas to cover all surfaces in each orientation. This avoids any feature recognition tasks during machining process. Only one cutting operation is performed in each orientation that considerably reduces the number of operations and planning complexities. However, a small cutting tool requires more passes, which will result in longer machining times. Besides, the removal of thin layers is inefficient especially when the bulk of the material needs to be removed during early phases of machining.

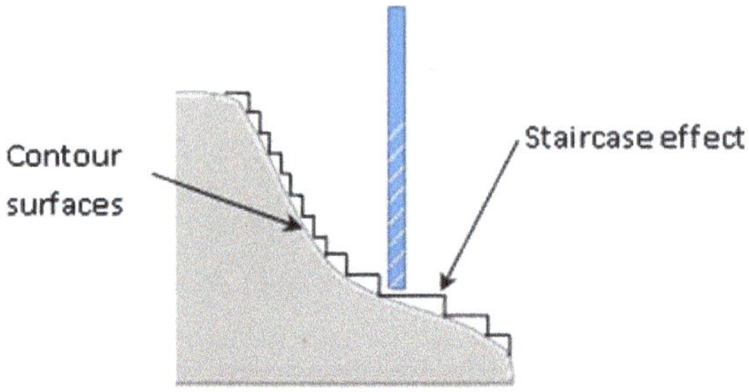

Figure 3. Staircase effect presented on contoured surfaces.

These considerations tend to restrict the capabilities of CNC machining even if some level of automation is embedded in the process planning.

Due to the inefficiency caused by a single cutting tool, later developments suggest the use of different tool sizes to handle roughing and finishing operations (Frank, 2007). As a result, roughing operations utilize larger tool diameters selected based on the size of the workpiece. Hence, high volumes of material can be removed during roughing operations whilst leaving sufficient material for finishing processes. Meanwhile, the approach of selecting the smallest tool diameter remains the same for executing finishing operations. This allows the cutting tool to reach all features presented on the part. Both roughing and finishing operations employ the same cutting areas. Thus, within one cutting orientation, two different toolpaths are developed and this directly increases the number of machining operations. Additionally, tool changes are required in each orientation and can be carried out through automated systems built into the CNC machine. This combination of different tool sizes based on roughing and finishing operations succeeds in enhancing the application of CNC machines in producing parts rapidly.

The introduction of different cutting tool sizes has resulted in a further study that attempts to assist in the selection of a proper combination of end mill tool sizes for roughing and finishing operations. Different tool sizes possess different abilities in removing the material and shaping the part. Large tools are capable of removing more material in shorter time, but are restricted in accessing small cutting regions. To the contrary, small tools have much improved accessibility but very inferior rate of material removal. Therefore, an appropriate combination of tool sizes is necessary to machine the workpiece effectively. The region that is accessible to different tool sizes can be calculated through the Tool Access Volume (TAV) method. Then, the analysis is extended to identify appropriate tool selection and sequencing by using a method known as Relative Delta Volume Clearance (RDVC) (Lim, Corney, Ritchie, & Clark, 2001). Based on information gained from these analyses, tool sizes and a set of cutting orientations can be identified by applying a genetic algorithm technique (Renner, 2008). Accordingly, the machining times are substantially reduced by using carefully selected combinations of tool sizes.

A recent study has proposed another approach to further improve the machining operations in CNC-RM (Osman Zahid, Case, & Watts, 2014). It was successfully carried out by manipulating the orientations for roughing operations. Instead of performing roughing and finishing operations using the same set of orientations, rough cuts were extracted and performed through another set of orientations. This set of orientations was based on a combination of four cutting orientations, 0–90–190–270°. A customized programme was developed to simulate the cutting process and propose efficient cutting directions. This method reduced the required cutting tool depth during roughing operations for avoidance of tool breakage. Since roughing and finishing operations are based on different sets of orientations, tool changes occur only once throughout the process. The time spent on roughing operations frequently increases, but it was found that this resulted in a reduction of total machining time.

Over the years, much research has been aimed at reducing machining time by manipulating cutting parameters as a part of process planning (Bouzid, 2005; Lavernhe, Tournier, & Lartigue, 2008; Palanisamy, Rajendran, & Shanmugasundaram, 2007). A good surface finish with low machining time can be obtained through the proper selection of machining parameters (Bharathi Raja & Baskar, 2012). The prominent parameters to be controlled are cutting speed, feed rate and depth of cut. Based on a multiple regression model, a previous study indicated that the depth of cut was the most influential parameter in determining the level of surface roughness achieved (Agarwal, 2012). Therefore, it is undeniable that cutting parameters are among the major factors that influence surface quality and machining time. However, in CNC-RM applications, the aim is to rapidly generate machining operations that reflect the planning and execution stages. Optimizing cutting parameters would require the handling of a large number of variables, and this tends to complicate the planning tasks and limits the level of automation that can be adopted in the planning stage. Therefore, the study conducted here integrates different end mill tools and observes the implications for machining time and surface quality. However, efficient cutting parameters suggested from previous studies could still be incorporated during the planning stage. Based on specific materials and tool sizes, a set of efficient parameters could be incorporated using decision-making tools in the development of machining plans.

Generally, the developments in selection of cutting tools and machining operations have enhanced the ability of CNC machining for rapid processes. The time required to machine parts has been substantially reduced with the introduction of roughing and finishing operations, appropriate tool size combinations and different rough cut sets of orientations. All these have enhanced the application of end mill tools in CNC machining, particularly for rapid manufacturing. However, far too little attention has been paid to the geometries of the end mill tool while developing the machining operations. Hence, there are no clear guidelines on how to combine the different tool geometries in the processing steps of CNC-RM. Nevertheless, several studies have strongly addressed the significance of cutting tool geometries on the level of quality achieved in the machining process (Patel, 2010; Reddy & Rao, 2005). Even if cutting can proceed at a very small depths of cut, neglecting the tool geometries tends to constrain the abilities of CNC machining in producing high quality parts. Therefore, considering the rapid manufacturing application, integrating different types of end mill tools is thought to be an approach to fulfil the quality requirements of end-use products.

3. Methodology

In attempts to implement CNC machining for rapid manufacturing processes, finishing tool selection is strictly defined based on the smallest diameter single end mill tool. This approach minimizes the tasks required in the planning phase and realizes the rapid generation of machining operations. Nevertheless, there is an implication for part quality as a single end mill tool is not capable of catering for a wide range of part surfaces. Therefore, integrating different end mill tools is a possible way of resolving the problem, but the guidelines must be clearly defined so that process planning can be handled effectively. Several important tasks are identified including separation of cutting regions and the types of end mill tool selected.

3.1. Cutting tools and surface classification

In the CNC-RM approach, finishing operations are executed through predetermined orientations and aim to shape the material completely. Hence, this process cuts the material with very small depths of cut to obtain a fine surface finish. Previous developments have implemented different cutting tools to finish cut parts built by welding process through a layer deposition method (Akula & Karunakaran, 2006). However, this approach utilized various type of tools and several surface categories, and seems impractical as it tends to complicate the planning tasks. However, some modifications to minimize and eliminate irrelevant setups will help to adopt this method in CNC-RM processes.

From one particular cutting direction, surfaces of the part can be classified into flat and non-flat surfaces. A flat surface is defined as a surface that is perpendicular to the cutting tool direction whereas the rest of the surfaces are categorized as non-flat surfaces. These classifications allow specific cutting tools to be assigned to machine the part effectively. Therefore, one orientation will perform up to two cutting operations if both types of surface are present. The first areas that will be cut are the flat surfaces using a flat end mill cutter. The process continues by machining non-flat surfaces using a ball nose end mill. Based on Figure 4, assuming cutting starts at 0° cutting orientation, the tool is engaged with the workpiece from ZC direction. In this particular direction, the dark areas represent flat surfaces that are perpendicular to the cutting tool direction. The first finishing operation is performed on this surface using a flat end mill tool. Then, the CNC machine replaces the current tool with a similar-sized ball nose end mill. This tool is used to machine the other surfaces shown as light grey areas on the part.

Generally, there are three common end mill tools (Figure 5) that differ based on the shapes of the tool tip. These are flat, bull nose and ball nose end mills. In this study, only two tools are selected to be utilized in finishing operations. This selection is expected to greatly reduce the complications for process planning as only two variables are present instead of three. A flat end mill is capable of machining plane surfaces with a minimum scallop size compared to other tools (Elber, 1995; Ryu, Choi, & Chu, 2006). Therefore, this is exactly the right tool to cut the flat surfaces classified on parts. The second selection utilizes a ball nose tool to cater for non-flat surfaces. Based on previous studies, the capabilities of this tool in handling sculptured surfaces are proven and high-quality surface finish can be produced (Chen, Huang, & Chen, 2005; Chen & Shi, 2008; Vijayaraghavan, Hoover, Hartnett, & Dornfeld, 2008). Moreover, this tool can be easily guided to engage with part surfaces and produces simpler NC code for machining (Chen & Shi, 2008). Meanwhile, a bull nose end mill is capable of removing high volumes of material with efficient machining times (Patel, 2010). But, due to the tool tip geometry, there is limited access to certain part features and this can result in

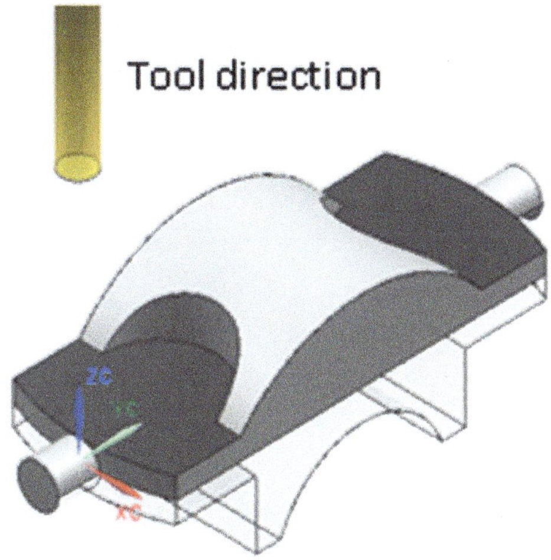

Figure 4. Classification of flat and non-flat surfaces in one cutting direction.

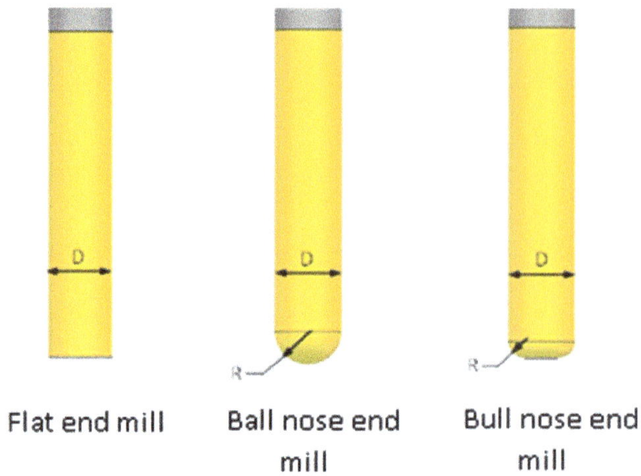

Figure 5. Three common shapes of end mill tool (Engin & Altintas, 2001).

high volumes of excess material. This limitation is shown in Figure 6. Therefore, in this study, only flat and ball nose end mills are selected for cutting tool integration in the CNC-RM processes.

3.2. Machining and verification processes

The introduction of different end mill tools requires a revision of the machining parameters, especially the cutting levels. Similar to roughing operations, finishing operations continue until the central axis of the cylindrical workpiece is reached. However, this is only applicable if a flat end mill tool is used in the operation. In the case where both tools are used, the ball nose cutter requires extended cutting levels to completely remove the remaining material. This is critically important for a part that contains closed regions that only permit tool access in limited directions. Figure 7 illustrates the situation where there is excess material due to there being insufficient cutting levels. Therefore, the cutting levels for a ball nose tool are extended by at least half the cutting tool diameter from the centre of the workpiece. Other parameters such as cutting feed and spindle speed are dependent on the diameter of the end mill tool. The finishing

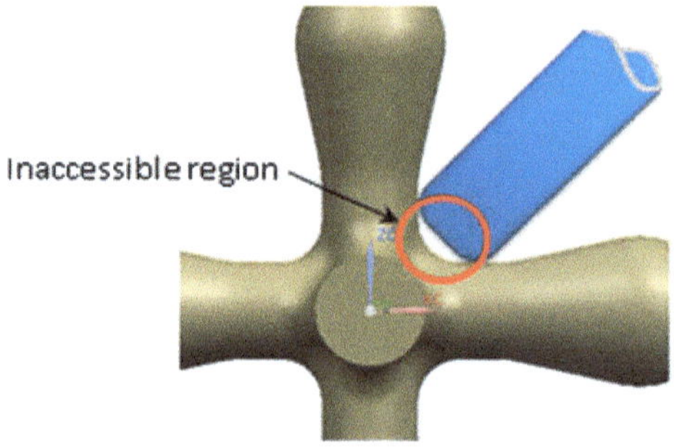

Figure 6. Restricted accessible regions for bull nose end mill tool.

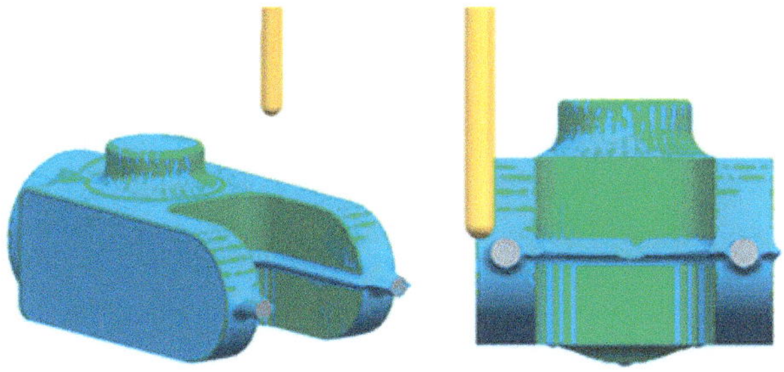

Figure 7. Remaining uncut materials left by ball nose end mill tool.

operations in this study employ the same tool diameter regardless of the type of tool used. Hence, both share the same value of spindle speed and feed rate. In this study, spindle speed and feed rates were determined based on the size of the cutting tools. The parameters are proposed by the CAD/CAM software itself and are not necessarily optimum cutting parameters. Depending on the models, depths of cut ranged between 1 and 1.5 mm for rough machining and 0.1 and 0.3 mm for finishing operations.

The simulation analysis was carried out with NX 7.5 software using the manufacturing application. A series of machining operations were created and the software used to estimate machining time to fabricate the parts. In order to calculate the excess volume left, the analysis was extended by using CGTech VERICUT ® 7.2.3 (CGTech, 2012). Recently, this software has been interfaced to NX which provides convenient access to run the simulation once operations have been created. There are several pre-processing steps while calculating the excess volume. First, the machining operations are translated into a cutter location source file (.clsf). Then, the file is run on VERICUT® software to simulate the cutting operations. Once completed, the 'X-caliper' tool is activated to

Table 1. Results for drive shaft and knob model.

Approach Criteria		Drive shaft		Knob	
		Approach 1	Approach 2	Approach 1	Approach2
Machining time	h:min:s	04:13:20	04:02:35	03:50:09	03:52:40
Finishing time		02:38:30	02:28:39	02:44:49	02:27:16
Non-cutting time		00:16:58	00:16:08	00:09:11	00:10:20
Roughing time		01:17:52	01:17:48	01:15:04	01:15:04
Cutting orientations		90–180–280–0–32–180–0°	90–180–280–0–32–180–0°	90–180–280–0–180–45–315°	90–180–280–0–180–45–315°
Machined volume	mm^3	150,273.23	150,432.83	112,946.97	113,203.60
Current part volume		53,244.05	53,084.45	21,660.63	21,403.91
Excess volume		514.82	355.22	527.99	271.27

obtain volumetric information. This includes stock volume, machined volume and current part volume. Prior to that, the machined part volume is calculated by a 'measure bodies' tool available in the NX system. All this information is gathered and the excess volume is calculated by subtracting the current part volume from the machined part volume. Finally, the results obtained are used to evaluate the effectiveness of integrating end mill tools in the machining processes.

4. Results and discussion

The simulation studies were conducted using different models including drive shaft, knob, salt bottle and toy jack models. Figure 8 illustrates the models adopted for the simulation studies. To visualize the implications of the suggested approach, similar cutting operations relying totally on the use of a flat end mill tool were developed. Therefore, Approach 1 was based on a process that utilized a single flat end mill whereas Approach 2 represented the method developed in this study that integrates different cutting tools. In this study, machining parameters were determined based on the size of the cutting tools. Since there are two types of operations (roughing and finishing), two sets of cutting parameters were involved. Meanwhile in the finishing operations, two different tools are introduced (flat and ball nose end mill). However, the machining parameters are similar since only one size of cutting tool is used. It is important to note that all the machining parameters are constant between Approach 1 and 2. Thus, the improvements gained from this study totally relied on the implication of different end mills during finishing operations. The study is not meant to validate the optimum cutting parameters, but the values could be adopted in the planning stage.

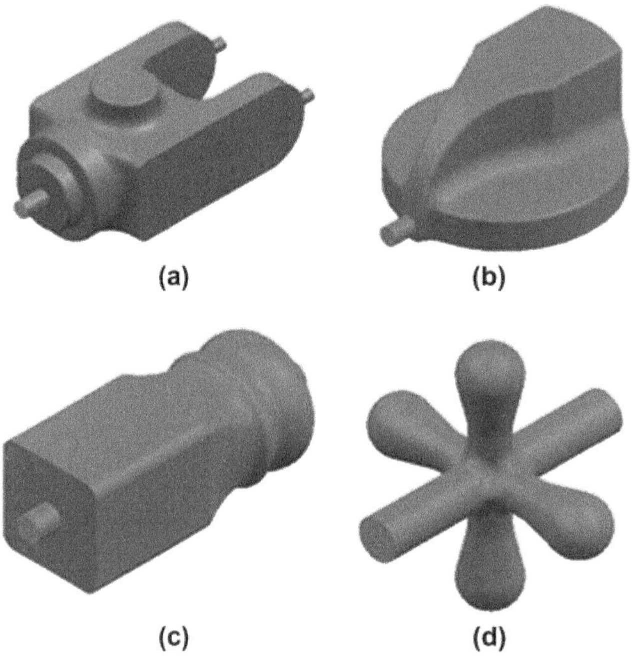

Figure 8. (a) Drive shaft, (b) knob, (c) salt bottle and (d) toy jack model.

There are two main evaluation criteria that are based on machining time and excess volume. Tables 1 and 2 summarize the results produced from simulation studies. Machining time is recorded in (h:min:s) format and volume information is in mm^3. The amount of material removed from the workpiece is presented as machined volume in the table. Part and stock volumes are calculated earlier based on the virtual models. It is apparent from these tables that there are considerable differences between Approaches 1 and 2. Out of the four models, three indicate a reduction of cutting time by implementing the approach developed in this study. Interestingly, all models achieved low excess volumes once different cutting tools were used in finishing operations. Assigning specific tools to particular surfaces, the ranges of excess volume reduce from 0.9–3.3% down to 0.7–1.6% of the total part volume.

4.1. Implications for machining time

Data gathered in the tables clearly indicate the benefits of implementing different cutting tools in CNC-RM processes. The majority of the models studied show a decreasing trend of machining time compared to a single tool approach. Finishing cutting times contribute most of the savings. This signals the effectiveness of cutting tools removing material on assigned surfaces. Depending on part geometries, cutting times differ from 2 min up to 11 min. For example, results for the drive shaft model in Table 1 indicate that the current method in Approach 2 manages to reduce machining time by about 11 min compared to Approach 1. Nonetheless, some models show only a small reduction of machining time such as the toy jack model. In this model, only ball nose tools were used as only non-flat surfaces were present on the part. For a medium production run, this reduction might still be important in minimizing cost and time. Oppositely, the knob model indicated a slightly longer machining time compared to Approach 1. The variety of flat and non-flat surfaces present on the part are possibly the main factors influencing cutting time.

Table 2. Results for salt bottle and toy jack model.

Approach Criteria		Salt bottle		Toy jack	
		Approach 1	Approach 2	Approach 1	Approach 2
Machining time	h:min:s	04:32:02	04:23:26	03:25:22	03:21:42
Finishing time		02:38:44	02:29:33	02:16:04	02:12:44
Non-cutting time		00:11:57	00:12:28	00:14:06	00:13:44
Roughing time		01:41:21	01:41:26	00:55:12	00:55:12
Cutting orientations		0–90–190–270–45–180–270°	0–90–190–270–45–180–270°	50–140–240–320–49–140–228–320°	50–140–240–320–49–140–228–320°
Machined volume	mm^3	76,676.80	76,845.85	109,662.15	109,792.30
Current part volume		34,499.39	34,330.33	7769.86	7639.71
Excess volume		417.56	248.5	252.74	122.59

Table 3. Excess materials distribution based on studied models.

Drive shaft		Knob	
Approach 1	Approach 2	Approach 1	Approach 2
514.82 mm^3 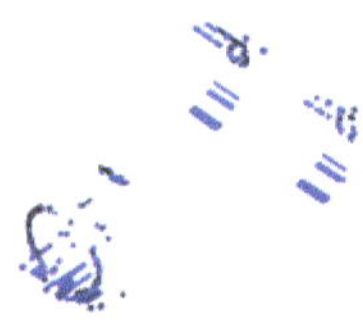	355.22 mm^3 	527.99 mm^3 	271.27 mm^3 
Salt bottle		Toy jack	
Approach 1 417.56 mm^3 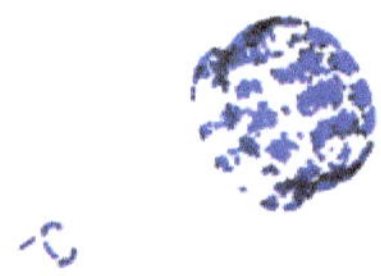	Approach 2 248.5 mm^3 	Approach 1 252.74 mm^3 	Approach 2 122.59 mm^3 

The simulation studies adopted different sets of orientations for roughing and finishing operations as proposed from previous research (Osman Zahid et al., 2014). Therefore, integrating different end mill tools did not add to the number of orientations, but it increased the number of operations and tool changes. The presence of flat and non-flat surfaces in one cutting orientation will require both end mill tools and one tool change operation. Taking the salt bottle as example, there are two orientations consisting of flat and non-flat surfaces. Hence, four finishing operations are required within the two cutting directions. Meanwhile, in a single tool approach (Approach 1), only one finishing operation is needed in each orientation and this is quite straightforward compared to Approach 2. However, considering the effectiveness of material removal, it is worthwhile integrating the cutting tools.

4.2. Volume of excess materials

Previous studies have validated the capabilities of ball nose cutters on sculptured surfaces. Hence, assigning the tools based on flat and non-flat surfaces manages to minimize the excess volume on parts. Surprisingly, simulation results consistently indicate that excess volume was reduced for all models studied. Some models show a considerable reduction. For instance, excess volume for the knob model decreases dramatically from 527 mm^3 down to 271 mm^3. In CNC machining, uncut materials can be reduced by modifying cutting parameters in finishing operations including depth of cut, number of passes, feed rates and speed. However, in the CNC-RM approach, all these parameters remain constant except for the type of cutting tools used. Therefore, the achievements of this study show that the cutting tool combination is the main factor that influences the results.

Furthermore, the implications can be visualized by analysing the material distribution diagram as shown in Table 3. Comparing Approach 1 and 2, it can be seen that most of the excess material on non-flat surfaces of the part have been reduced. This indicates that the ball nose cutter effectively minimizes the staircasing effect. Keeping the flat end mill cutter to machine flat surfaces is useful in overcoming the weakness of ball nose tools.

5. Limitations and future work

Using multiple end mill tools in finishing operations manages to greatly reduce the excess volume left on the machined parts. The results of simulation studies indirectly portray the benefits of the approach in terms of surface quality. But, observation of the cutting simulations has revealed a limitation in shaping the sacrificial support of the part. If the support structure is connected to a flat vertical surface, a ball nose end mill faces a difficulty in completely shaping the structure. An excess volume (fillet shape) is created on the edge of the support and part. This problem is shown in Figure 9. On the other hand, a flat end mill is capable of forming the sharp edge, but is limited to areas that are perpendicular to the cutter direction. Excess volume will still be present in the rest of the connection area. Currently, the only way to minimize the problem is by using the smallest possible end mill tool during finishing processes. Larger tools will increase the excess volume and cause more work at the post-processing stage.

In real applications, several variations can be expected on the method proposed in this study. First, the rate of material removal is highly dependent on the size of cutting tools adopted in roughing and finishing processes. The implications can be seen based on accessibility and machining time. Selection of the smallest cutting tool in finishing operations will guarantee the accessibility, but machining time will probably be extended. Using optimization tools like TAV and RDVC that have been discussed earlier could be helpful in reducing the variations. Another factor that will cause variation between simulation and real machining is process interventions. In the simulation study, the estimated cutting time represents the complete operation including cutting and non-cutting time, but times for tool changes and workpiece rotations were not considered. If both are fully automated, it would have less impact on the cutting time. However, if the operations are carried out manually, which requires the machine to stop, then the estimated time is likely to be extended. For this reason, it is highly recommended that CNC-RM operations are executed in an automated manner.

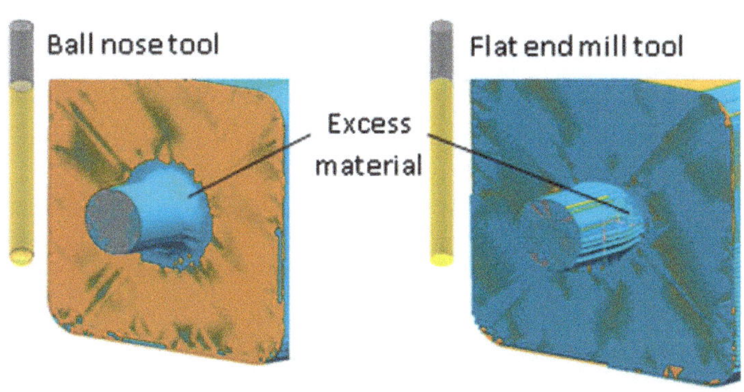

Figure 9. Excess material formed at the edge of sacrificial support.

The integration of cutting tools has benefited the production stage of the CNC-RM, but, it affects the planning stage as more variables need to be handled. Previously, process planning in CNC machining restricted implementation in RM as it involves many manual tasks and is highly dependent on human skill (Anderberg, Beno, & Pejryd, 2009; Relvas & Simoes, 2004). However, a recent development in CAD/CAM has introduced several features that can be used to effectively handle the process planning tasks (Agrawal, Soni, & Dwivedi, 2013). Therefore, an assisting tool can be developed to integrate the cutters while constructing the machining operation in the system and make it adaptable to any kind of part features. On the other hand, the identification of flat and non-flat surfaces within the cutting orientations needs to be performed correctly to ensure the right end mill tools are selected. Flat surfaces are not only the horizontal and vertical surfaces of the CAD model, but, depending on cutting direction, inclined surfaces can also be considered as flat surfaces if they are perpendicular to the direction of tool movement. Considering these issues, the medium of communication between the user and the system must be preserved to assist decision-making during process planning. To realize this, customized programming tools and computerized systems can be implemented in the CNC-RM process plan.

6. Conclusion

This simulation study has revealed the advantages of using different cutting tools in finishing operations particularly for CNC-RM processes. The present study was designed to improve part quality by reducing uncut material left after completion of machining. Generally, using different tools to cater for flat and non-flat surfaces is an effective way to reduce excess volume. Furthermore, machining times also can be reduced depending on part geometries. In order to visualize the implications, the results were compared to a previous method that adopted a single flat end mill tool throughout the cutting operations. Distinctly, the comparisons show the method developed in this study is capable of enhancing part quality and appearance. Further studies could focus on integrating this approach into CNC-RM process planning. The surface classification task needs to be handled carefully so that it will not complicate the planning task. Also, further verification is still required through machining experiments so that real parts can be fabricated and analysed completely. In the simulation studies, the excess volume was a key indicator to determine the surface quality of the part. However, in real machining operations, the surface quality needs to be examined through roughness analysis.

Disclosure statement

No potential conflict of interest was reported by the authors.

References

Agarwal, N. (2012). Surface roughness modelling with machining parameters (speed, feed and depth of cut) in CNC milling. *MIT International Journal of Mechanical Engineering, 2*, 55–61.

Agrawal, A., Soni, R. K., & Dwivedi, N. (2013). Development of integrated CNC-RP system through CAD/CAM environment. *International Journal of Mechanical and Production Engineering Research and Development, 3*, 1–10.

Akula, S., & Karunakaran, K. (2006). Hybrid adaptive layer manufacturing: An Intelligent art of direct metal rapid tooling process. *Robotics and Computer-Integrated Manufacturing, 22*, 113–123.

Anderberg, S., Beno, T., & Pejryd, L. (2009). CNC machining process planning productivity – A qualitative survey. *Proceedings of the International 3rd Swedish Production Symposium*, Göteborg, Sweden, 228–235.

Bharathi Raja, S., & Baskar, N. (2012). Application of particle swarm optimization technique for achieving desired milled surface roughness in minimum machining time. *Expert Systems with Applications, 39*, 5982–5989.

Bouzid, W. (2005). Cutting parameter optimization to minimize production time in high speed turning. *Journal of Materials Processing Technology, 161*, 388–395.

CGTech. (2012). *VERICUT 7.2: Release notes.*

Chen, J., Huang, Y., & Chen, M. (2005). A study of the surface scallop generating mechanism in the ball-end milling process. *International Journal of Machine Tools and Manufacture, 45*, 1077–1084.

Chen, T., & Shi, Z. (2008). A tool path generation strategy for three-axis ball-end milling of free-form surfaces. *Journal of Materials Processing Technology, 208*, 259–263.

Davim, J. P. (2001). A note on the determination of optimal cutting conditions for surface finish obtained in turning using design of experiments. *Journal of Materials Processing Technology, 116*, 305–308.

Elber, G. (1995). Freeform surface region optimization for 3-axis and 5-axis milling. *Computer-Aided Design, 27*, 465–470.

Engin, S., & Altintas, Y. (2001). Mechanics and dynamics of general milling cutters: Part I: Helical end mills. *International Journal of Machine Tools and Manufacture, 41*, 2195–2212.

Frank, M. C. (2003). *The development of a rapid prototyping process using computer numerical controlled machining* (PhD thesis). The Pennsylvania State University.

Frank, M. C. (2007). Implementing rapid prototyping using CNC machining (CNC-RP) through a CAD/CAM interface. *Proceedings of the Solid Freeform Fabrication Symposium*, Autin, TX, 112–123.

Frank, M., Joshi, S. B., & Wysk, R. A. (2002). CNC-RP: A technique for using CNC machining as a rapid prototyping tool in product/process development. *Proceedings of the 11th Annual Industrial Engineering Research Conference*, Orlando, FL, 19–22.

Frank, M. C., Wysk, R. A., & Joshi, S. B. (2004). Rapid planning for CNC milling - A new approach for rapid prototyping. *Journal of Manufacturing Systems, 23*, 242–255.

Frank, M. C., Wysk, R. A., & Joshi, S. B. (2006). Determining setup orientations from the visibility of slice geometry for rapid computer numerically controlled machining. *Journal of Manufacturing Science and Engineering, 128*, 228–238.

Lavernhe, S., Tournier, C., & Lartigue, C. (2008). Optimization of 5-axis high-speed machining using a surface based approach. *Computer-Aided Design, 40*, 1015–1023.

Li, Y., & Frank, M. C. (2006). Machinability analysis for 3-axis flat end milling. *Journal of Manufacturing Science and Engineering, 128*, 454–464.

Lim, T., Corney, J., Ritchie, J., & Clark, D. (2001). Optimizing tool selection. *International Journal of Production Research, 39*, 1239–1256.

Osman Zahid, M. N., Case, K., & Watts, D. (2014). Optimization of roughing operations in CNC machining for rapid manufacturing processes. *Production & Manufacturing Research, 2*, 519–529.

Palanisamy, P., Rajendran, I., & Shanmugasundaram, S. (2007). Optimization of machining parameters using genetic algorithm and experimental validation for end-milling operations. *The International Journal of Advanced Manufacturing Technology, 32*, 644–655.

Patel, K. (2010). *Web based automatic tool path planning strategy for complex sculptured surfaces* (Master thesis). University of Waterloo.

Reddy, N. S. K., & Rao, P. V. (2005). Selection of optimum tool geometry and cutting conditions using a surface roughness prediction model for end milling. *The International Journal of Advanced Manufacturing Technology, 26*, 1202–1210.

Relvas, C., & Simoes, J. (2004). Optimization of computer numerical control set-up parameters to manufacture rapid prototypes. *Proceedings of the Institution of Mechanical Engineers, Part B: Journal of Engineering Manufacture, 218*, 867–874.

Renner, A. (2008). *Computer aided process planning for rapid prototyping using a genetic algorithm* (Master thesis). Iowa State University.

Ryu, S. H., Choi, D. K., & Chu, C. N. (2006). Roughness and texture generation on end milled surfaces. *International Journal of Machine Tools and Manufacture, 46*, 404–412.

Vijayaraghavan, A., Hoover, A., Hartnett, J., & Dornfeld, D. (2008). Improving end milling surface finish by workpiece rotation and adaptive toolpath spacing. *International Journal of Machine Tools & Manufacture, 49*, 89–98.

Permissions

All chapters in this book were first published in PMR, by Taylor & Francis Online; hereby published with permission under the Creative Commons Attribution License or equivalent. Every chapter published in this book has been scrutinized by our experts. Their significance has been extensively debated. The topics covered herein carry significant findings which will fuel the growth of the discipline. They may even be implemented as practical applications or may be referred to as a beginning point for another development.

The contributors of this book come from diverse backgrounds, making this book a truly international effort. This book will bring forth new frontiers with its revolutionizing research information and detailed analysis of the nascent developments around the world.

We would like to thank all the contributing authors for lending their expertise to make the book truly unique. They have played a crucial role in the development of this book. Without their invaluable contributions this book wouldn't have been possible. They have made vital efforts to compile up to date information on the varied aspects of this subject to make this book a valuable addition to the collection of many professionals and students.

This book was conceptualized with the vision of imparting up-to-date information and advanced data in this field. To ensure the same, a matchless editorial board was set up. Every individual on the board went through rigorous rounds of assessment to prove their worth. After which they invested a large part of their time researching and compiling the most relevant data for our readers.

The editorial board has been involved in producing this book since its inception. They have spent rigorous hours researching and exploring the diverse topics which have resulted in the successful publishing of this book. They have passed on their knowledge of decades through this book. To expedite this challenging task, the publisher supported the team at every step. A small team of assistant editors was also appointed to further simplify the editing procedure and attain best results for the readers.

Apart from the editorial board, the designing team has also invested a significant amount of their time in understanding the subject and creating the most relevant covers. They scrutinized every image to scout for the most suitable representation of the subject and create an appropriate cover for the book.

The publishing team has been an ardent support to the editorial, designing and production team. Their endless efforts to recruit the best for this project, has resulted in the accomplishment of this book. They are a veteran in the field of academics and their pool of knowledge is as vast as their experience in printing. Their expertise and guidance has proved useful at every step. Their uncompromising quality standards have made this book an exceptional effort. Their encouragement from time to time has been an inspiration for everyone.

The publisher and the editorial board hope that this book will prove to be a valuable piece of knowledge for researchers, students, practitioners and scholars across the globe.

List of Contributors

Yadong Fang
Department of Mechanical and Electrical Engineering, Xi'an Technological University, Xi'an, China

Reina Angkiriwang, I Nyoman Pujawan and Budi Santosa
Department of Industrial Engineering, Sepuluh Nopember Institute of Technology, Kampus ITS Sukolilo, Surabaya 60111, Indonesia

Mahdi Bagheripoor and Hosein Bisadi
Department of Mechanical Engineering, Iran University of Science and Technology, Tehran, Iran

Victor Chukwunweike Nwokocha and Iganatius Ani Madu
Department of Geography, University of Nigeria Nsukka, Nigeria

Sherif Mostafa and Nicholas Chileshe
School of Natural and Built Environments, University of South Australia, Adelaide 5000, Australia

Sang-Heon Lee
School of Engineering, University of South Australia, Mawson Lakes 5095, Australia

Jantanee Dumrak
Global Management Program, Torrens University Australia, Adelaide, SA 5000, Australia

Hassan Soltan
Production Engineering and Mechanical Design Department, Faculty of Engineering Mansoura University, Mansoura 35516, Egypt

R.K. Tyagi
Department of Mechanical Engineering, Amity School of Engineering and Technology, Amity University, Noida, India

Jalel Ben Hmida, Jim Lee and Xinchun Wang
Systems Engineering, University of Louisiana at Lafayette, P.O. Box 44170, Rougeou, Hall Room # 244, Lafayette, LA 70504, USA

Fathi Boukadi
Department of Petroleum Engineering, University of Louisiana at Lafayette, P.O. Box 44408, Madison Hall, Room # 134, Lafayette, LA 70504, USA

Ihsan K. Al Naimi and Kasim M. Daws
Mechanical Department, College of Engineering, Baghdad University, Baghdad, Iraq

Moneer H. Al Saadi
Welding Department, University of Kerbala, Baghdad, Iraq

Niels Bay
Department of Mechanical Engineering, Technical University of Denmark, Lyngby, Denmark

Douglas H. Grabenstetter and John M. Usher
Department of Industrial and Systems Engineering, Mississippi State University, MS 39762, USA

Narges Banaeian, Hossein Mobli and Mahmoud Omid
Department of Agricultural Machinery Engineering, Faculty of Engineering and Technology, University of Tehran, Karaj, Iran

Izabela Ewa Nielsen
Department of Mechanical and Manufacturing Engineering, Aalborg University, Aalborg, Denmark

Suha Karim Shihab, Zahid A. Khan, Aas Mohammad and Arshad Noor Siddiquee
Department of Mechanical Engineering, Jamia Millia Islamia (A Central University), MMA Jauhar Marg, New Delhi 110025, India

Chigbogu Godwin Ozoegwu
Department of Mechanical Engineering, Nnamdi Azikiwe University, PMB 5025, Awka, Nigeria

Muhammed Nafi s Osman Zahid, Keith Case and Darren Watts
Mechanical and Manufacturing Engineering, Loughborough University, Loughborough, Leicestershire LE11 3TU, UK

Index

A

Abrasive Jet Machining, 99-102, 107-109

Abrasive Jet Machining Process, 107

Aluminum Alloys, 36, 38, 128, 142

Analytic Hierarchy Process (ahp), 63

Automotive Applications, 181

Axial Velocity, 37

C

Computational Engineering, 52

Computer Managed Maintenance System (cmms), 81

Computer Numerical Controlled (cnc), 216

Conceptual Model, 16-18, 30, 32-34

Contact Resistance, 128-129, 134-135, 139, 141-143

Continuous Manufacturing Systems, 111-112, 114, 126

D

Delphi Method, 161, 167, 178-179

Demand Uncertainty, 18-19, 21, 23, 25-26, 33, 35

E

Electrical Resistance, 128, 134-135, 140

Element Analysis, 1, 47-48, 197

Engineer-to-order (eto), 144

Engineering Environment, 144-145, 149-150, 154, 157-159

Environmental Criteria, 161-162, 164, 168-169, 171-172, 179

Estimation Algorithm (ea), 150

F

Finite Element Method, 36, 48, 182, 198

Flexible Manufacturing Systems (fmss), 113

Flicker-noise Spectroscopy, 208, 214

Food Industry, 161, 163-165, 167, 170, 173, 176, 179-180

G

Goal-setting Theory, 51

Green Supplier Selection, 161-164, 167-168, 170-173, 176, 178-179

Green Supply Chain Management (gscm), 162

Greenhouse Gas (ghg), 163

Grey Relational Analysis (gra), 168

H

Hard Steels, 181

Heat-affected Zone (haz), 135

Heuristic Algorithms, 111, 114

Hot Strip Rolling Process, 36

House Of Waste (how), 62, 64

Human Resource Management (hrm), 87

I

Inhomogeneous Shear Strain, 188, 203

J

Just-in-time (jit), 87, 113

K

Kinematic Analytical Approach, 37, 47

L

Labour Productivity, 51, 54

Lead Time Reduction, 21, 27-28, 31

Lean Maintenance (lm), 62-63, 76

Lean Thinking, 62-64, 66, 68, 71, 74, 77, 81-82, 93, 95-96, 98

Linear Programming (lp), 112

Longitudinal Stresses, 37

Low Capital Intensity, 50, 54-56, 59

Low-cost Automation, 1, 3-4

M

Maintenance Process, 62-63, 67-69, 71, 77-78, 82-83, 85-90, 92-93, 97

Maintenance Wastes, 62, 64, 68, 89-90, 93

Manufacturing Industries, 50, 52-53, 55-56, 59, 61, 94, 194

Maynard's Operation Sequence, 1, 5

Mean Absolute Lateness (mal), 155

Mean Tardy (mt), 146

Meta-mechanism Combination, 1, 9-10

Micro-machining Process, 99
Microstructural Properties, 36, 43
Mill Tools Integration, 216
Mixed Integer Linear Programming (milp), 113
Model Predictive Control (mpc), 113
Motion Analysis, 1-2, 4, 6, 8-10, 13
Motion Study, 1
Multi-objective Decision-making (modm), 63
Multi-variable Simulation Model, 36
Multiple Linear Regression, 150

O
Operation Process Optimization, 2
Orthogonal Turning Process, 207-208
Overall Equipment Effectiveness (oee), 67
Overdue Shipment Rate (osr), 112

P
Packaging Supplier, 161, 172-173
Plasma Etching, 99, 101, 106, 108-110
Plasma Etching Process, 106
Plasma Spray Technology, 99, 108
Proactive Approaches, 15
Process Uncertainty, 18, 23, 26
Production Line, 1-5, 13, 63
Production Management, 34, 52, 60, 95, 159-160
Production Scheduling, 111-112, 116, 122, 126, 159
Proportion Tardy (pt), 146
Pull-backward Heuristic Procedure (pb), 118-119
Push-forward Heuristic Procedure (pf), 118, 120

Q
Quality Constraints, 111-112, 115, 117-118

R
Rapid Manufacturing, 216, 220, 229
Raw Material, 23, 161, 164, 172-174
Resistance Spot Welding, 128, 142-143
Roll Force, 36-38, 41, 43-49

S
Statistical Process Control (spc), 113
Strategic Management, 51, 60-61, 94
Subcontracting Constraints, 50-56, 59
Supply Chain Flexibility, 15-17, 19-22, 25, 27-30, 32-35
Supply Chain Uncertainty, 15, 17, 25-26
Surface Pretreatment, 128, 130

T
Thickness Distribution, 36
Tight Dimension Tolerance, 36
Torque Fluctuations, 36, 43, 47
Total Productive Maintenance (tpm), 62, 65
Total Quality Management (tqm), 87
Traditional Sequencing, 145
Triangular Fuzzy Number (tfn), 168

U
Uncertainty Typology, 15-16, 22, 25, 30-31
Unconventional Machining Processes, 99-100

V
Value Stream Mapping, 62-63
Value Stream Mapping (vsm), 63
Velocity Shear Instability, 99-100, 107-110

W
Wave Attenuation, 207-208, 211-214
Welding Equipment, 130
Workpiece Material, 105-106, 130, 139, 190, 193, 208v